The Continental Permian in Central, West, and South Europe

NATO ADVANCED STUDY INSTITUTES SERIES

Proceedings of the Advanced Study Institute Programme, which aims at the dissemination of advanced knowledge and the formation of contacts among scientists from different countries

The series is published by an international board of publishers in conjunction with NATO Scientific Affairs Division

A	Life Sciences	Plenum Publishing Corporation
B	Physics	London and New York
C	Mathematical and Physical Sciences	D. Reidel Publishing Company Dordrecht and Boston
D	Behavioral and Social Sciences	Sijthoff International Publishing Company Leiden
E	Applied Sciences	Noordhoff International Publishing Leiden

Series C - Mathematical and Physical Sciences

Volume 22 - The Continental Permian in Central, West, and South Europe

The Continental Permian in Central, West, and South Europe

Proceedings of the NATO Advanced Study Institute
held at the Johannes Gutenberg University, Mainz, F.R.G.,
23 September - 4 October, 1975

edited by

H. FALKE

Geological Institute, Johannes Gutenberg University, Mainz, F.R.G.

D. Reidel Publishing Company

Dordrecht-Holland / Boston-U.S.A.

Published in cooperation with NATO Scientific Affairs Division

Library of Congress Cataloging in Publication Data

Nato Advanced Study Institute, Johannes Gutenberg University, 1975.
    The Continental Permian in central, west, and south Europe.

    (NATO advanced study institutes series: Series C, Mathematical and physical sciences; v. 22)
    'Published in cooperation with NATO Scientific Affairs Division.'
    Bibliography: p.
    Includes index.
    1. Geology, Stratigraphic — Permian — Congresses. 2. Geology — Europe — Congresses.
I. Falke, Horst, 1909 -   II. North Atlantic Treaty Organization. Division of Scientific Affairs.
III. Title. IV. Series.
QE674.N37 1975   551.7′56′094   76 - 18138
ISBN-13: 978-94-010-1463-2   e-ISBN-13: 978-94-010-1461-8
DOI: 10.1007/978-94-010-1461-8

Published by D. Reidel Publishing Company
P.O. Box 17, Dordrecht, Holland

Sold and distributed in the U.S.A., Canada, and Mexico
by D. Reidel Publishing Company, Inc.
Lincoln Building, 160 Old Derby Street, Hingham, Mass. 02043, U.S.A.

C O N T E N T S

PREFACE

     Geological research on the continental Permian in
West, Central, and South Europe has been neglected in
former days. Nevertheless, interest in it increased in
the last twenty years and became important owing to
the problems connected with continental sedimentation
and also because of the important discoveries of oil,
gas, and uranium deposits. Therefore, detailed investigations in all countries where continental Permian
appears have been made with the help of boreholes.

     These investigations included sedimentation basins
in which the above-mentioned deposits are not expected.
Consequently, the numerous results received in connection with still unanswered questions led to the idea
to invite the specialists to a meeting. The first
meeting was held in Pisa 1966, and a further one in
Vienna 1969. Both conferences exclusively dealt with
the deposits of the Verrucano province, both in the
Alps and Apennines. With respect to the importance of
such meetings a further one was intended to be arranged
in Mainz dealing with the whole continental Permian in
West, Central, and South Europe. With financial help
of the NATO Scientific Affairs Division this meeting
could be arranged as a NATO Advanced Study Institute.
Besides a few exceptions nearly all important European
occurrences have been discussed and an outlook into
North Africa Permian deposits was given as well as
into the Dunkard Formation USA, which is important for
comparison with European deposits. Beyond that, questions
of general interest of that time were discussed as for
example climate and the red colouring of the sediments
as well as summarizing the tectonic and volcanic events
concerning plate tectonics.

     In order to guarantee further extension of these
results the lectures are published in the present pro-

ceedings for which I am very much obliged to the publishers. But my thanks are especially extended to the
NATO Brussels which sponsored this meeting and made the
Advanced Study Institute possible. Further help was
granted by the Auswärtige Amt and Deutsche Forschungs-
gemeinschaft, Bonn and the Akademische Auslandsamt,
Mainz and is gratefully acknowledged. Finally, only the
desire and hope remains that the published papers will
activate further investigations.

Mainz, December 1975   H. F a l k e

INTRODUCTION

Falke (1972) divided the continental Permian in West, and South Europe into three provinces:

1) into the Northern or Subvariscan province;
2) into the central or variscan province;
3) into the Verrucano province.

The first-mentioned province extends from Great Britain through the area of the North Sea and the North German plain into Poland. This province is first of all marked by a great NW-SE striking sedimentation basin north of the central European mountains filled with volcanic material and sediments of special genesis, as for instance a salt sequence.

The variscan province marks a region which reaches from West Poland via Czechoslovakia, Central and South Germany into France and with some reservations into Spain. That province is characterized by appearance of many basins of different size aligned in the direction of the particular predominante Variscan structures and by a sequence which mostly can be divided into a lower section = "Autunian" = Lower "Rotliegend", and an upper section = "Saxonian" = Upper "Rotliegend".

Finally, there is the Verrucano province to which especially belong the Alps and Apennines. Here the occurrences have suffered a more or less strong metamorphism owing to the alpine tectonic movements.

The following contributions deal with the particular provinces

H. Falke (ed.), The Continental Permian in Central, West, and South Europe, 1. All Rights Reserved.
Copyright © 1976 by D. Reidel Publishing Company, Dordrecht-Holland.

NORTHERN END OF EUROPEAN CONTINENTAL PERMIAN - THE OSLO REGION

Chr. Oftedahl

Department of Geology, The University of Trondheim,
7034 Trondheim-N.T.H., Norway

ABSTRACT. The Permian sedimentary and volcanic stratigraphy of
the Oslo graben is reviewed. An analysis of the faulting leads to
a three-fold division with displacements more than 1000, 100, and
10 m respectively. Fault directions point to both dextral and
sinistral movements if considered an en echelon pattern. Doming
of the surroundings is considered insignificant to the east,
possible to the west. Continuation of the Oslo graben fault
pattern to the south is discussed: The Mittelmeer-Mjösen zone
can be deleted.

VOLCANIC AND SEDIMENTARY ROCKS

The Oslo region is an area of around 8000 km^2, consisting of
Cambrosilurian sediments, Permian volcanic and plutonic rocks
(1500 and 5000 km^2, respectively), see fig.1. It is bounded against
the Precambrian all around, and its supracrustals are preserved
because of the subsidence, amounting to more than 1000 m on the
average, with a downdrop estimated to 2-3 km, possibly 4 km along
some parts of the master faults. In the following a geologic
history of its surface is attempted, followed by a discussion of
the tectonic history and its relation to Central Europe.

Within the downfaulted graben area the Precambrian basement
is seen in a few places. On a sub-Cambrian peneplane rests a
Cambrosilurian sequence of shales and limestones with a maximum
thickness of 1000 m, followed by a late Silurian continental sandstone series, also with a maximum thickness 1000 m. The sequence
was folded in the Devonian, and a peneplane developed in early

H. Falke (ed.), The Continental Permian in Central, West, and South Europe, 2-13. All Rights Reserved.
Copyright © 1976 by D. Reidel Publishing Company, Dordrecht-Holland.

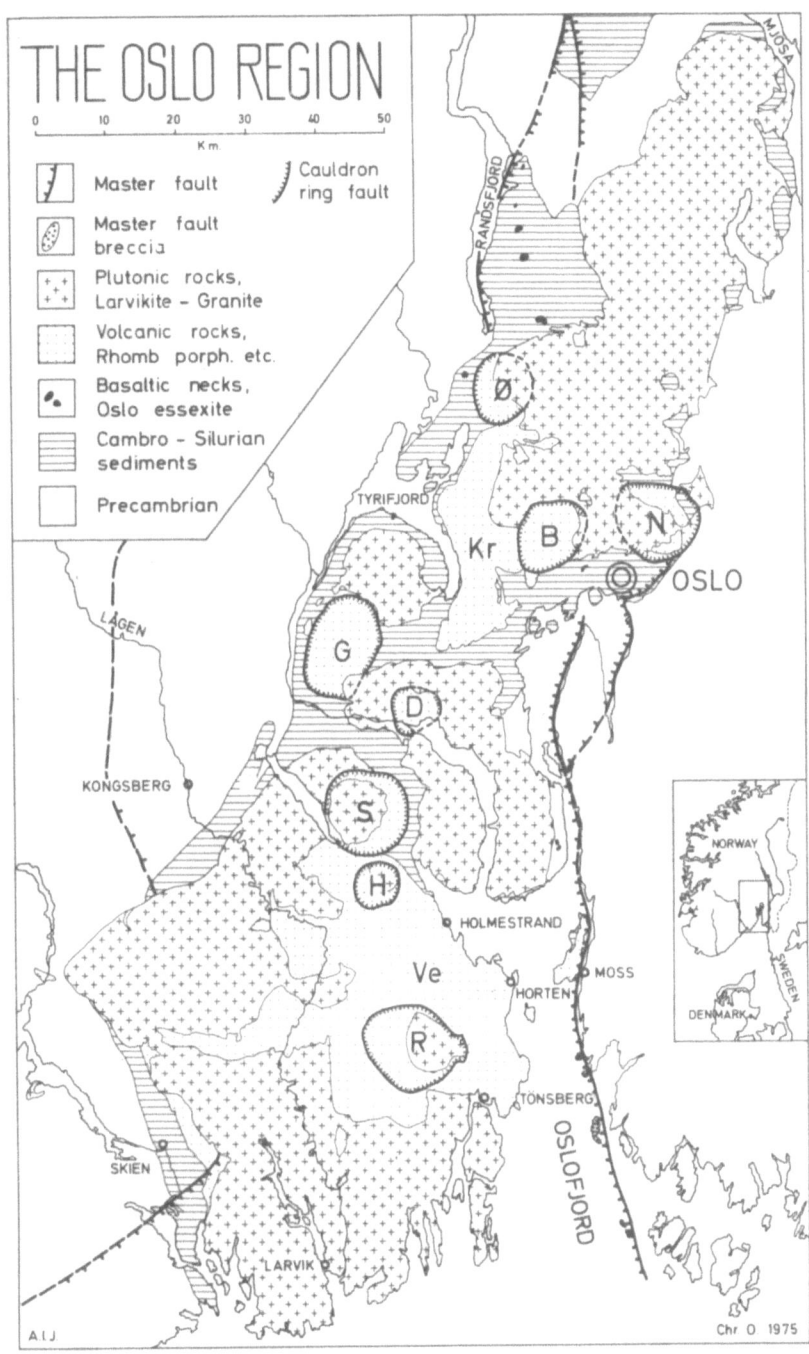

Fig.1. Abbreviations appear from the text.

Permian. On this peneplane rests a thin basal series, followed
by plateau lavas of basaltic and latitic rocks (rhomb porphyries)
with trachytic and rhyolitic rocks admixed in the later part of
the lava plateau development. After the development of most of
the lava plateau a plutonic phase started, giving plutons of
monzonitic rocks (larvikite and kjelsåsite), syenitic rocks (e.g.
nordmarkite) and granitic rocks.

The basal Permian series is called the Asker group by Dons
and Györy [1]. It is highly varying in development and has a
maximum thickness of 50 m west of Oslo, but may attain a greater
thickness in the southern end of the Oslo region. The group
consists of redbeds at the base (Kolsås Formation) followed by grey
conglomeratic sandstone and quartz conglomerate. The youngest is
the Skaugum Formation of partly pyroclastic sediments, with fossils
of a flora and a fresh water fauna which are now considered to belong to the lowermost Lower Permian (prof. G. Henningsmoen, oral
information). The quartz conglomerate is the most persistent
formation of the basal series and may be the only member present
in many places [2,3]. I interpret the redbeds as belonging to a
series which may have covered most of south-eastern Norway, and
the sandstone with quartz conglomerate as formed on delta plains
during the start of the graben subsidence.

The igneous rocks of the Oslo region are known from old days
essentially by the work of W.C.Brögger. A survey of the region
with reference to all older literature is given by Oftedahl [4].

The volcanic rocks now occur in two large areas and some
smaller areas, because the stoping of deepseated magma has removed
most of the lava plateau within the graben in the present erosion
level. The surprisingly detailed stratigraphy of the big northern
area (the Krokskogen area) appeared in 1:100.000 geologic maps by
W.C.Brögger and J.Schetelig in 1917, and for the central Vestfold
area to the south in similar maps from 1926. Contributions to the
stratigraphy also derive from a number of cauldrons. The four most
prominent (Bærum, Glitrevann, Sande and Drammen) were described
in detail earlier [5] and four others are being studied, with
preliminary results announced (Naterstad [6] for Nittedal, B.T.
Larsen for Øyangen [7], and Oftedahl [8] for Ramnes and Hillestad
cauldrons). The lava plateaus and the cauldrons are dominated by
rhomb porphyries and basaltic rocks, with one exception, the Ramnes
cauldron. Only here occurs a thick succession of ignimbritic
rhyolites, with a total maximum thickness of 1400 m [9, 10]. In
table 1, the stratigraphy of the Oslo region is presented in a
compressed form.

The mapping of the central Vestfold plateau has shown the
author that the Oslo volcanism was characterized by very easy-
flowing basalts and rhomb porphyries. Each rhomb porphyry type

Table 1

Stratigraphy of the Oslo volcanics

Krokskogen	Vestfold	Ramneś cauldron
		R_1-R_{11}: Mostly ignimbrites
		RP_1, B, T, and R flows
	and B_3-B_5	
	$RP_{13}-RP_{26}$, with T_2-T_4	
B_3	B_3 and T_1	
RP_1-RP_{11},with B_2	RP_1-RP_4, local flows	
B_1	B_1	

specially numbered can be distinguished from other types by
characteristic shape, size and packing of the plagioclase phenocrysts. Each type makes up a plane parallel plate which quite
often consists of a great number of single flows. Both upper and
lower contacts of the lava flow show a surprisingly flat landscape. Development of small river valleys or canyons has never
been found. Thus each flow simply filled the lowermost areas of
the underlying flow in a very faintly undulating landscape.

The contact between single flows or types may be either just
a knifesharp contact between two faintly vesicular lavas or may
be marked by a thin sedimentary horizon. Most ordinary are pockets
of red sandstone 1-10 cm thick. Continuous horizons are 10-100 cm,
very rarely several meters. These sediments are mixtures of two
components, volcanic erosion products and wind-blown quartz-sand
from outside the graben area. A strongly red baked sediment
appears when the quartz dominates. Otherwise the volcanic débris
occur as all mixtures from mud via silt and sand to gravel and
head-sized boulders, both rounded and angular. Only in the
cauldrons during start of the caldera subsidence thicker sedimentary beds are observed. Thus early caldera lake sediments
amount to 30-40 m in the Bærum caldera and to about 100 m in the
Nittedal caldera [2].

The only direct proof of the existence of a comprehensive
lava plateau outside the present Oslo region is found in the
Sevaldrud breccia [11] occurring 12 km west of the western master
fault in the northern Oslo region. This breccia occurring in the
Precambrian contains blocks of Precambrian gneisses, Cambrosilurian
sedimentary rocks, and Permian volcanics, and both basalts and
rhomb porphyries of various types. Another indication from the

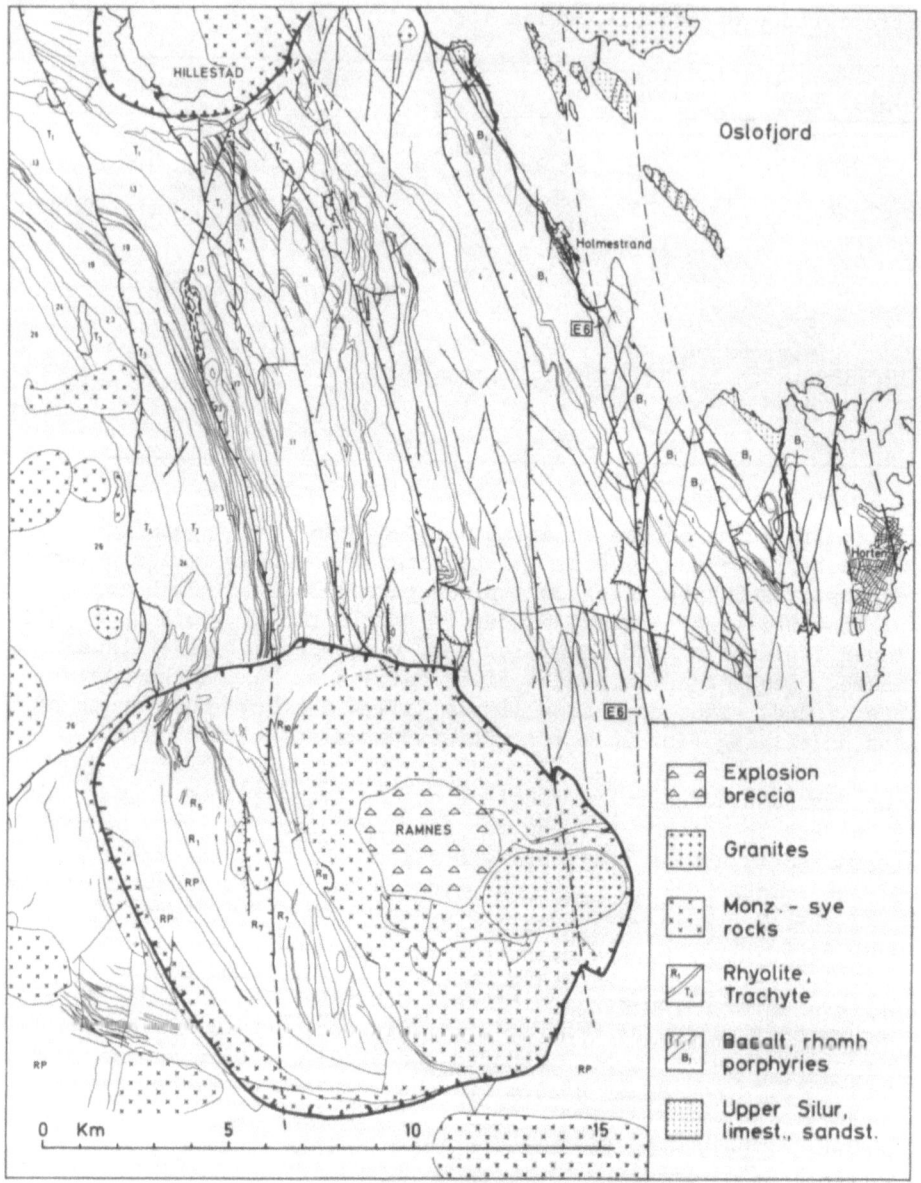

Fig.2.   Geologic map of the central Vestfold lava plateau, with
         the Hillestad and Ramnes cauldrons, and the most important
         points of the lava stratigraphy.

east side of the region is derived from a fault breccia, exposed
on a number of small islands in outer Oslofjord (fig.1). The
gravelly sandstone with conglomerate beds contain rhomb porphyry
blocks not rarely of 1 m size, and locally of house size. Other
boulders are made up by basalts, trachyte, rhyolites and upper
Silurian sandstone. There is little doubt that the material came
down the precipice of the master fault from the east as supposed
by Störmer [12], thus testifying of a comprehensive lava sequence
east of the southern Oslo region. Other but less certain indicators are the rhomb porphyry dikes which occur north of the
region, down the south-western coast of Norway and down the Swedish
Skagerak coast.

The described features and several others support the general
thesis of the Oslo region as a great lava plateau covering perhaps
most of southeastern Norway and possibly considerable parts of
adjacent Sweden. The lava emissions must have built up a very
flat but faintly domal lava plateau with little erosion between
eruptions, following rapidly upon each other. An earlier estimate
was 1000-10.000 years between each lava type [13]. One may even
be tempted to assume eruption intensity of present day Iceland:
one eruption per five years. For estimation of the total lava
volumes the question of subsidence during eruptions is critical.
So far no reports support with certainty the possibility that the
lava eruptions were nearly balanced in volume by subsidence. It
seems reasonable to assume the start of graben faulting with the
basal quartz conglomerate and a fair amount of subsidence during
the eruption period. But the volumes overflowing the present day
graben margins must have been considerable, estimated to order of
size 15-30%, rather than 1-5%. Since southern central Norway
seemingly remained dry land from Permian to present day, the
erosion products of this lava plateau should be present in
sedimentary deposits south of the Oslo region.

A much discussed problem is whether doming of the Precambrian
occurred close to the master faults, as is obvious for the
Rhinegraben and the East African rifts. The relief east of the
graben area is very even, with ridge tops approaching a plane of
less than 100 m in elevation. Norwegian geologists have long
supposed that this plane is the sub-Cambrian peneplane which
erosion barely has passed through. On the west side of the
northern graben area the relief goes to 600 m immediately west of
the master fault and to 1500 m at a distance of 45 km. Thus the
west side may have been elevated 1500 m or more, indicating a
marked asymmetry, like the Rhinegraben. But this requires that
the present relief is post-Permian, an assumption which is quite
uncertain.

TECTONIC FEATURES

Fault pattern

    It is evident from the general map of the Oslo region (fig.1)
that the region as it now appears does not show master faults on
both sides of the graben like for instance the Kenya rift in East
Africa. Whereas the main axis of the region trends to the NNE, a
few major faults run roughly N-S. The fault starting around Oslo
and running south on the east side of the Oslofjord has a displacement of more than 1000 m at Moss (the Moss master fault) and 2-3 km
further south. The big fault through lake Randsfjord must have
been a similar western master fault. Possibly these master faults
were connected, but the connection is now obliterated by the plutons
north of Oslo.

    The map sheets of W.C.Brögger and J.Schetelig printed 1917
to 1926 also show a next generation of faults, although never
drawn as fault lines. These faults appear on maps of the
Krokskogen area [14], and the Vestfold area [15]. The section of
the Vestfold lava plateau shows a regular antithetic step faulting
with displacement of 100-300 m on normal faults (fig.2). The
Krokskogen area is somewhat similar. On the whole this second
generation follows the master faults, trending a little west of
north.

    A third generation of small-scale faulting is apparent from
the detailed mapping of lava stratigraphy in the Vestfold lava
plateau. Displacements on this system range from 10 to 30 m.
Quite surprisingly the prevalent direction makes a sharp angle
with the two first generations of faults and parallels the main
axes of the Oslo region. This small-scale faulting shows up only
where exposures are good and where a detailed stratigraphy exists.
It is somewhat hypothetical and erratic in the eastern part of the
lava plateau between the Hillestad and Ramnes cauldrons (see fig.2)
but much more certain and regular in the western part. In fig.2
these faults are indicated in the eastern part but not in the
western where they might destroy the detailed stratigraphy. The
faulting also occurs in the upper Silurian limestone islands at
Holmestrand [16] and is shown on fig.2.

    To explain the major faulting (the two first classes) H.Cloos
[17] assumed that the fault pattern is an en echelon pattern
("Fiederspalten") which then necessitate a displacement of the
western block to the SSW. In his analysis of the fault pattern
indicated by the conglomerate island of the Oslofjord, Störmer
[18] assumed an en echelon faulting with a dextral strike slip,
opposite to that of Cloos. From observations however, strike slip

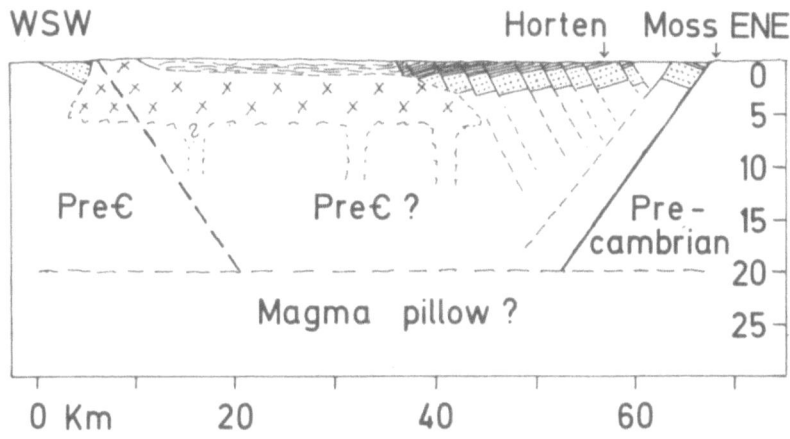

Fig.3. Hypothetical section across the southern Oslo graben, with first and second order faults.

is proved to exist only at one fault traversing the Ullernåsen volcanic neck in Oslo [19], with a 200 m dextral slip.

What is certain, however, is that the faulting had a tension component in an east-west direction, with development of normal faulting in three size classes. If the Fiederspalten hypothesis is applied, the present knowledge of the Oslo region fault systems seem to indicate a sinistral slip for the first and second order faulting, and a dextral slip for the third order faults. From resemblance with central Europe Störmer [20] assumed that the faulting of the Outer Oslofjord conglomerate could be Late Mesozoic or even Tertiary in age, a good guess even today. This could then be the age of the third class faults.

One solution of this dilemma lies in assuming a very complex tectonic history, with sinistral and dextral slips, as well as simple tension, like that supposed for the Rhinegraben by Laubscher [21]. The principal fault pattern of the two first fault classes appear in fig.3, which is drawn in analogy with the pattern for the Rhinegraben by Laubscher. The asymmetry of the second class faults across the graben is striking. The eastern master fault is certain, but the western is hypothetical, because in the west the stoping plutons removed the possible faults. The faulting hypothetically stops at the top of a Rhinegraben type pillow at depth around 20 km. From a recent explosion seismic program Tryti [22] confirmed the earlier suggestions about such a pillow [23].

Aki et al [24] found P-wave anomalies around the northern end

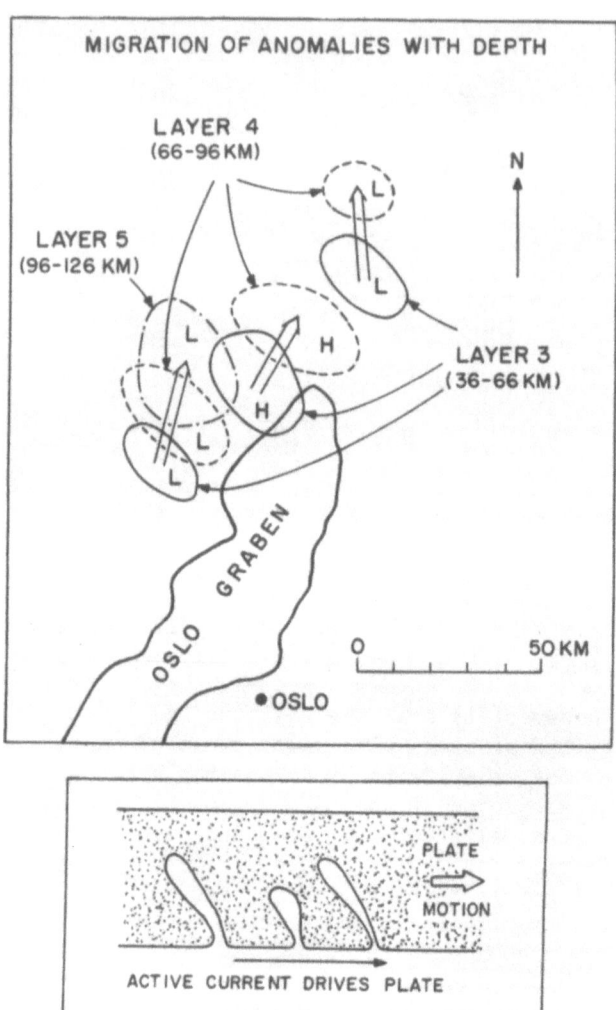

Fig.4.   Velocity anomalies and their interpretation [24].

of the Oslo region, in the area of the Norsar seismic array.   The
anomaly patterns of seismograms from special earthquakes were
analyzed in three dimensions, with the crust and upper mantle
divided into layers and each layer into blocks.   With a complex
analysis an anomaly pattern was derived from the area, see fig.4.
The authors conclude that the positive and negative anomaly
channels sloping to the NNE are explained by an active current
in the asthenosphere dragging the lithospheric plate along.   This
northerly drag west of the Oslo region means a dextral movement,
opposite that required by the first and second order faults.

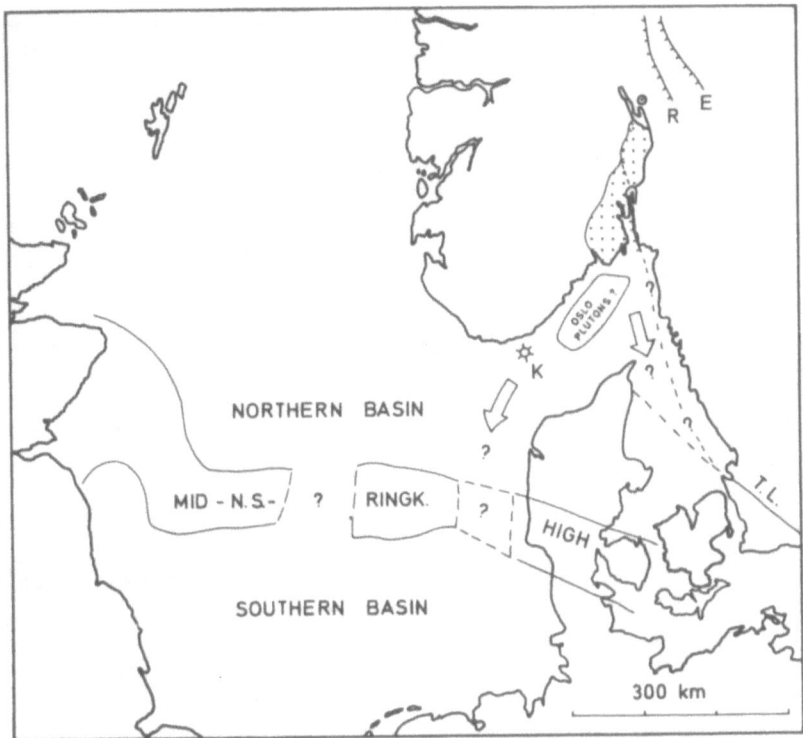

Fig.5. Sketch map of the North Sea (from Ziegler [27] and
        Southern Norway. K: Kristiansand Permian volcano.
        R and E: Rendal and Engerdal Permian(?) faults.
        T.L.: Tornquist line.

Southern continuation of the Oslo graben.

    It is very tempting to assume that the Permian fault pattern
of the Oslo graben continues in a way to the south to link up
with the Rhinegraben and the graben of the Rhone valley to form
a trans-European rift valley system like those of East Africa.
The term "Die Mittelmeer-Mjösen Zone" was coined by H.Stille in
1925 and has been much used in the past 50 years. Results from
last years' geophysical investigations and oil prospecting in the
North Sea have shown that the term should rather be deleted.
Schönenberg [25] has already doubted its validity.

    Gravity and aereal magnetic anomaly studies of the North Sea
and Skagerak up to the Oslo region suggest a continuation of the
Oslo magmatism towards the central North Sea along the coast of
southern Norway. A detailed study by Åm [26] of the strong magnetic
and gravity anomaly in the sea 25 km south of southernmost Norway

is interpreted as both a Permian and a Tertiary basalt plug nearly
superimposed on each other. For the area between the Kristiansand
volcano and the southern tip of the present Oslo region, Åm
supports earlier gravimetric studies suggesting a continuation
of the Permian plutonism, as shown on fig.5. This figure also
shows the North Sea situation during Lower and Middle Permian
(Rotliegendes). The Ringköbing-Fyn-High continued from Denmark
across the North Sea, to divide the area into a northern basin
with continental to shallow-marine sandstones and shales, including
volcanism, and a southern basin with sandstone and evaporites [27].
As emphasized by Ziegler, the rifting of the Viking-central graben
of the North Sea started only in Triassic to be fully developed
in the Jurassic.

    Another possibility for the southern continuation lies in
linking the master fault of southern Oslofjord with the Tornquist
line, indicated as operative in the Ziegler maps from Late Devonian
through Carboniferous to Middle Permian, even though the Danish-
Polish Furrow developed in the Triassic. The possibility of the
Oslo magmatism continuing this way got a faint support with a
find from the Danish island Fyn. Petroleum drilling around 1950
was stopped when hitting assumed Precambrian, an arcose with
pebbles like nordmarkite from the Oslo region [28]. A third
possibility which cannot be completely excluded, is to link the
Oslo region with a more westerly fault system. Ziegler states
that the Viking graben possibly had its earliest traces in the
northern North Sea during the Permian. Ballard and Uchupi [29]
in treating the Fundy fault system off the coast of Maine, drew
the late Devonian-Permian Cabbot fault system from coastal U.S.
and Newfoundland to Ireland and Scotland, with proposed dextral
strike slip like that of Aki et al. From Scotland with new red
continental sandstone in the Midland valley graben east-going
faults could link up the start of the Viking graben with the
Oslo graben. Contradicting strike slip direction makes this less
probable. On the other side Storetvedt [30] proposed a 200-300 km
sinistral slip on the Great Glen fault, with a late/post Caledonian age (or better Carboniferous/Permian?), a slip which
coincides with that required by the large scale Oslo graben
faulting.

REFERENCES

1.   J.A.Dons, and E. Györy, Norsk Geol. Tidsskr. 47, 1967, 62.
2.   Naterstad, News from the Oslo region research group, 1,
     1971, 34.
3.   E. Rohr-Torp, Norges geol. unders., 300, 1973, 56.
4.   Chr. Oftedahl, Norges geol. unders., 208, 1960, 298.
5.   Chr. Oftedahl, Vid.Akad. Skr. Oslo, 3, 1953.

6. J.Naterstad, News from the Oslo region research group, $\underline{\underline{1}}$, 1971, 29.
7. B.T.Larsen, News from the Oslo region research group, $\underline{\underline{2}}$, 1971, 8.
8. Chr. Oftedahl, Geol. Rundschau, $\underline{\underline{57}}$, 1967, 203.
9. E.S. Jensen, Unpubl. thesis, Copenhagen Univ. 1964.
10. Chr. Oftedahl, Geol. Rundschau, $\underline{\underline{57}}$, 1967, 212.
11. W.C.Brögger, Norsk Geol. Tidsskr., $\underline{\underline{11}}$, 1931, 286.
12. L. Störmer, Norsk Geol. Tidsskr., $\underline{\underline{15}}$, 1935, 82.
13. Chr. Oftedahl, Vid.Akad. Skr. Oslo, $\underline{\underline{3}}$, 1952, 17.
14. O.Holtedahl and J.A. Dons (ed.), Geological guide to Oslo and district, Universitetsforlaget, Oslo, 1966.
15. Chr. Oftedahl, Geol. Rundschau, $\underline{\underline{57}}$, 1967, 15.
16. J. Kiær, Vid.-Selsk. Skr. Christiania, 1906.
17. H. Cloos, Fortschritte Geol. u. Pal., $\underline{\underline{7}}$, 1928, 297.
18. L. Störmer, Norsk Geol. Tidsskr., $\underline{\underline{15}}$, 1935, 101.
19. J.A. Dons, Vid.Akad. Skr. $\underline{\underline{2}}$, 1952, 76.
20. L. Störmer, Norsk Geol. Tidsskr., $\underline{\underline{15}}$, 1935, 105.
21. H.P. Laubscher, Graben Problems (UMP Report 27), Schweizerbart, Stuttgart, 1970, 82.
22. J.A. Tryti, Unpubl. thesis, Univ. of Bergen, 1975.
23. I.B. Ramberg and S.B. Smithson, Tectonophysics, $\underline{\underline{11}}$, 1971, 419.
24. H. Aki, A. Christofferssen and E. Husebye, 1975. In press. (submitted to Journal of Geophys. Res.).
25. R. Schönenberg, Einführung in die Geologie Europas, Rombach, Freiburg, 1971, 262.
26. K. Åm, Norges geol. unders., $\underline{\underline{287}}$, 1973, 20.
27. P.A. Ziegler, AAPG Bull., $\underline{\underline{59}}$, 1975, 1079.
28. P. Holmsen, Medd. Dansk Geol. For., $\underline{\underline{14}}$, 1958, 5.
29. R.D. Ballard and E. Uchupi, AAPG Bull., $\underline{\underline{59}}$, 1975, 1064.
30. K. Storetvedt, Geol. Mag., $\underline{\underline{111}}$, 23.

A REVIEW OF THE LOWER PERMIAN IN AND AROUND THE BRITISH ISLES

Denys B. Smith

Institute of Geological Sciences, Princes Gate,
London SW7, England

1. INTRODUCTION

In an earlier account [1] the author described the pattern of
sedimentation during the Lower Permian in land areas of the
British Isles but was able to include only limited information
from the surrounding shelf seas. However, the pace of offshore
exploration has been so rapid that it is now possible to place
the Lower Permian sediments of the present land areas within
their wider context, and it is with this aspect that the present
paper is mainly concerned.

The information from the British Shelf seas comes from both
indirect and direct sources, and is of varied quality. The
indirect evidence is provided mainly by marine geophysical surveys
and includes almost all the information now available for undersea
areas north and west of the British Isles; it is generally
imprecise and suffices mainly to indicate the broad outlines of
basins where Lower Permian rocks may lie. In contrast, the newly-
released borehole data on Lower Permian rocks of the southern
North Sea [2] provides a wealth of precise information for a large
area to the east of the British Isles, and more limited information
from the Irish Sea and Northern North Sea basins provides a clue
to the Lower Permian rocks of further extensive shelf areas. The
new data show, as had long been suspected, that Permo-Triassic
sedimentary basins lie close to much of the British coast and that
the Lower Permian rocks preserved on land record only a small part
of the contemporary history of the region. The major discovery is
of the existence of extensive continental sebkhas and saline playa
deposits in deeper parts of several of the offshore basins. As in
the basins on land, inland drainage appears to have been dominant

H. Falke (ed.), The Continental Permian in Central, West, and South Europe, 14-22. All Rights Reserved.
Copyright © 1976 by D. Reidel Publishing Company, Dordrecht-Holland.

and the floors of some basins probably lay well below sea level.

2. PATTERNS AND HISTORY OF SEDIMENTATION

Little new information has become available since the pattern of
Lower Permian sedimentation in present land areas of the British
Isles was outlined [1, 3] and this aspect need be restated only
briefly. Broadly, following Hercynian folding, uplift and block
faulting, the area was subjected for perhaps sixty million years
to subaerial erosion and continental sedimentation, at the end of
which the initial high relief had become greatly subdued and large
areas reduced to a rolling desert peneplane. Reddening of strata
beneath this surface is almost ubiquitous, exceptionally reaching
depths of 580m but generally penetrating 5 to 20m. Evidence of
the immature stages of the peneplanation cycle is now found only
in the west of the British Isles, where continuing subsidence in
a number of inland basins and fault-bounded troughs enabled coarse
late Carboniferous and early Permian breccias to be preserved;
doubtless breccias were at one time much more extensive, especially
in the relatively stable eastern areas, but were swept away as the
Hercynian mountains were progressively removed. In the larger
basins, breccias passed downslope into fanglomerates and these in
turn into sands and silts. Volcanicity was a major feature only
in west Scotland, where the late Carboniferous and early Permian
scene was dominated in places by the outpourings of more than
sixty vents, and in south-west England where volcanism of similar
age may have been briefer and less extensive.

The history of late Palaezoic continental sedimentation in
the British Isles was generally similar to that in inland basins
in arid and semiarid regions today, and although it is clear that
the various basins evolved at rates determined mainly by local
factors, almost all have a broadly similar sedimentary sequence.
Overlying a sharp unconformity generally spanning at least one
stage, this sequence consists of early coarse marginal breccias
and fanglomerates (grading laterally into sands and silts in the
larger basins) which pass upwards by alternation into mainly wind-
blown sands that accumulated to thicknesses exceeding 300m under
the influence of prevailing easterly and east-north-easterly winds.
This generally simple sequence was varied where the formation of
thick piles of volcanic rocks led to repeated rejuvenation and
renewed influx of coarse detritus, and rejuvenation may also have
accompanied continuing earth movements in some basins.

Useful fossils are rare in the late-Palaeozoic continental
rocks of the British Isles and there are many basins where no
fossils whatever have been recorded. Determination of the age of
these deposits is therefore difficult, and in most cases they can
be shown only to be post Westphalian C or D, pre Upper Permian or
Scythian. Where present, fossils are either plants, as in the

Mauchline Basin of Scotland and in the English Midlands, or
reptilian remains and footprints in several basins. The plants
from the Mauchline Basin have been provisionally referred to the
early Autunian [3], and occur near the base of the sequence.
Footprints occur at several localities in south-west Scotland [4]
and indicate a general late Lower Permian age for strata high in
the local sequences, but the most valuable footprint evidence is
from low in the late Palaeozoic sequence in the English Midlands
where footprints have been assigned firmly to the upper Autunian
[5]. Pelycosaur remains from the same area have also been assigned
to the Autunian [6]. The only other specific evidence of age is
in the form of radiometric age determinations of about 280 m.y.
for lavas near the base of the continental red-beds in south-west
England [7. 8].

Where Upper Permian marine strata are absent, it is probable
that continental beds at the top of the various sequences are of
Upper Permian age; a similar age must also be accorded to the
uppermost few metres of continental sediments in north-east
England where these were redistributed during the Zechstein
transgression.

3. SEDIMENTARY BASINS NORTH AND WEST OF THE BRITISH ISLES

Geophysical surveys of shelf areas off the north and west coast
of Scotland and in the Irish Sea, Celtic Sea and Western
Approaches have revealed the presence there of a series of
sedimentary basins of similar style to those in which late
Palaeozoic rocks occur on adjoining land areas; some, indeed,
are continuous with known basins on land and reasonable assessments
of the nature of their contents may be made. The basins are
generally inferred to have been initiated in Devonian or Carboniferous time, although some are thought to be mainly Tertiary.
The geophysical characteristics of rocks in several of the older
basins are similar to those of Permo-Triassic rocks in basins on
land, most of which contain some Lower Permian strata, and these
basins are shown in Fig. 1 in the belief that they too probably
contain Lower Permian rocks. If this is correct, sandstones and
fanglomerates probably predominate, with finer-grained playa
deposits in deepest parts of the larger basins; it must be
emphasized, however, that the actual contents of most of these
basins remain unknown.

Of the western basins so far identified, drilling results
have been published only for the East Irish Sea Basin where four
deep boreholes reveal up to 280m of barren continental red
sandstones passing into mudstones and siltstones in a central
area [3, 9]. The sandstones are probably continuous with the
Collyhurst Sandstone, which is 715m thick at Formby in the south-

east of the basin [10].

Western offshore sedimentary basins for which no drilling
results have been published include a number that extend onshore
and here the nature of their contents may partly be judged from
the marginal outcrops. Most of these basins lie off western
Scotland, where the higher marginal strata and some beds sampled
at sea have yielded Upper Permian or Triassic miospores [11].
Coarser lower beds in the marginal outcrops have so far yielded
no useful fossils, and hence their age cannot be firmly established.
However, the writer tentatively assigns most of these beds, and
their inferred thicker offshore equivalents to the Lower Permian
in the belief, based on comparisons with other basins where
stratigraphic control is better, that this was the peak phase of
breccia formation and accumulation. Thus, for example, the
alluvial Stornaway Formation [12] of the North Minch Basin, Lewis,
perfectly fits the Lower Permian scene in north-west Europe but
in several respects is atypical of the Triassic to which it and
many similar deposits in Scotland traditionally have been ascribed.

Amongst western and northern basins where the presence of
Lower Permian sedimentary rocks is least certain are those west
of the Orkney and Shetland Isles, in the Southern Irish and
Celtic Seas, and in the Western approaches. Basins west of the
Orkney and Shetland Isles are known mainly from research by Bott
and his co-workers, summarized by Bott [13]; several are thought
to be bounded by faults on one or more sides, and to have been
affected by synsedimentary earth movements. On the basis of
gravity profiles, up to 4 km of Palaeozoic or early Mesozoic
sedimentary rocks are thought to be present in at least one basin.
Farther south, the position of sedimentary basins in the southern
Irish Sea, Celtic Sea and Bristol Channel have been established
by a variety of methods including gravity and seismic surveys
[14, 15] and limited sea bottom sampling, and the presence of
Lower Permian rocks appears to be a reasonable probability.
Basins in the south-west approaches and English Channel are best
known from seismic surveys [16, 17, 18], from which up to 1.5 km
of sedimentary rocks of seismic velocities similar to those of
known Permo-Triassic and possibly Jurassic rocks have been inferred.

4. SEDIMENTARY BASINS EAST OF THE BRITISH ISLES

East of the British Isles, substantial thicknesses of Lower
Permian sedimentary rocks occur in both the Northern and Southern
North Sea basins but are thin and patchy on the intervening
Mid-North Sea High.

In the Northern North Sea Basin, these rocks are both oil-
and gas-bearing in suitable structures, but have been removed

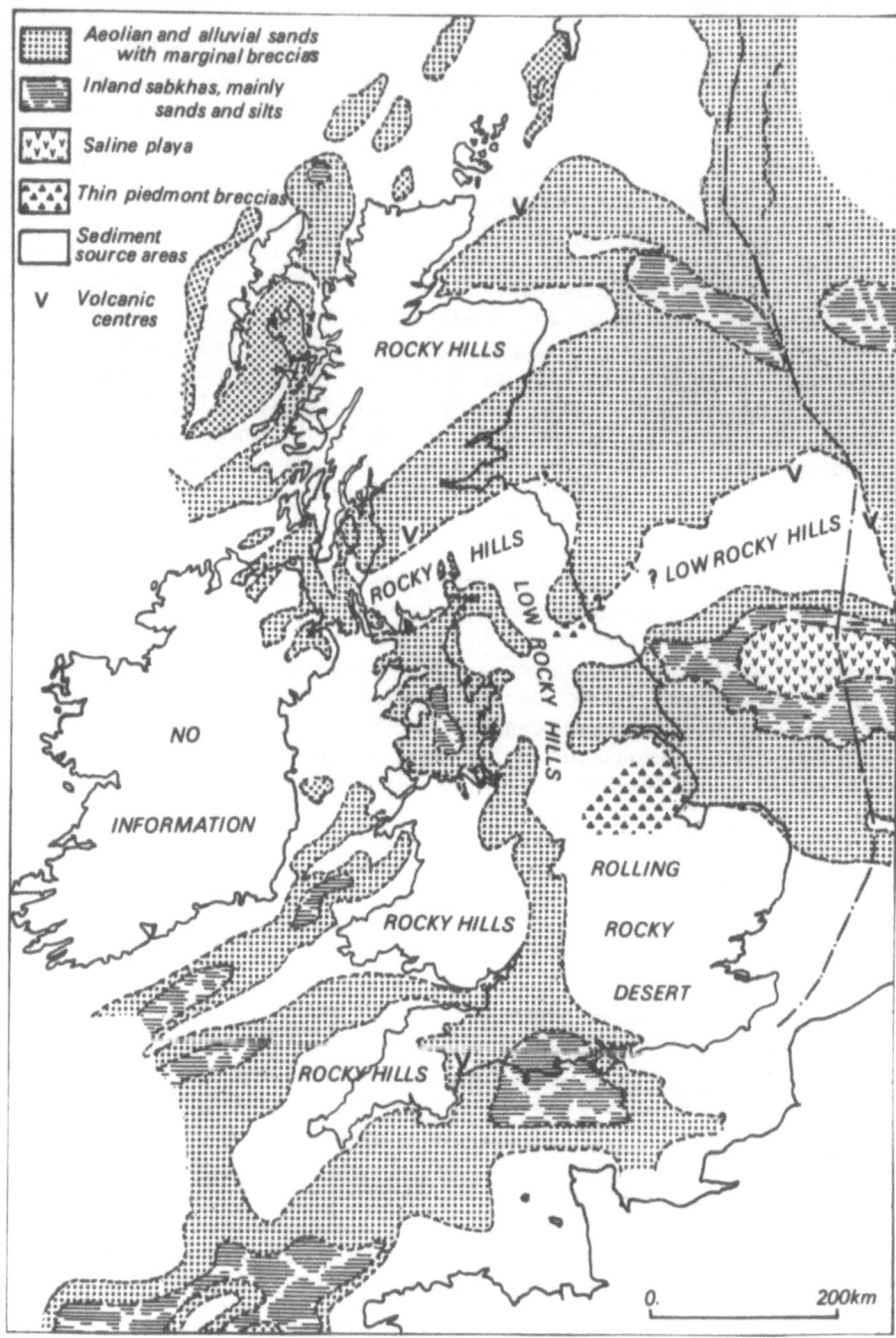

Fig. 1. Late Lower Permian palaeogeography of the British Isles and surrounding areas.

from many areas by Jurassic and later erosion. Few details of
their overall distribution in this basin are available, but
boreholes near the southern margin have proved up to 454m of
Lower Permian continental sandstones and thin interbedded 'shales'
[19, 20]. According to Pennington [20], the sandstones in the
upper 183m of the Lower Permian sequence in Well 30/24-2 are
mainly red fine-grained to very fine-grained friable subarkoses,
and comprise an alternating sequence of fluviatile and aeolian
rocks whose sediment came from the south-west. Pollen from a
thin shale 8.5m below the Zechstein in Well 30/24-3 is regarded
as a typical Zechstein assemblage [20] and may stem from the
initial Upper Permian flooding of the basin. Ziegler [21] shows
two large areas in this basin where the marginal sands pass into
continental sebkha deposits and such rocks may also be present
elsewhere in the basin. He further depicts three volcanic centres
located near basin-margin faults, including one in the area of
the Argyll Field where Lower Permian basalt has been recorded by
Pennington [20].

The Lower Permian and early Upper Permian sedimentary rocks
(Rotliegendes) of the Southern North Sea Basin include the main
natural gas reservoirs, and for this reason have been probed by
many boreholes and intensively studied. Few details have been
published, but excellent summaries by Glennie [22] and Marie [23]
have provided a wealth of data showing that this basin extends
eastwards into Northern Germany and that the continental
Rotliegendes rocks are thickest - more than 300m - in a west-east
belt some 150 km off the English coast. Most Rotliegendes rocks
in this basin are red, but grey predominates at the top of the
formation where up to 50m of sediment was redistributed during
the Zechstein transgression.

From his study of the lithology and sedimentary structures
of the red Rotliegendes rocks, Glennie [22] inferred deposition
in a wide range of desert sedimentary environments, and similar
conclusions were later reached by Blanche [24]. Their work
revealed that both wind and running water played major roles in
the transport and deposition of sediment, especially in marginal
parts of the basin where great spreads of aeolian, sheet-flood
and fluviatile ("wadi") sediments are complexely interbedded, and
that generally finer sediments in the deepest part of the basin
accumulated on extensive inland sebkhas and in saline playas;
here, salts carried in solution eventually crystallized
interstitially as nodular gypsum and as displacive and perhaps
bedded (subaqueous) halite. The mainly sandy rocks comprise the
Leman Sandstone Formation, whilst argillaceous, silty and
evaporite strata of playa and sabkha facies comprise the
Silverpit Formation [25].

Glennie [22] on the evidence of dip measurements, inferred

that most of the aeolian sands of the Leman Formation near the
basin margin accumulated on westward-migrating barchans, but that
traverse dunes were probably more common nearer to the centre of
the basin. Release of data on the approximate distribution of
the several facies enabled Marie [23] to construct a series of
palaeogeographic maps which show that the slightly elevated land
around the south-western margin of the basin constituted a major
source of sediment and was drained by several northward-flowing
water courses. Irreversible reduction of porosity and permeability
resulting from the compaction and diagenesis of Rotliegendes
sediments during deep burial were summarized by Blanche [24], and
the diagenesis of grey sediments at the top of the sequence has
been investigated by Pryor [26].

5. SUMMARY

New information from the continental shelf shows that most of the
Lower Permian rocks in land areas of the British Isles lie at or
near the margins of basins which lie mainly off the coast. Details
of the strata in many of the newly-proved basins are unknown or
have not been released, but where information is available it is
clear that the patterns of sedimentation recognized on land extend
far into the undersea basins; additionally, extensive inland
sebkhas and at least one major saline playa were present. By
combining the information from both land and undersea areas, it is
now possible to attempt a reconstruction of the late Lower Permian
palaeogeography of the whole region (Fig. 1), although it must be
remembered that, especially in some undersea areas in the west,
this reconstruction is little more than speculative.

REFERENCES

1. D. B. Smith, The British Isles, in: Rotliegend; Essays on
   European Lower Permian, ed. by H. Falke, Brill, Leiden, 1972.
2. Petroleum and the continental shelf of north-west Europe,
   ed. by A. W. Woodland. Applied Science Publishers Limited,
   Barking, England.
3. D. B. Smith, R.G.W. Brunstrom, P. I. Manning, S. Simpson and
   F. W. Shotton, A correlation of the Permian rocks in the
   British Isles, Special Report No. 5, Geological Society,
   London, 1974.
4. G. A. Hickling, British Permian footprints, Mem. Proc.
   Manchester lit. phil. Soc. 53, 1, 1909.
5. H. Haubold and W.A.S. Sarjeant, Tetropodfahrten aus den
   Keele und Enville Groups (Permokarbon: Stefan und Autun)
   von Shropshire und South Staffordshire, Grossbrittannien,
   Z. geol. Wiss., B95, Berlin 1, 1973.

6. Roberta L. Paton, Lower Permian Pelycosaurs from the English Midlands, _Palaeontology_, 17, 541, 1974.
7. J. A. Miller, K. Shibata and M. Munro, The potassium-argon age of the lava at Killerton Park, near Exeter, _Geophys. J._, 6, 394, 1962.
8. J. A. Miller and P. A. Mohr, Potassium-argon measurements on the granites and some associated rocks from south-west England, _Geol. J._, 4, 105, 1964.
9. V. S. Colter and K. W. Barr, Recent developments in the geology of the Irish Sea and Cheshire basins, 61-73 in: A. W. Woodland (Ed.), q.v. [2], 1975.
10. P. E. Kent, A deep borehole at Formby, Lancashire, _Geol. Mag._, 85, 253, 1948.
11. J. A. Chesher, C. E. Deegan, D. A. Ardus, P. E. Binns and N.G.T. Fannin, I.G.S. drilling with M. V. Whitethorn in Scottish waters 1970-71, _Rep. Inst. geol. Sci._, No. 72/10, 1972.
12. R. J. Steel and A. C. Wilson, Sedimentation and tectonism (?Permo-Triassic) on the margin of the North Minch Basin, Lewis, _Jl geol. Soc. Lond._, 131, 183, 1975.
13. M.H.P. Bott, Structure and evolution of the North Scottish Shelf, the Faeroe Block and the intervening region, 105-115 in: A. W. Woodland (Ed.), q.v. [2], 1975.
14. M. R. Dobson, W. E. Evans and R. W. Whittington, The geology of the south Irish Sea, _Rep. Inst. Geol. Sci._ No. 73/11, 1973.
15. A. J. Lloyd, R.J.G. Savage, A. H. Stride and D. T. Donovan, The geology of the Bristol channel floor, _Phil. Trans. Roy. Soc._, A, 274, 595, 1973.
16. W.B.R. King, The geological history of the English Channel, _Q. Jl geol. Soc. Lond._, 85, 77, 1954.
17. D. Curry, D. Hamilton and A. J. Smith, Geological and shallow subsurface geophysical investigations in the Western Approaches to the English Channel, _Rep. Inst. Geol. Sci._ No. 70/3, 1970.
18. F. Avedik, The seismic structure of the Western Approaches and the south Armorican continental shelf, and its geological interpretation, 29-43 in: A. W. Woodland (Ed.), q.v. [2], 1975.
19. T. P. Brennand and F. R. van Veen, The Auk Oil-field, 275-283 in: A. W. Woodland (Ed.) q.v. [2], 1975.
20. J. J. Pennington, The geology of the Argyll Field, 285-291 in: A. W. Woodland (Ed.), q.v. [2], 1975.
21. W. H. Ziegler, Outline of the geological history of the North Sea, 165-190 in: A. W. Woodland (Ed.), q.v. [2], 1975.
22. K. W. Glennie, Permian Rotliegendes of north-west Europe in light of modern desert sedimentation studies, _Bull. Amer. Ass. Petrol. Geol._, 56, 1048, 1972.
23. J.P.P. Marie, Rotliegendes stratigraphy and diagenesis, 205-21 in: A. W. Woodland (Ed.), q.v. [2], 1975.

24. J. B. Blanche, The Rotliegendes Sandstone Formation of the
     United Kingdom sector of the Southern North Sea Basin,
     Trans. Instn. Min. Metall., 82, B85, 1973.
25. G. H. Rhys, A proposed standard lithostratigraphic
     nomenclature for the southern North Sea and an outline
     structural nomenclature for the whole of the (UK) North
     Sea, Rep. Inst. Geol. Sci., No. 74/8, 1974.
26. W. A. Pryor, Petrology of the Permian Yellow Sands of
     northeastern England and their North Sea equivalents,
     Sedimentary Geology, 6, 221, 1971.

OUTLINE OF THE ROTLIEGEND (LOWER PERMIAN) IN THE NETHERLANDS

H. Albert van Adrichem Boogaert
Subsurface Department,
Netherlands Geological Survey,
Spaarne 17,
HAARLEM, The Netherlands.

ABSTRACT. The Rotliegend in the Netherlands is placed in an Upper
Rotliegend Subgroup, which consists of two formations:
The Slochteren Formation (sandstone and minor conglomerate) in
the southern and central part of the country, and the Silverpit
Formation (shale, siltstone and evaporites) in the northern part
of the country, with its main extension offshore in the North Sea.
   In the transitional zone of these two formations four members can be distinguished from top to bottom: the Ten Boer Member
(shale and siltstone), the Upper Slochteren Member (sandstone),
the Ameland Member (shale and siltstone) and the Lower Slochteren
(sandstone). A description of the type section of the Ameland
Member is given in this paper. A locally present fifth member,
consisting of thin, gray marine sandstone and conglomerate, is
provisionally called the gray upper member.
   The Rotliegend sediments were derived from the Variscan mountain chain and the London-Brabant Massif situated SE respectively
SW of the depositional basin. Deposition took place under arid to
semi-arid conditions. The Slochteren sandstones are of fluviatile
(wadi), aeolian and sebkha facies; shale and evaporites were deposited in a desert lake in the central part of the basin.

1. INTRODUCTION

   The discovery well Slochteren 1 of the Groningen gasfield,
drilled by the Nederlandse Aardolie Maatschappij B.V. in 1959,
proved the importance of the Rotliegend as a major reservoir.
Since then the Rotliegend sandstone has been subjected to an
intense exploration and evaluation program by a great number of
oil companies in the Netherlands, Germany and the United Kingdom,

H. Falke (ed.), The Continental Permian in Central, West, and South Europe, 23-37. All Rights Reserved.
Copyright © 1976 by D. Reidel Publishing Company, Dordrecht-Holland.

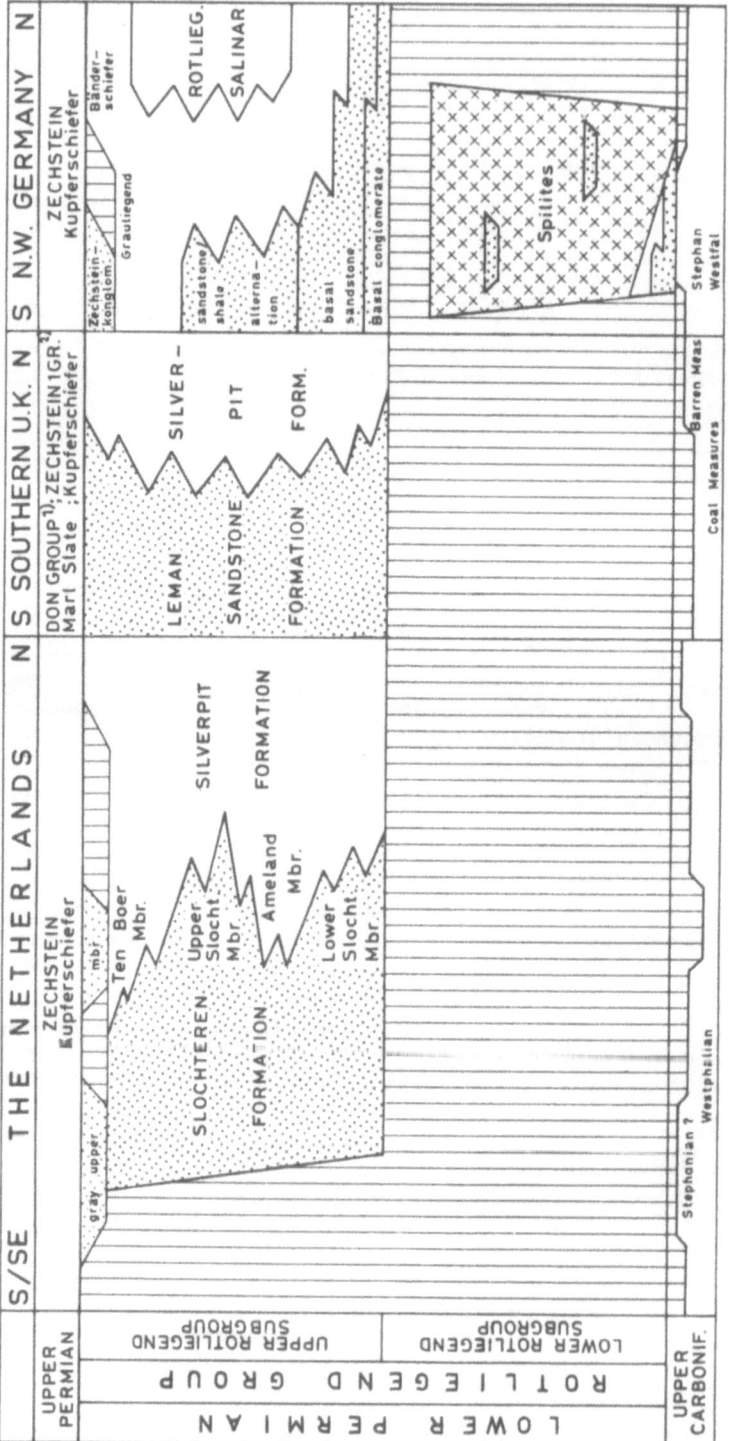

Table I. Correlation diagram of the Rotliegend of the Netherlands and adjacent areas.
(for the U.K. column: 1) after Smith et al., 1974; 2) after Rhys, 1974)
coarse-grained sediments; x x x volcanics

onshore as well as offshore. For competitive reasons only part of
the vast amount of information obtained from this exploration
effort has been published to date.

With the valued permission of the majority of the oil companies operating on the Netherlands onshore to disclose part of
the older exploration data, the Netherlands Geological Survey is
able to contribute to the picture which shows the development of
the Rotliegend in this country. Special thanks for giving permission to publish the GR-log and a lithological interpretation
of the Rotliegend are due to Petroland B.V. for data of the
L/7-1 well, and to Drs. J.A. van Hoeflaken who gave permission
to use data of the L/10-5 well from a forthcoming comprehensive
publication on the L/10 - gasfield of Placid International Oil Ltd.
comprising additional details on the L/10-5 well.

2. ROTLIEGEND STRATIGRAPHY

2.1. Definition of lower and upper boundaries

The Rotliegend comprises a sequence of predominantly reddish,
continental, siliciclastic and evaporitic rocks. It overlies more
or less disconformably the Upper Carboniferous, of which the surface ranges in age from Westphalian A to Stephanian (van Wijhe
& Bless, 1974).

The Carboniferous can be distinguished from the Rotliegend
by its dark gray colours (in its upper part weathered to purplish) and its greater degree of compaction. Furthermore, as de
Booy (1968) had already observed, the sandstones in the higher
part of the Upper Carboniferous contain a remarkable amount of
mica, a mineral rare in the Rotliegend sandstones.

The Rotliegend is conformably or paraconformably overlain
by the marine Zechstein. One must, however, realize that:
a. the marine Zechstein transgression has reworked the upper part
of the Rotliegend sands (Glennie, 1972).
b. locally the Zechstein transgression has deposited a relatively
thin, basal sandstone or conglomerate bed, known in Germany as
the "Zechsteinkonglomerat", containing reworked Carboniferous
and Rotliegend pebbles, among which ventifacts occur (Fabian &
Müller, 1962).
c. a marine influence in the highest part of the Rotliegend has
to be accounted for, as has been proved in Germany (e.g. Plumhoff,
1966).

Colour is not in all cases conclusive to distinguish the
continental Rotliegend from the marine Zechstein, because locally
part of the Rotliegend sandstones are not red or pink but yellowish or greyish, and the upper most part may have been bleached
due to percolation of reducing formation water from the overlying
Zechstein.

This means that in some areas it can be difficult to establish the exact boundary between the continental and the marine
strata. Falke (1972, p.45) mentions the same kind of problems
in Germany. Hence the base of the Zechstein has been defined in

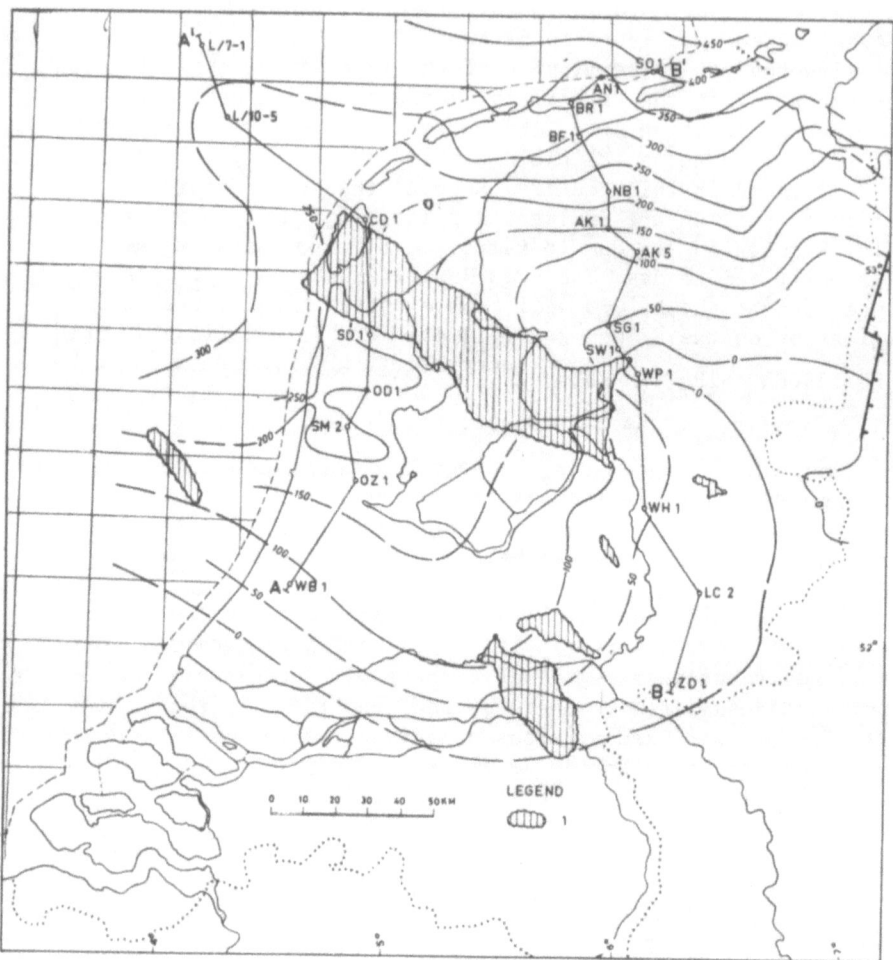

Fig. 1. Isopach map of the Upper Rotliegend Subgroup. Complemen-
ted with data from Heybroek (1974) and Lutz et al. (1975).

Contour interval = 50 m
1   areas where the Rotliegend has been truncated by later erosion
A———A' line of section of Fig. 3
B———B' line of section of Fig. 4
For the legend to the abbreviations of the names of the wells,
see Table II.

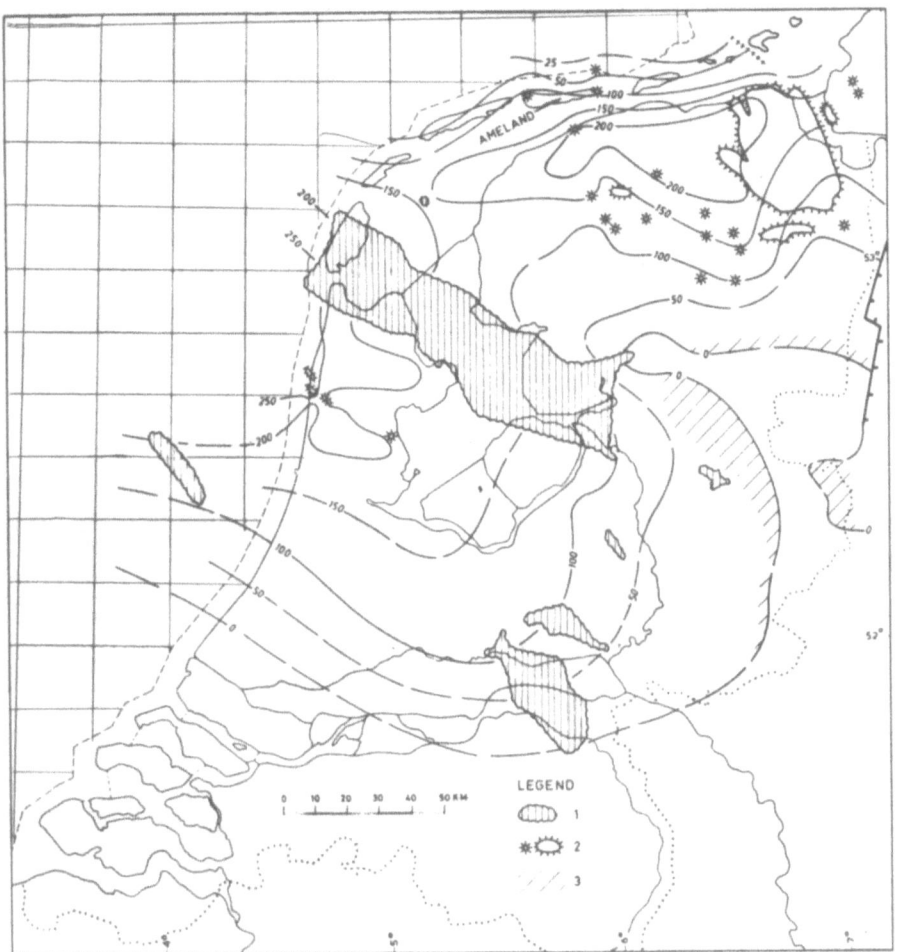

Fig. 2. Sandstone/conglomerate isolith map of the Slochteren
Formation.

Contour interval = 50 m
1 areas where the Rotliegend has been truncated by later erosion
2 onshore (incl. territorial waters) gas wells and fields productive from the Slochteren Formation.
3 main distribution of the gray upper member in the central-eastern part of the Netherlands

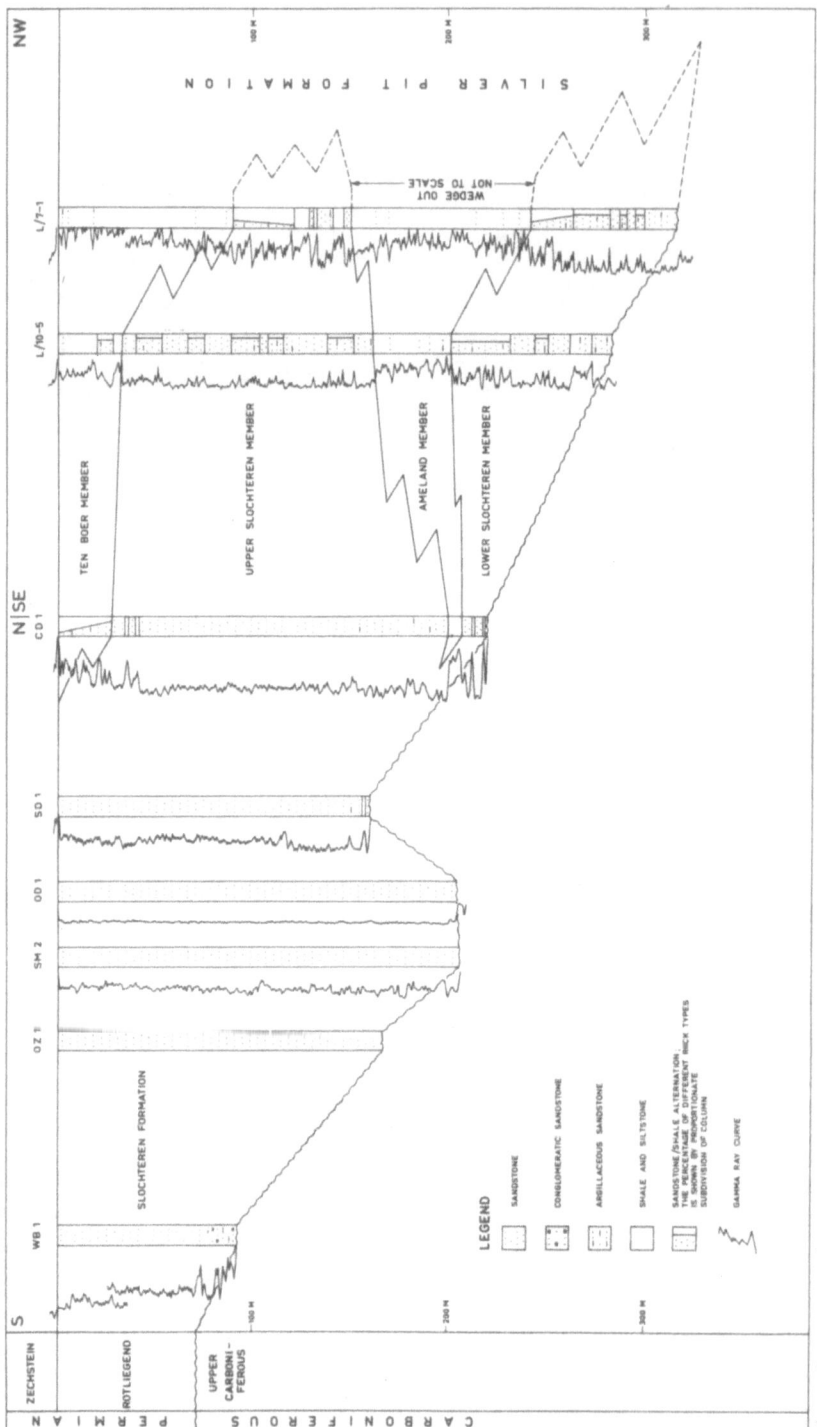

Fig. 3. Stratigraphic section of the Rotliegend from Woubrugge 1 to L/7-1.
For location of the section, see Fig. 1.

this paper, for practical reasons, by the base of the Kupferschiefer (Coppershale), a readily recognizable, wide-spread
marker bed. It is a quarter to one meter thick bed of shale,
giving an unmistakably high reading on the gamma-ray log.

In areas where a thin marine sandstone or conglomerate
underlying the Kupferschiefer can be distinguished with a fair
degree of confidence, that unit has been classified as a separate
informal member of the Rotliegend, although genetically it belongs
to the basal part of the transgressive Zechstein sequence.

2.2. Lithostratigraphy

The sequence of rocks delineated in the previous section
is considered as a group and is called the Rotliegend group, following the recommendation of Rhys (1974).

Volcanic and volcano-clastic strata present in NW Germany,
and which have been assigned to the Lower Rotliegend, have never
been encountered in the Netherlands (van Amerom, 1972; Lutz et al.,
1975). They appear to be restricted to synsedimentary fault controlled depressions. It is suggested here to place these deposits
in a Lower Rotliegend Subgroup. The Rotliegend sequence overlying
these volcano-clastic deposits with a slight unconformity, which
is attributed to the Saalian phase (Falke, 1972), in consequence
constitutes the Upper Rotliegend Subgroup.

The correlation between Germany and the Netherlands, pointed
out by Brouwer (1972), indicates that the Rotliegend in the Netherlands has to be assigned to the Upper Rotliegend Subgroup. Its
distribution and thickness is shown on Fig. 1.

This isopach map shows that the Rotliegend is absent in the
southern and south-eastern part of the Netherlands. The zero
contour represents approximately the depositional edge. The thickness increases progressively from the south to the north and northwest, interrupted by a WNW striking zone of lesser thickness in
the northern half of the country. Areas where the Rotliegend has
been truncated by later erosion (mainly Late Kimmerian) are marked
with a vertical hatching.

Inspecting the Upper Rotliegend Subgroup one distinguishes
two rock bodies, being a sandstone/conglomerate body in the south
and a shale/siltstone/evaporite body in the north (see Table 1 and
the sections on Figs. 3 and 4). The sandstone/conglomerate body
is called the Slochteren Formation, upranking its position from
a member, as originally defined by Stäuble and Milius (1970) in
the Groningen area, to a formation. The conglomerate and conglomeratic sandstone form only a subordinate part of the Slochteren
Formation. The distribution of this formation can be seen on the
sandstone/conglomerate isolith map depicted on Fig. 2. It shows
that south of a west-east oriented line, running midway the
northern half of the country, the Rotliegend consists almost exclusively of sandstone with some conglomerate. The main sandstone
development occurs in the NW mainland and in the NE provinces.

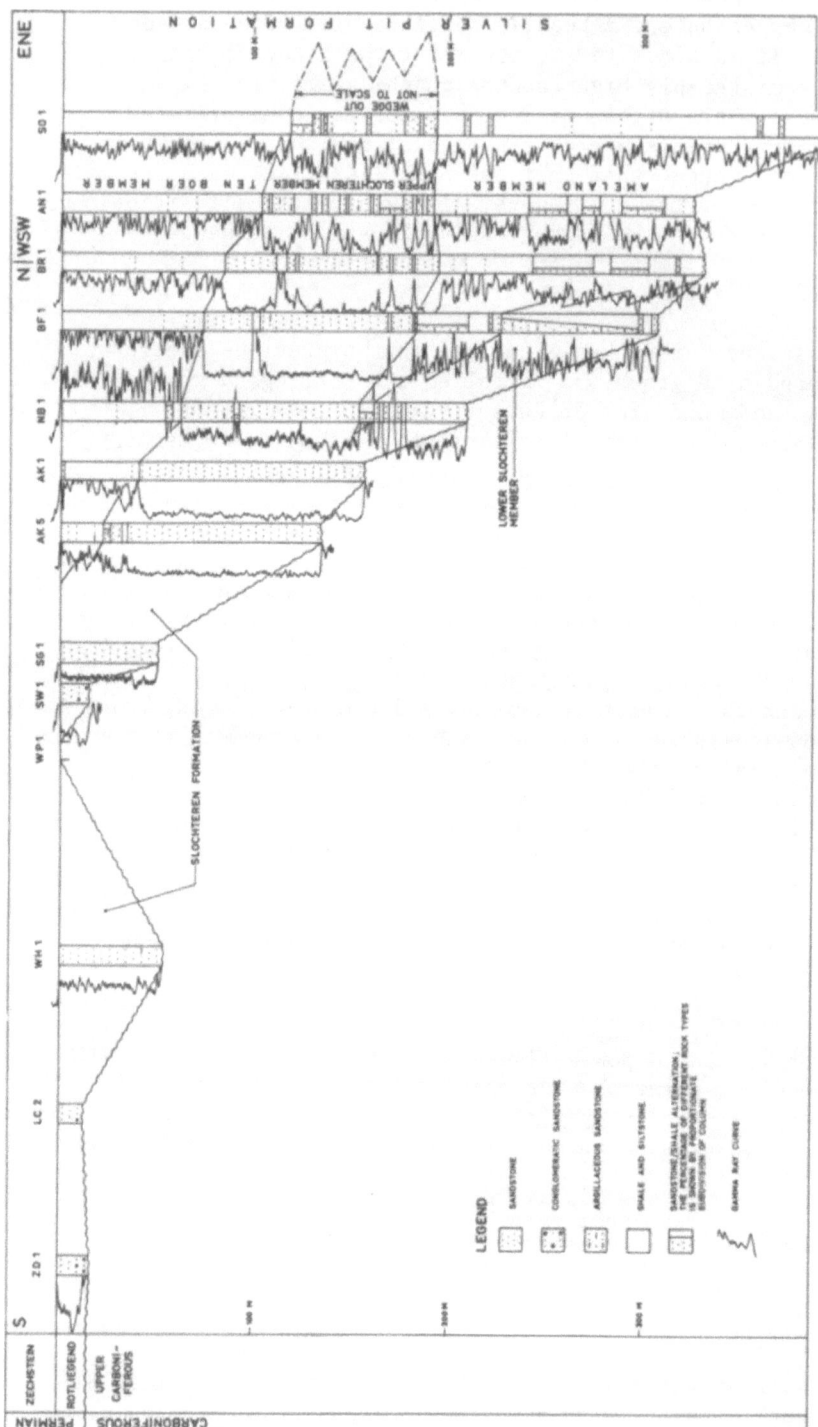

Fig. 4, Stratigraphic section of the Rotliegend from Zeddam 1 to Schiermonnikoog-Zee 1. For location of the section, see Fig. 1,

Just north of the mainland the sandstone thickness decreases rapidly. Onshore gasfields productive from the Slochteren Formation are indicated on the map.

The fine-clastic/evaporite unit corresponds with the Silverpit Formation as defined by Rhys (1974). As Trusheim (1971) has pointed out, the shale/evaporite sequence in the Rotliegend should not be called "Haselgebirge", because this is a name for a tectonically induced salt/clay/sulfate breccia.

The transition between the Slochteren and the Silverpit Formations occurs in a E-W trending zone, crossing the northernmost part of the Netherlands. Here one can distinguish four lithological units, of which two are sandy and two are shaly. They are classified as members of the above-mentioned formations (table 1, and figs. 3 and 4). From the top to bottom they are:

4. an upper shale/siltstone member, called the Ten Boer Member (Stäuble & Milius, 1970);
3. an upper sandstone member, called the Upper Slochteren Member;
2. a lower shale/siltstone member, called the Ameland Member;
1. a lower sandstone member, called the Lower Slochteren Member.

abbreviation	well	company
AK 1	Oudega-Akkrum 1	Chevron
AK 5	Akkrum 5	Chevron
AN 1	Ameland-North 1	Mobil
BF 1	Blija-Ferwerderadeel 1	NAM
BR 1	Buren 1	BP/Gulf
CD 1	De Cocksdorp 1	Petroland
LC 2	Lochem 2	NAM
NB 1	Noord-Bergum 1	NAM
OD 1	Obdam 1	BP/Gulf/Mobil
OZ 1	Oostzaan 1	NAM
SD 1	Slootdorp 1	Petroland
SG 1	Sonnega-Weststelling-werf 1	NAM
SM 2	Schermer 2	Amoco
SO 1	Schiermonnikoog-Zee 1	Chevron
SW 1	Steenwijkerwold 1	NAM
WB 1	Woubrugge 1	NAM
WH 1	Wijhe 1	NAM
WP 1	Wanneperveen 1	NAM
ZD 1	Zeddam 1	NAM
	L/7-1	Petroland
	L/10-5	Placid

Table II. Legend to abbreviations of wells on Figs. 1 to 4.

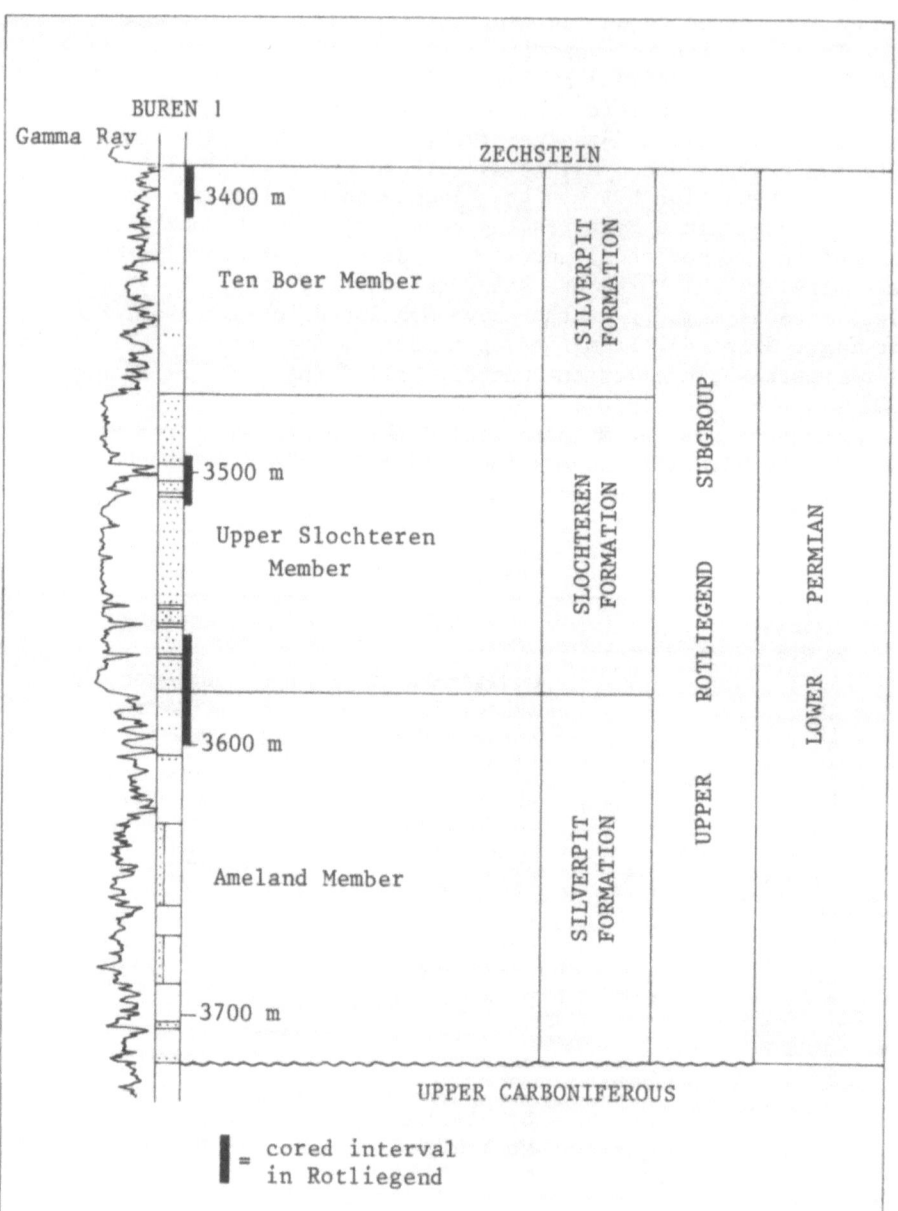

Fig. 5. Type section of the Ameland Member in the well Buren 1
For legend, see Fig. 3.   (BP/Gulf).

Because of the intertonguing of the Slochteren and Silverpit Formations and the shaling out of the sandstones towards the
north, the formation and member boundaries are often somewhat
arbitrary.

The Ameland Member of the Silverpit Formation, dividing the
Slochteren Formation into two parts, is called after the island
of Ameland along the northern coast of the Netherlands.

As type section the well Buren 1 has been chosen, drilled by
B.P./Gulf on this island in 1964. The gamma-ray log of the type
section of the Ameland Member, interval 3581.5 - 3716.5 m, is
shown in Fig. 5. Cores covering this interval range from 3581.5
to 3603 m. Slabs of these cores are present in the core house of
the Netherlands Geological Survey. The top of the member is defined by the occurrence of a predominantly red, shale/siltstone
sequence underlying the massive sandstones of the Upper Slochteren Member. The base is defined by occurrence of dark grey to
purplish, fine clastics of the Upper Carboniferous. The sediments
of the Ameland Member consist of red to red brown shales and siltstones, often sandy, with thin sandstone beds intercalated. Small
disseminated anhydrite nodules are common.

To the south and the west of the type area the Ameland Member
is not directly underlain by the Carboniferous, but by the well
developed sandstones of the Lower Slochteren Member (e.g. in
Blija-Ferwerderadeel 1 and L/10-5 on Fig. 3 and 4). The Upper
Slochteren Member reaches further to the north than the Lower
Slochteren Member in the Wadden Island area. In the NW offshore
area it is the other way round.

A separate member of the Slochteren Formation, which can be
distinguished locally, is the thin marine sandstone and conglomerate bed underlying the Zechstein Kupferschiefer, previously mentioned in chapter 2.1. and provisionally called the gray upper member.

This till now informal member does not correlate with the
German Grauliegend, which consists of bleached Rotliegend deposits
(Falke, 1972, p.54), but with the Zechsteinkonglomerat (see Brueren,
1959 and Fabian & Müller, 1962). It has been decided not to use
this German name for this member because:
a. in the classification presented in this paper it is confusing
to have a member in the Rotliegend with the name of the overlying
stratigraphic unit;
b. this member is not always developed as a conglomerate, as it
may consist of a sandstone or a conglomeratic sandstone. It mainly occurs in the central-eastern part of the country. Further
study is necessary to define its mineralogical composition, reworked elements, and exact geographical distribution.

A schematic view of the stratigraphic classification as outlined in this paper is presented on Table 1. A tentative correlation with the United Kingdom offshore Rotliegend and the NW
German sequence is incorporated.

Following Smith et al. (1974), the NW European Rotliegend

(incl. the Zechsteinkonglomerat) has provisionally been placed in
the Lower Permian, pending further correlations with the Russian
type sections of the Permian. Up to now no palaeontologic, paly-
nologic or radiometric age determinations of the Upper Rotliegend
Subgroup in the Netherlands are available. In view of this, and
of the observations of Visscher et al. (1974) no attempt has been
made to assign the Upper Rotliegend to a chronostratigraphic
stage (formerly it used to be assigned to the Saxonian).

2.3. Depositional environment

     Deposition of the Rotliegend in N.W. Europe took place under
arid to semi-arid conditions. As Glennie (1972) has demonstrated
clearly, many similarities with sedimentation in a modern desert
environment can be observed.

     The Rotliegend sediments were derived from the Variscan moun-
tain chain in the SE and the London-Brabant Massif in the SW, and
were transported by ephemeral streams to the north, forming large
alluvial fans comprising ill-sorted sandstones, conglomerates and
minor shale intercalations (wadi deposits). A large desert lake
with a fluctuating water level existed in the centre of the depositional basin, with its axis situated to the north of the Netherlands in the North Sea between Yorkshire and central-north Germany,
Anhydritic clays and several rock salt sequences were deposited
in the central part of the basin (Trusheim, 1971; Rhys, 1974).

     The prevailing E to NE trade winds during the Permian gave
rise to the formation of extensive dune complexes of sand, blown
out of the wadi deposits. The aeolian sandstones occur in the
areas between alluvial fans and are partly interbedded with wadi
and interdune sebkha deposits. Sebkha deposits marginal to the
desert lake were laid down in the transitional zone between this
lake, and the wadi and dune area. They consist of fine grained,
anhydritic sandstones and siltstones, intertonguing with basinal
shales, and are rapidly pinching out towards the north. The general facies distribution is shown in Fig. 6 (from Lutz et al.,
1975).

     Blanche (1973) and Marie (1975) give a more detailed descrip-
tion of the Rotliegend sandstones in the U.K. sector of the North
Sea, with special reference to porosity and permeability, that
applies to the Netherlands area as well.

     The aeolian sandstones show the best reservoir properties.
The porosity and permeability of the other types of sandstone
are reduced by their primary clay content, and early carbonate
and anhydrite cements. Deep subsequent burial of the Rotliegend
sandstone resulted in intergranular growth of authigenic clay
minerals (Stalder, 1973).

     The formation of fibrous illite, apart from other diagenetic
processes, decreases the permeability considerably, increasing
the total specific surface of the rock and consequently the water
saturation to such an extent that any gas present in the rock

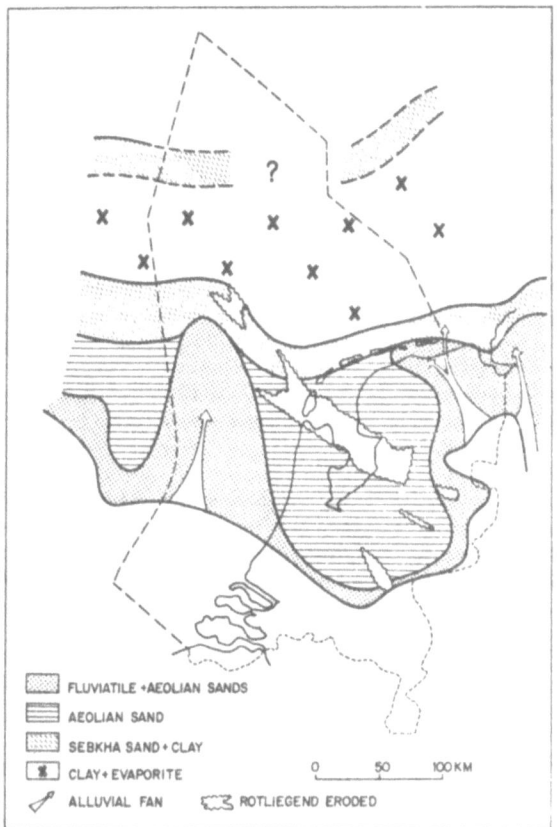

Fig.6. Facies distribution of the Rotliegend in the Netherlands
and the Netherlands part of the North Sea.
After Lutz et al. (1975). Published with permission of Applied
Science Publishers Ltd.

may not be producible at economic rates. The maximum depth of
burial of a producible Rotliegend reservoir appears to lie at
about 3500 m. However, in case of an early filling of the reservoir with hydrocarbons or the presence of a seal preventing the
escape of formation fluids this depth might be as great as 5500 m
(Lutz et al., 1975).

ACKNOWLEDGEMENTS

The author is indebted to the Director of the Geological
Survey for allowing the publication of this paper. He gratefully
acknowledges his colleagues who assisted with valuable advice and
critical reading of the manuscript.

REFERENCES

Amerom, H.W.J. van, 1972 - Das karbonische Alter der Tiefbohrung
    Wanneperveen 1, eine Revision. Geologie en Mijnbouw., vol.
    51, pp.491-495.

Blanche, J.B., 1973 - The Rotliegendes Sandstone Formation of the
    United Kingdom sector of the Southern North Sea Basin. Trans.
    Instn. Min. Metall. (Sect. B. Appl. Earth Sci.), vol.82,
    pp.B85-89.

Booy, T. de, 1968 - Mineral assemblages in Permo-Carboniferous
    sediments in the eastern Netherlands. Verh. Kon. Ned. Geol.
    Mijnbouwk. Genootsch., geol. serie, vol.25, pp.21-23.

Brouwer, G.C., 1972 - The Rotliegend in the Netherlands. In:
    Rotliegend, Essays on European Lower Permian, ed. by H. Falke.
    Brill, Leiden, pp.34-42.

Brueren, J.W.R., 1959 - The stratigraphy of the Upper Permian
    "Zechstein" formation in the eastern Netherlands,
    Atti del Convegno di Milano su I Giacimenti Gassiferi dell'
    Europa Occidentale, Accademia Nazionale dei Lincei, Roma,
    vol.I, pp.243-274.

Fabian, H.J. & G. Müller, 1962 - Zur Petrographie und Alterstellung
    präsalinarer Sedimente zwischen der mitteren Weser und Ems.
    Forschr. Geol. Rheinld u. Westf., vol.3, part 3, pp.1115-1140.

Falke, H., 1972 - The continental Permian in North- and South
    Germany. In: Rotliegend, Essays on European Lower Permian,
    ed. by H. Falke. Brill, Leiden, pp.43-113.

Glennie, K.W., 1972 - Permian Rotliegendes of NW Europe interpre-
    ted in the light of modern desert sedimentation studies.
    Amer. Ass. Petr. Geol. Bull., vol.56, pp.1048-1071.

Heybroek, P., 1974 - Explanation to tectonic maps of the Netherlands.
    Geologie en Mijnb., vol.53, pp.43-50.

Lutz, M., J.P.H. Kaasschieter & D.H. van Wijhe, 1975 - Geological
    factors controlling Rotliegend gas accumulations in the Mid-
    European basin. Proc. 9th World Petroleum Congress. Preprint
    Panel Disc. 2, 14 pp. Applied Science Publ., Essex,

Marie, J.P.P., 1975 - Rotliegend stratigraphy and diagenesis. In:
    Petroleum and the Continental Shelf of NW Europe, vol. 1:
    Geology (A.W. Woodland, ed.). Applied Science Publ., Essex,
    pp. 205-210.
Plumhoff, F., 1966 - Marines Oberrotliegendes (Perm) im Zentrum
    des nordwestdeutschen Rotliegendbekkens. Neue Beweise und
    Folgerungen. Erdöl und Kohle, vol.19, pp.713-720.

Rhys, G.H., 1974 (compiler) - A proposed standard lithostratigra-
    phic nomenclature for the southern North Sea and an outline
    structural nomenclature for the whole of the (UK) North Sea.

Rep. Inst. Geol. Sci., No. 74/8, 14pp.

Smith, D.B., R.G.W. Brunstrom, P.I. Manning, S. Simpson & F.W. Shotton, 1974 - A correlation of Permian rocks in the British Isles. Journ. geol. Soc. London, vol.130, Special Rep. No. 5, 45pp.

Stalder, P.J., 1973 - Influence of crystallographic habit and aggregate structure of authigenic clay minerals on sandstone permeability. Geologie en Mijnb. vol.52, pp.217-220.

Stäuble, A.J. & G. Milius, 1970 - Geology of Groningen gasfield, Netherlands. Amer. Ass. Petr. Geol. Mem. 14, pp.359-369.

Trusheim, F., 1971 - Zur Bildung der Salzlager im Rotliegenden und Mesozoicum Mitteleuropas. Beih. geol. Jb., vol.112, 51pp.

Visscher, H., M.G. Huddleston Slater-Offerhaus & T.E. Wong, 1974 - Palynological assemblages from "Saxonian" deposits of the Saar-Nahe Basin (Germany) and the Dôme de Barrot (France) - An approach to chronostratigraphy. Review Palaeobot. and Palynol., vol.17, pp.39-56.

PROBLEMS OF THE CONTINENTAL PERMIAN IN THE FEDERAL
REPUBLIC OF GERMANY

H. Falke

Geologisches Institut, Johannes Gutenberg-
Universität, 65 Mainz, W. Germany

ABSTRACT. The occurrences of Continental Permian in
the Federal Republic of Germany are generally well
understood but some problems remain. These are outlined
and discussed in the present paper.

INTRODUCTION

        The continental Permian of the Federal Republic of
Germany can only be discussed in connection with the
adjacent countries to the East and West. The present
writer distinguishes two main areas for the distribution
of the continental Permian of central and western
Europe (Falke 1972):
1. A Northern or Subvariscan province which extends
from Great Britain to Poland and includes the North Sea
area, the Netherlands and the North German plain. This
was called the Central European depression by Katzung
(1972). The predominent tectonical structures show this
province to be trending NW-SE. It is crossed by a
Rhenish structural trend (NNE-SSW) in the area of the
Ems and Weser (Fig. 1). It contains the northern part
of the FRG.
2. A Southern or Variscan province extending from France
to western Poland, and which contains the southern part
of the FRG. Its areas of deposition of continental Permian show NE-SW aligned troughs, corresponding to
Variscan structural trends.

        These Permian sediments are called "Rotliegend",

H. Falke (ed.), The Continental Permian in Central, West, and South Europe, 38-52. All Rights Reserved.
Copyright © 1976 by D. Reidel Publishing Company, Dordrecht-Holland.

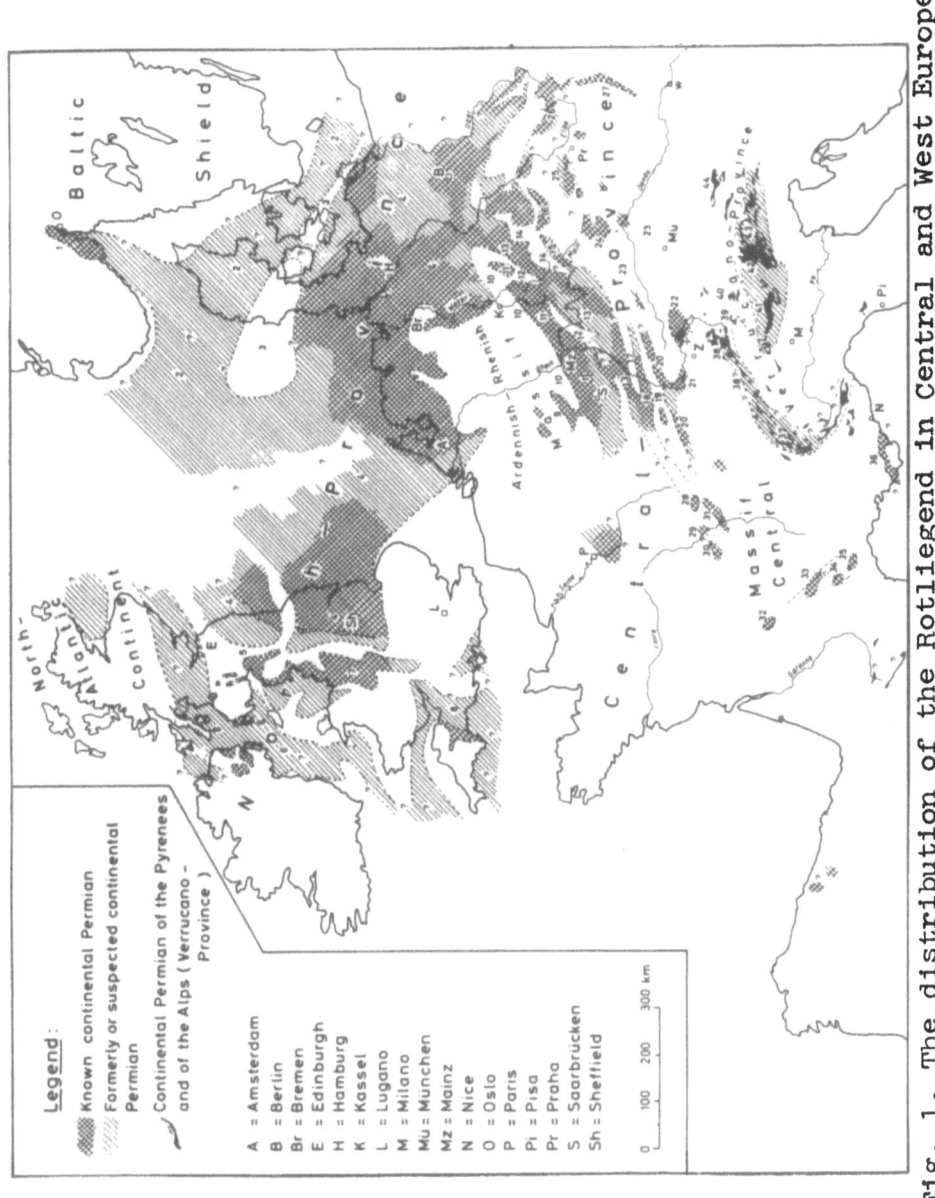

Fig. 1. The distribution of the Rotliegend in Central and West Europe.

a term derived from the word "Totliegendes" which was
used by the miners in central Germany for the red and
barren deposits underlying the copper-bearing marls of
the "Zechstein" (marine Upper Permian) which was and
is exploited in this region.

THE FIXING OF THE LOWER AND UPPER BOUNDARIES OF THE "ROTLIEGEND"

In both provinces mentioned a sharp contact is
found with the underlying beds where the "Rotliegend"
is unconformable (with a marked stratigraphic break)
to Carboniferous and Lower Palaeozoic rocks. In all
other cases the lower boundary is more or less difficult to establish exactly. In the Ems Basin of the
northern province a continuous sedimentation exists
between the Stephanian and Lower "Rotliegend". Index
fossils are lacking and there are no marked changes in
sedimentary conditions apart from a higher content of
micas in the Lower Permian (Hedemann 1972). Consequently it is not so easy to draw the boundary although in
part of this region the presence of a conglomerate
provides a conventional base to the Rotliegend. Also
in the Saar-Nahe region of the southern or Variscan
province a conglomerate, the so-called "Dirminger"
Conglomerate, is used for establishing the boundary
with the underlying Stephanian C-deposits. The position
of this conglomerate is determined by the first occurrence of the Callipteris conferta in the overlying strata.
Callipteris conferta and Callipteris naumanni are Permian markers according to a recommendation of the
Heerlen Congress of 1927, but there is still some controversy as to whether the occurrences of Callipteris
or the abundant appearance of Callipterids is to be
regarded as indicative of an early Permian age. For
example, Doubinger and Bouroz (1974) place the Igornay
Beds of the Autun Basin, the type-locality for the
"Autunian" = Lower "Rotliegend", in their Stephanian D,
despite the presence in these beds of rare represen-
tations of Callipteris conferta and Callipteris nicklesi.
In their opinion it is not the first appearance, but
the abundant occurrence of these plants which is decisive for the drawing of the boundary. It should be
added here that Bouroz and Doubinger rely mainly on the
distribution of microfossils for the recognition of the
Lower Autunian boundary. In the Saar-Nahe trough the
Dirming Conglomerate mentioned above may not always
occur at one and the same stratigraphic level. According to Drozdzewski (1969) between this conglomerat

and the Stephanian/Rotliegend boundary there may be a
distance up to 90 m. The behaviour of this horizon in
time and space must be further elucidated by future
investigations. The microflora in this area, as studied
by Helby (1966) and Reimann (1973), shows the importance of the disappearance of Lycospora and Thymospora
and the abundant occurrence of saccate forms such as
Illinites unicus and Cheileidonites in the vicinity of
Lower Rotliegend boundary. However, the microfloral
boundary lies at a lower level than that established
by macrofloral remains, the first mentioned being at the
horizon of the Breitenbach coal seam.In consequence of
these differences it should be clarified which boundary
should be preferred either the micro- or macroflora boun-
dary.As a result of these uncertainties it is necessary
to examine yet again the boundary between Carboniferous
and Permian in the northern Schwarzwald (Black Forest)
and in the Naab trough (Fig. 1).

At the top of the "Rotliegend" the boundary is
easily drawn where sediments of marine "Zechstein" -
"Thuringian" = Upper Permian are found as overlying
deposits. This is the case in Northern Germany and in
the north-eastern part of southern Germany. It is particularly easy where the Kupferschiefer (copper shale)
can be used as a boundary horizon. In this case the
underlying "Weissliegend", a bleached sandstone of
partly aeolian origin and consequently also the Corn-
berg sandstone (both in the north-eastern part of South
Germany) still belong to the Lower Permian. However,
the origin and stratigraphic position of the latter
are always still in discussion as the equivalent sediments in central Germany. Recent investigations place
the "Weissliegend" in East Hessen into the uppermost
Lower Permian (Schäfer 1969). In this connection the
tetrapod-tracks used by Haubold and Katzung (1972, 1975)
have recently furnished an important contribution to
the age determination. One must also mention here that
some authors have supposed the presence of a stratigraphic gap between the "Zechstein" and the "Rotliegend".
This gap certainly exists where the Zechstein is missing,
as in the Saar-Nahe region, the Baden-Baden trough and
the Naab trough; all areas where the Bunter sandstone
of Triassic age is the immediately overlying deposit.
Here an important question exists with regard to the
length of the time gap and whether the youngest strata
of the terrestrial Permian facies may be a time equivalent of the Zechstein. This has in fact been suggested for the Kreuznach beds in the Nahe trough (Falke
1966). According to Visscher (1973) this gap is very

extensive because he still puts the higher section of
the Rotliegend of the Saar-Nahe region into the
Autunian that is the lower section of the Lower Permian
named to the type-locality of the Autun Basin in France.
Further investigations will be necessary to establish
whether or not the highest Rotliegend is the terrestrial
equivalent of the marine Zechstein. No fossils of "Zechstein" age have been found as yet in the disputed sediments, and only the kind of deposits has been used as
an argument for a Zechstein equivalent. As such have
been used occurrences of dolomite or carnelian dolomite
at certain horizons, traces of evaporites and the
existence of aeolian sands in form of dunes as for
example the Kreuznach beds (Falke 1972). However, these
arguments can only be used with reservation because
they may be identified in the highest Rotliegend of
Hessen just below the overlying marine Zechstein
(Kowalcyk and Prüfert 1974). Therefore, it would be of
decisive importance if certain plants could be found
e.g. <u>Ullmannia frumentaria</u>. In this respect it should
be noted that assemblage of fossils of Lower and Upper
Permian ages might be found in the uppermost sediments
of the Upper Rotliegend near Lieth in Schleswig Holstein
(Backhaus 1964). Furthermore, there have been observed fo-
raminifers in the shales of the salt sequence of the
northern province which can be regarded as a precursor
of the Zechstein sea (Plumhoff 1966). Therefore, it
would be necessary to draw the boundary with the first
occurrence of these fossils. At present the boundary
is placed below the "Zechsteinkonglomerat" which underlies the "Kupferschiefer", a most important marker
horizon near of the base of the "Zechstein".

THE PROBLEMS OF THE SUBDIVISION OF THE ROTLIEGEND

In the Federal Republic of Germany the Rotliegend
is subdivided into the Lower and Upper Rotliegend a
subdivision which equates approximately to the division
of the continental Permian in France into Autunian and
Saxonian. This subdivision is aided by the Saalian unconformity. However, more than one unconformity occurs
at its locus typicus, the "Halle"-Porphyry complex in
central Germany (von Hoyningen-Huene 1960 a, 1960 b,
Katzung 1968). Several unconformities also exist in
the Thuringian Forest (Lützner 1960, Andreas, Enderlein,
Michael 1966). Therefore, it is better to speak of
Saalian movements than of a Saalian phase when local
unconformities are found. The Saalian phase proper is
used for drawing the boundary between Lower and Upper

Rotliegend. Obviously, it reflects only one of several movements and it is very much in order in each particular case to state whether the unconformity is that corresponding to the Saalian phase or merely one of the Saalian movements. Apart from this it can be missing as for example in the Saar-Nahe Basin. In this respect the results of the investigations by Haubold and Katzung (1972, 1975) are of interest. They propose the Thuringian Forest as the type locality for the base of the Upper Rotliegend because the unconformity here coincides with a change in the kind of tetrapod tracks. They maintain that the various parts of the Rotliegend can be distinguished on the basis of tetrapod tracks. Apart from the limited nature of these palaeontological arguments it seems necessary to do additional work on the tetrapod tracks before one can rely on them as index-fossils.

In the northern part of FRG i.e. in the Subvariscan province, analogous conditions obtain to those in central Germany, viz. an unconformity is found as the upper limit of a volcanic sequence which mainly corresponds to the Lower Rotliegend. This unconformity is regarded tentatively as representing the Saalian phase. Summarizing, it can be said that the determination of this boundary by the Saalian phase is fraught with difficulties, particularly in the Saar-Nahe Basin where the only unconformities present are local in occurrence and linked to the updoming of porphyry massifs in this area. In this respect it must be mentioned the fault zone on the southeast side of the Hunsrück within the Nahe Syncline. It is trending NE-SW and after recent geophysical measurements it has still removed the "Moho" level. Present researches have shown that repeated movements have taken place: 1. during the Upper Carboniferous time because of Stephanian sediments which break off on this fault zone as demonstrated by boreholes, 2. on the boundary between Lower and Upper Rotliegend, proved by basic intrusions which are bound on this fault zone and have originated during the "Grenzlager"-volcanism, and 3. after the sedimentation of the Upper Rotliegend because its sediments are still removed. It is very important to clear up the further relations in order to determine too if the Saar-Nahe Basin is a "Graben" or not.

Apart from this they are faults probably older than the Upper Rotliegend but these indications have to be investigated more exactly in the future. In view of these difficulties the change in the sedimentation

pattern is used for drawing the boundary between Lower
and Upper Rotliegend in the Saar-Nahe Basin. This
change is influenced by the climate and the relief.
The climate became progressively drier as is indicated
by the occurrences of fanglomerates, frequently poorly
sorted sediments in some alluvial fans, the presence
of evaporites (especially in the northern province)
and by dunes. The tectonic movements which took place
during this time resulted in a rejuvenation of the topographic relief which is reflected in the formation of
coarse-grained sediments near the base of the Upper
Rotliegend. These sedimentary characteristics are in
evidence in all the basins of both provinces. However,
it should be investigated whether this change in sedimentation pattern occurred at the same time in the
single areas of deposition. As already knwon but pointed
out especially by Visscher (1974) there is no difference
between the plantfossils which have been found very
occasionally in a few Upper Rotliegend localities such
as near Sobernheim in the Nahe Syncline (Geib 1950,
Visscher 1974) and those of the Lower Rotliegend. Therefore, Visscher places the Upper Rotliegend of the Saar-
Nahe region in the Autunian, which with help of the
Saalian phase is regarded as corresponding to the Lower
Rotliegend whilst the Saxonian is generally equated to
the Upper Rotliegend. However, the identity of floral
remains for Lower and Upper Rotliegend is a well known
fact and these plants have consequently never been used
for determining the boundary between these two stages.
After investigations of tetrapod tracks found near
Martinstein and Nierstein in the Upper Rotliegend sediments Haubold and Katzung (1975) put the lower part of
the Wadern Group of the Saar-Nahe Basin in the Autunian
and consequently also the underlying Grenzlager Group,
which is composed mainly of volcanic rocks with a few
intercalations of tuffs and other sediments. This decision must again be tested, but it is attractive since
it would remove the apparent anomaly that the Saar-Nahe
region is the only Rotliegend basin where the main
volcanic activity corresponds to the Upper Rotliegend.
Apart from the tetrapod tracks mentioned, additional
information will be required in order to solve this
problem. One may point out here that the rhyolite
intrusions of the Kreuznach Massif (Lorenz 1973) belong
to the Grenzlager volcanism that forms the base of the
Upper Rotliegend. According to the radiometric dating
by Orliac and Schäfer (1969) it has been shown that
they have an age between 272 and 278 mill. years. This
would mean that these massifs originated in early Rotliegend time. Since these volcanic rocks penetrated

Lower Rotliegend sediments, one is forced to conclude
that the latter formed during Carboniferous times. This
result would be in accordance with the opinion of
Visscher (1972) who placed part of the Autunian in the
Carboniferous, but contradicts not only the first
appearance of Callipteris but also the entire floral
development of Lower Rotliegend resp. Autunian times.
Further investigations are clearly necessary. A formal
decision would be desirable on the preferred use of
either Autunian or Lower Rotliegend and Saxonian or
Upper Rotliegend as internationally recognized stages.

An attempt has been made to subdivide the Lower Rotliegend in the Halle-trough with the aid of plant fossils,
and to recognize an Autunian$_1$ and Autunian$_2$ (Remy and
Kampe 1962). In the Thuringian Forest a subdivision into
Upper and Lower Autunian was proposed by Haubold and
Katzung (1975) on the basis of tuffs, tetrapod tracks
and associated macrofloral remains including Walchia
germanica and, particularly, Callipteris scheibei. With
regard to the plants this subdivision is in accordance
with that given by Doubinger (1974) for some French
occurrences. The absence of adequate investigations
such a subdivision is not yet applicable in southern
Germany. It would be an important contribution for the
urgently required correlation of the sequences of the
sediments in the single basins. Such palaeontological
researches would be of great importance for the Saar-
Nahe Basin because they could help to decide whether
the Lower Rotliegend succession in this basin is suitable for stratotypic purposes. It is known that the
type Autunian at Autun is developed rather differently
from the Autunian in other basins, thus invalidating
this conceptual type area. This question may be discussed further in the next chapter.

SOME PROBLEMS REGARDING THE ORIGIN OF THE DEPOSITS

With few exceptions the Rotliegend of the northern
province is only known from boreholes. The corresponding data are generally secret. However, some of the
logs are occasionally released and the present proceedings contain two papers which complete the evidence
available thus far and which prove, once again, that
the Upper Rotliegend sediments were deposited under
semi-arid to arid conditions. It seems that no typical Lower Rotliegend facies has been found in the western section
of this large basin. As far as is known to the present

writer, the German part of this Subvariscan province
only yielded a sequence of grey to varicoloured sediments with Autunian in the Lippspringe borehole
(Fabian 1974). East of the Ems Basin the main part of
the Lower Rotliegend consists of predominantly volcanic
rocks with basic or alternating basic and acid material
(e.g. in the eastern part of this depression, at Flech-
tinger Höhenzug, Mecklenburg where intercalated tuffs
and sediments also occur). Here they attein a thickness of more than 1000 m. The "diabases" have been
frequently affected by spilitisation (Eckhardt 1968).
This metamorphism has not yet been interpreted satisfactorily, i.e. if Eckhardt also believes that a K or
Na supply is not necessary in the present case.

After the Saalian unconformity comes the Upper
Rotliegend. It consists of an alternation of conglomerates, reddish brown sand-, silt- and claystones. Southeast of Hamburg and extending northwestwards into the
North Sea a salt sequence is intercalated. According
to Trusheim (1971), this sequence consists of 4 saltbearing sections separated by thick mudstone successions.
Thinner mudstones also occur interbedded with the salt
layers. In contrast to a normal salt deposit of marine
origin no carbonate precipitation seems to have taken
place and anhydrite precipitation was rare. Therefore,
extraordinary conditions occured and these have not
yet been explained satisfactorily. Especially the
question of the origin of the salt solution has not
yet been solved. Different opinions exist about the
origin of the salt. Brunstrom & Walmsley (1969) "pre-
fer the theory that sodium chloride free from other
salts was derived from subaerial erosion of older evaporites, especially emergent salt plugs". In this
connection, according to Trusheim (1971), two possibilities exist for the origin of the salt. It may have come
from the South by squeezing out of salt layers within
the Rhenish Schiefergebirge as a result of the Variscan
folding. However, no indication has been found of formerly saltbearing deposits in these Devonian sediments.
Alternatively, he supposed that the salt-solution might
have come from the North by "suberosion" of diapires.
No such diapires have been found in this area. Apart
from these hypotheses it is possible that the salt-solution originated from the weathering of minerals,
particularly feldspars, in the sediments deposited in
this area. It might have been transported by groundwater into the depressed basin centre where the salt
was then precipitated by an evaporation under a semi-
arid to arid climate. According to Richter-Bernburg

(1953), a basin of predominantly continental character
existed into which the seawater could only restrict
invade and its content of $CaCO_3$ and $CaCO_4$ was almost
precipitated before in other places within the satu-
ration-shelve. The boreholes have not yet proved this
area, and it can therefore not be decided if this
opinion may be right. In any case marine intercalations
are present in the uppermost series of the Upper Rotliegend as demonstrated by Plumhoff (1966) and Backhaus
(1964). These marine invasions might be also interpreted as a possible indication that this depression in
which the saltlake originated was located at greater
depth than sea-level so that an inundation could easily
take place. It is expected that boreholes in the North-
Sea may contribute to a solution of this problem.

In the Variscan or Central province the many problems exist as well. As in many basins the sediments in
the Saar-Nahe trough show a cyclic sedimentation which
starts with coarse-grained, red deposits and ends with
grey fine-grained strata. It is not yet clear which
causes are responsible for the origin of these cycles
i.e. whether they are climatic or tectonic. It is
especially interesting that the coarse-grained layers
are combined with red coloured deposits which consist
predominantly of shales (resp. mudstones) which secon-
darily reddened, at least in part, the intercalated
sandstones. Walker (1967 a, b, 1973), could demonstrate
that the red colour of some sediments is of secondary
origin. It would be very desirable to investigate this
in the Saar-Nahe basin, in order to determine the
primary or secondary origin of the red beds.

Work should also be done on the single processes
of sedimentation. It is quite probable that many of the
sediments are of deltaic origin, although present day
analogues are unknown. Such studies would furnish an
essential contribution to the correlation of the sediments. Negendank (1972) has described turbidities, but
it would be important to obtain further evidence. Such
investigations are of great importance in order to
establish the environment as is proved by the paper
of Stapf (1972) who has described the distribution and
the composition of the carbonate rocks in the Lower
Rotliegend of the Saar-Nahe region. These carbonate
rocks, occurring generally at the end of a cycle are
sometimes together with a coal seam. All these data
would help in distinguishing the various facies within
a basin and in correlating between basins. Naturally,
the fossil contents have to be taken into consideration

in order to provide further possibilities for a comparison between basins. There are still not enough indicators to compare for example the succession of the Saar-Nahe Basin with those of the Naab-Trough or the Thuringian Forest. This results may sound pessimistic because they prove that the present difficulties of correlation are eventually unsurmountable as the existing studies have already gone a long way towards elucidating the processes of sedimentation.

In the Upper Rotliegend, e.g. in the Saar-Nahe Basin, it is especially important to define exactly the sedimentation patterns in the fanglomerates which here occur at the base of this sequence as a result of the rejuvenation of the relief. The sedimentological features in single outcrops seem to indicate that the clasts are deposited under different conditions regarding the quantity of water influencing the sedimentation. Similar investigations must be realized in the overlying beds which become increasingly finer towards the top of the sequence. These sediments show in some outcrops that they are often poorly sorted and sometimes laminated. The dolomite horizons at the top of the Rotliegend also have to be mentioned. These can no longer be considered as possible indicators for a "Zechstein" age of these beds since Lietz (1967) demonstrated dolomite banks in the strata immediately underlying the Zechstein of Hesse. Results of investigations about these dolomite and carnelian-dolomite horizons of this section will probably be published in the near future which also inform on the origin of this deposits.

Summarizing, it may be stressed again that a number of Rotliegend problems of the Federal Republic of Germany still remain to be solved. The solution of these problems is also of importance outside this area.

REFERENCES

ANDREAS, D., ENDERLEIN, Fr., MICHAEL, J. (1966): Zur Entwicklung des Rotliegenden im Thüringer Wald auf Grund neuer Kartierungsergebnisse.- Ber.deutsch.Ges.geol.Wiss., A., Geol.Paläont., 11, p. 119-130, Berlin.
ANDREAS, D., HAUBOLD, H. (1973): Erste Information über die Richtgrenze Unteres/Oberes Autun (Unteres Perm, Unterrotliegend) im Niveau der Goldlauterer Schichten des Thüringer Waldes.-

Z.geol.Wiss.Berlin, 1, p.509-514, Berlin.
ANDREAS, D., HAUBOLD, H., KATZUNG, G. (1975 a): Zur
    Grenze Stefan/Autun (Karbon/Perm).-Z.geol.
    Wiss.Berlin, 3, p.699-716, Berlin.
ANDREAS, D., HAUBOLD, H. (Berlin 1975): Die biostrati-
    graphische Untergliederung des Autun (Unteres
    Perm) im mittleren Thüringer Wald - DDR.-
    Schriftenr.geol.Wiss, Berlin, 3, p. 4-86,
    Berlin.
BACKHAUS, E. (1964): Die Lamellibranchiaten aus den
    roten permischen Schichten von Lieth bei
    Elmshorn (Holstein).- Geol.Jb., 82, p.103-
    130, Hannover.
BOUROZ, A. et DOUBINGER, J. (1974): Les relations entre
    Stéphanien supérieure et l'Autunien d'après
    le contenu de leur stratotype.- C.R.Acad.Sc.
    Paris, 279, p.1745-1748, Paris.
BRUNSTROM, R.G.W. & WALMSLEY, P.J. (1969): Permian eva-
    porites in North Sea basin.- Bull.Amer.Ass.
    Petrol.Geol., 53, p. 870-883, Tulsa.
DOUBINGER, J. (1974): Etudes palynologiques dans l'Au-
    tunien.- Review of Palaeobotany and Palyno-
    logy, 17, p. 21-38, Amsterdam.
DROZDZEWSKI, G. (1969): Sedimentation und Struktur des
    nordöstlichen Saarbeckens.- Oberrhein.geol.
    Abh., 18, p. 77-117, Karlsruhe.
ECKHARDT, F.-J. (1968):-Vorkommen und Petrogenese
    spilitisierter Diabase im Rotliegenden im
    Weser-Ems Gebiet.- Geol.Jb., 85, p. 227-264,
    Hannover.
FALKE, H. (1966): Zur Geochemie der Schichten der
    Kreuznacher Gruppe im Saar-Nahegebiet.-
    Geol.Rdsch., 55, p. 59-77, Stuttgart.
FALKE, H. (1972): The continental Permian in North and
    South Germany.- in Falke: "Rotliegend. Essays
    on European Lower Permian." Inter.Sediment.
    Petrograph.Ser., Vol.XV, p. 43-113, Brill,
    Leiden.
GEIB, K.W. (1950): Über eine pflanzenführende Schicht-
    folge in den Waderner Schichten des ro bei
    Sobernheim im Nahebergland.- Notizbl.hess.
    L.Amt Bodenforsch., VI, p. 193-200., Wiesbaden
HAUBOLD, H., KATZUNG, G. (1972): Die Abgrenzung des
    Saxon..- Geol., 21, p. 849 - 863, Berlin.

HAUBOLD, H., KATZUNG, G.(1972a): Das Typusgebiet der
    Autun/Saxon Grenze im Thüringer Wald.- Ber.
    deutsch.geol.Wiss., A, Geol.Paläont., 17,
    p.883 -910, Berlin.
HAUBOLD, H., KATZUNG, G. (1975): Die Position der

Autun/Saxon-Grenze (Unteres Perm) in Europa und Nordamerika.- Schriftenr.geol.Wiss.Berlin, 3, p. 87-138, Berlin.

HEDEMANN, H.A., FABIAN, H.J., FIEBIG, H., RABITZ, A. (1972): Einführung in die Geologie des Gastlandes. 1. Das Karbon in marin-paralischer Entwicklung.- C.R. 7.Intern.Congrès Intern. Stratigraph.Géol.Carbonifère, 1, p. 29-47, Krefeld.

HELBY, R. (1966): Sporologische Untersuchungen an der Karbon/Perm Grenze im Pfälzer Bergland.- Fortschr.Rheinl.u.Westfl., 13, p.645-704, Krefeld.

VON HOYNINGEN-HUENE, E. (1960a): Jungpaläozoische Krustenbewegungen im östlichen Harzvorland.- Geol.Jb., 9, p.759-767, Berlin.

VON HOYNINGEN-HUENE, E. (1960 b): Das Permokarbon im östlichen Harzvorland.- Freib.Forschungsh., C, 93, p.1-116, Berlin.

KAMPE, A., REMY, W. (1962): Ausbildung und Abgrenzung des Stephanian in der Halleschen Mulde.- Mber.Deutsch.Akad.Wiss., Berlin, 4, p.54-68, Berlin,

KATZUNG, G. (1968): Perm.- im"Grundriss der Geologie der Deutschen Demokratischen Republik", Akademie- Verlag, p. 199-218, Berlin.

KATZUNG, G. (1972): Stratigraphie und Paläogeographie des Unterperm in Mitteleuropa.- Geol., 21, p. 570-584, Berlin.

KOWALCZYK, G., PRÜFERT, J. (1974): Gliederung und Fazies des Perms in der Wetterau (Hessen).- Z.D.Geol.Ges., 125, p. 61-90, Hannover.

LIETZ, J. (1967): Der Grenzbereich Rotliegendes - Zechstein im Gebiet des Spessart und der Wetterau in Paläogeographischer Sicht.- Jber.u. Mitt.Oberrh.Geol.Ver., N.F. 49, p. 129-146,Stgt.

LORENZ, V. (1973): Zur Altersfrage des Kreuznacher Rhyolithes unter besonderer Berücksichtigung der Stratigraphie und Überschiebungstektonik in seiner südlichen Umrandung (Saar-Nahe Gebiet, SW-Deutschland).- N.Jb.Geol.Paläont. Abh., 142, p.139-164, Stuttgart.

LÜTZNER, H. (1961): Saalische Bewegungen im Rotliegenden des mittleren Thüringer Waldes.- Geol. Rdsch., 51, p. 560-566, Stuttgart.

LÜTZNER, H. (1964): Die saalische Phase im Gebiet von Ilmenau (Thüringer Wald).- Abh.Deutsch.Akad. Wiss.Berlin, Jg. 1964, p. 281-308, Berlin.

LÜTZNER, H. (1972): Lithostratigraphie und Paläotektonik des Rotliegenden der Schleusinger Rand-

zone (Thüringer Wald).- Ber.deutsch.Ges.
geol.Wiss., A, Geol.Paläontol., 17, p. 811-
834, Berlin.
NEGENDANK, J.F. (1970): Sedimentologische Erscheinun-
gen aus dem Unterrotliegenden des Saar-Nahe-
gebietes.- Abh.hess.L.Amt Bodenforsch., 56,
Horst Falke Festschrift, p. 151-162,Wiesbaden.
NEGENDANK, J.F. (1972): Turbidite aus dem Unterrotlie-
genden des Saar-Nahe-Gebietes.- N.Jb.Geol.
Paläont., Mh., Jg. 72, p. 561 - 572, Stutt-
gart.
ORLIAC, J., SCHÄFER, K. (1969): Das Rubidium-Strontium-
Alter unterpermischer Rhyolithe des Saar-
Nahe Gebietes.- N.Jb.Geol.Paläont., Mh.,
p. 246-252, Stuttgart.
PLUMHOFF, Fr.(1966): Marines Oberrotliegendes (Perm)
im Zentrum des nordwestdeutschen Rotliegend-
becken.- Neue Beweise und Folgerungen.-
Erdöl und Kohle, 19, p.713-720, Hamburg.
REIMANN, K.U. (1973): Ergebnisse mikrosporologischer
Untersuchungen an der Wende Karbon/Perm in
der Saar-Nahe-Senke. Diss.Mainz.
REMY, W. (1962): Die Existenz des Stefan und die
Stefan-Rotliegend-Grenze in paläobotanischer
Sicht.- C.R. 4. Congrès Stratigraph. Geol.
Carbonifère, Heerlen 1958, p. 599-602, Heerlen.
REMY, W. (1964): Zur Untergliederung des Stéphanien
und Autunien Grenze Stéphanien/Autunien.-
C.R. 5. Congrès Stratigraph. Géol. Carbonifère,
Paris 1963, p. 177-187, Paris.
REMY, W., KAMPE, A. (1961): Ausbildung und Abgrenzung
des Autunien in der Halleschen Mulde.- Mber.
Deutsch.Akad.Wiss.Berlin, 3, p.394-408,
Berlin.
RICHTER-BERNBURG, G. (1953): Die paläogeographischen
Voraussetzungen für die Bildung nordwest-
deutscher Salzlager.- Jb.Geograph.Ges.Han-
nover f1953, p. 166-182, Hannover.
SCHÄFER, K. (1969): Das Rotliegende der Treischfelder
Bohrungen in Osthessen.- Notizbl.hess.L.Amt
Bodenforsch., 97, p.152-194, Wiesbaden,
STAPF, K. (1972): Lithologische Untersuchung der Alten-
glaner Schichten im saarpfälzischen Unterrot-
liegenden mit besonderer Berücksichtigung
der Karbonatgesteine.- Diss. Mainz.
TRUSHEIM, F. (1971): Zur Bildung der Salzlager im Rot-
liegenden und Mesozoikum Mitteleuropas.-
Beihefte Geol.Jb., 112, p. 1-51, Hannover.
VISSCHER, H. (1973): The Upper Permian of western Europe.-
A palynological approach to chronostratigraphy.-

in: A. Logan and L.V.Hills "The Permian and Triassic Systems and their mutual boundary". Mem.Can.Soc.Petrol.Geol., 2, p. 200-219.

VISSCHER, H., HUDDLESTON SLATER-OFFERHAUS, Marieke G., and WONG, T.E. (1974): Palynological assemblages from "Saxonian" deposits of the Saar-Nahe basin (Germany) and the Dome de Barrot (France).- An approach to chronostratigraphy.- Review Palaeobotany and Palynology, 17, p.39-56, Amsterdam.

WALKER, Th.R. (1967): Formation of Red Beds in Modern an Ancient Deserts.- Geol.Soc.America Bull., 78, p. 353-368.

WALKER, Th.R. (1974): Formation of Red beds in Moist Tropical Climates: a Hypothesis.- Geol.Soc. America Bull., 85, 633-638.

PERMIAN BASINS IN THE BOHEMIAN MASSIF

Vlastimil M. Holub

Geological Survey Prague,Praha 1,
Malostranské nám. 19, ČSSR

ABSTRACT. A review of the stratigraphy, facies, palaeogeography, and tectonic development of the Permian of
the Bohemian Massif is given.

The Permian rocks of Bohemia and Moravia are of
terrestrial origin and belong to the Variscan (Central
European) province according to the division of H. Falke. Regionally they are represented in the following
areas: 1/ in the Sudetic region, 2/ in the Central and
western Bohemian area, and 3/ in the Boskovice and Blanice furrows.

The Carboniferous/Permian boundary lies partly in a
continuous sequence, but in some areas a sedimentary
break is developed.

The palaeogeography is defined as the share of sediments of the different sedimentary environments on the
composition of the sedimentary fill.

The character of the sedimentary area and of its
fill was mainly influenced by diastrophism and climate.
The filling of the basins had a cyclic character.
Diastrophic movements were connected with the final
evolution of the Variscan orogeny. Volcanic activity
was controlled also by cyclicity, and is manifested by
the alternation of basic to intermediate and acid volcanism. The Late Palaeozoic volcanism culminated in the
Autunian period, but in that time also was terminated.
The volcanic centres are associated with the zones of

H. Falke (ed.), The Continental Permian in Central, West, and South Europe, 53-79. All Rights Reserved.
Copyright © 1976 by D. Reidel Publishing Company, Dordrecht-Holland.

intense Variscan tectogenesis. The rate of sedimentation
was controlled by contemporaneous subsidence.

INTRODUCTION

The Permian rocks of Bohemia and Moravia represent
molassoid fills of the intermontane depressions of the
Bohemian Massif and are only of continental origin.
According to the division of Falke,1972, they belong
to the Variscan (Central European) province. They
originated in the postgeosynclinal phase of the Variscan
orogenic cycle. The development of these basins suggests
a gradual cratonization of the individual parts of the
Bohemian Massif. In the Autunian, the filling of the
depressions was terminated; the Saxonian and Thuringian
sediments are already of the platform type.

The contents of the basins mainly consist of clastic
sediments representing the periodic transport from the
denuded part of the massif. They are characterized by a
cyclic (rhytmic) sedimentation pattern caused by diastrophic and climatic influences. The sedimentation pattern
and their content of coal seams reflect the gradual humidity decrease of the climate precipitation intervals of
which continuously diminished and were less efective
during the Stephanian, and especially during the Permian.
During periods with greater humidity sediments, pyroclastics and coal seams were deposited, while during arid
periods of the Stephanian reddish-brown and red sediments
were formed (a variegated facies including the origin of
sediments of red-bed type). The Permian humid time-spans
were short, and the sediments formed during those times
occupied restricted local areas only and consist of beds
with bituminous marlstones and limestones. The climate
was not consistently arid, however, since even during
the Permian some small coal seams were formed too.

The most abundant Permian rocks in the Bohemian
Massif are clastic sediments - siltstones, claystones,
sandstones, arkosic sandstones and conglomerates; carbonate rocks of chemogenic and organogenic origin (limestones and dolomites) are of small extent, but are
characteristic of the sequence. Rare, but distinctive
of the Autunian, are caustobioliths, chiefly bituminous
pelites and scarcely also coal. Bituminous coal seams
are confined to the Upper Stephanian and the oldest lower Autunian and are of no economic value, except for
the deposits in the Boskovice Furrow. Volcanic rocks are
represented either by basic and intermediate types

(melaphyres), or by acid palaeorhyolites and their tuffs.

The lower Autunian sediments are represented in all Bohemian Permian basins. In the Stephanian C and lower Autunian periods the area of sedimentation had the largest extention of the all continental Late Palaeozoic sedimentation covers of the Bohemian Massif. During the upper Autunian, Saxonian and Thuringian the sedimentation area was reduced.

In this article, a review of the geology of the Permian basins of the Bohemian Massif is given. The main attention is given to palaeogeographical and tectogenetical developments of the sedimentary area, which were the main objects of our investigations in the last years. A detailed geographical and stratigraphical description of individual basins is given by Holub,1972, 1973 and 1976.

REGIONAL DIVISIONS OF BASINS AND OCCURRENCES

The Upper Stephanian and Permian sedimentary areas originated mostly in the north-eastern part of the massif, but they occur also in its northern, north-western and eastern parts (see Fig.1), with the exception of the Blanice and Boskovice furrows which extend into its southern core area.

Only a minor part of the sediments crops out since they are generally covered by younger formations, especially those of Upper Cretaceous and Tertiary ages. The areas of sedimentation are shaped either as isometric basins or as elongate depressions, the so called furrows. The latter are very narrow basins of sedimentation striking NNE-SSW.

With regard to the regional distribution, lithofacies and stratigraphy, the Stephanian C-Permian basins can be divided into the following three groups (see Fig.1):

I.Permo-Carboniferous basins in the Sudetic region in the north-eastern part of the Bohemian Massif (Fig.1, hatching 2) with Upper Stephanian and Permian deposits are as follows:

1. Intra-Sudetic Basin (=Lower Silesian B.) (Fig.1, 2/12) is an intermontane depression between the cristal-

line massifs of the Krkonoše Mts., the Orlické hory Mts.,
the Bystřické hory Mts. and Góry Sowie Mts. This depression desplays the greatest completeness of the stratigraphic column of the Late Palaeozoic basins, since its deposits range from the Dinantian through the Upper Carboniferous and Permian to the Triassic.

2. Krkonoše Piedmont Basin (Fig.1,2-11) originated
at the southern edge of the Krkonoše Mts. In the post-
Saalian times the new basin (the Trutnov-Náchod B.) came
into being between the Krkonoše and Orlické hory Mts.
crystalline masses. A geological cross-section through
the western part of the Krkonoše Piedmont Basin is shown
by Fig.2.

3.The Stephanian and Autunian sediments of the Krkonoše Piedmont Basin continue below deposits of the Cretaceous cover westwards across the Lusatian Fault up to the
crystalline elevation of Maršovice-Bezděz filling the
Mnichovo Hradiště Basin(Fig. 1, 2-10,Fig.2,Fig.3).

4. To this group also belong Permian occurrences in
the Orlické hory Mts. and on their slopes and at their
foot up to the Malonín uplift, on the Hořice and Zvičina
Ridges, occurrences near the Lusatian Fault - at Hodko-
vice nad Mohelkou and near Krásná Lípa.

II. Late Palaeozoic of the Central and western Bohemian
area (see Fig.1, hatching 1) containing Upper Stephanian
and lower Autunian sediments in the uppermost part of the
stratigraphic column only, from the structural point of
view, include: 1/ Plzeň B.(Fig.1,1/1), 2/Manětín B.(Fig.
1,1/2),3/Radnice B.(Fig. 1,1/3), 4/Žihle B.(Fig.1,1/4),
5/Rakovník B.(Fig.1,1/5),6/Kladno B. (Fig.1,1/6). These
basins as well as other small occurrences rest unconformably on the folded early Palaeozoic and Proterozoic
rocks of the Teplá-Barrandian region and also on granitoid
intrusives, and constitute present denudation relics,
where locally Stephanian and Autunian rocks are eroded.
They are partly or entirely devoid of a cover of younger
sediments.

7/ Roudnice B. (Fig.1,1/7), 8/ occurrence between
Děčín and Česká Lípa (Fig.1,1/8) and 9/ Mšeno B. (Fig.
1,1/9,Fig.3)containing sediments of geologically proved
Autunian age, are overlain by younger formations underlying the Bohemian Cretaceous Basin.

Minor occurrences in the Železné hory Hills can be

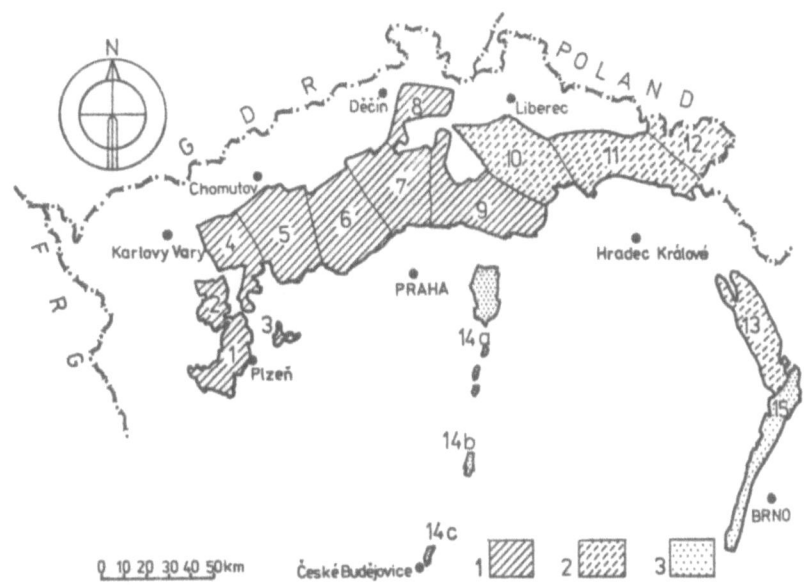

Fig. 1. Permo-Carboniferous continental basins in the Bohemian Massif.

1- Central Bohemian area : 1-1 Plzeň Basin, 1-2 Manětín B., 1-3 Radnice B., 1-4 Žihle B., 1-5 Rakovník B., 1-6 Kladno B., 1-7 Roudnice B., 1-8 occurrence between Děčín and Česká Lípa, 1-9 Mšeno B.

2- Sudetic area : 2-10 Mnichovo Hradiště B., 2-11 Krkonoše Piedmont B., 2-12 Intra-Sudetic B., 2-13 Orlické hory Piedmont B.

3- Furrows :14- Blanice Furrow : 14a-occurrence at Český Brod and Vlašim, 14b- occurrence at Tábor, 14c-occurrence at České Budějovice;
15- Boskovice Furrow.

added to this group.

The occurrences of the Permo-Carboniferous in the Krušné hory Mts. are assigned to this group too. In the Krušné hory Mts. region a number of Permian occurrences can be found; on the territory of Bohemia there is a small Permian basin near Brandov. In this region there are many bodies of volcanics of Permo-Carboniferous age, the most important of which is the Teplice Porphyry.

III. The third group includes the"furrows" (Furchen).
In Czechoslovakia, the term is aplied to markedly elongate, narrow and tectonically predisposed depressions
trending with the meridians. The best known furrows of
the Bohemian Massif are the Blanice (Fig.1,14), and
Boskovice (Fig.1,15) furrows. They belong to the structural systems trending N-S through the European epi-Varis-
can platform. They came into existence after the "Asturian phase" of the Variscan folding.

STRATIGRAPHY AND PALAEOGEOGRAPHICAL RECONSTRUCTION

Stratigraphical review and C/P boundary

        Permo-Carboniferous basins in the Sudetic area
(see Tab.1, left column) display the most complete
stratigraphical sequence of all the continental Late
Palaeozoic basins in the Bohemian Massif, where, besides
others, Stephanian, Autunian, Saxonian and Thuringian
deposits are represented. Some sedimentary breaks, which
were affected by posthumous movements of the Variscan
orogeny, appear in the basinal fill. In the Central and
western Bohemian area (see Tab. 1, middle column) the
Upper Stephanian (palaeontologically documented) and
lower Autunian (without marked organic remains) are
represented. In the Krušné hory Mts. the red-bed sequence without palaeontological records, considered as lower
Autunian and overlapping with disconformity Westphalian
sediments, is developed. The fillings of the furrows
(see Tab.1, right column) stratigraphically correspond
to Upper Stephanian and lower and upper Autunian.

        The dating and biostratigraphy of the Late Palaeo-
zoic continental basins of the Bohemian Massif rely
almost exclusively on the study of fossil floras which
are richly represented in the coal-bearing formations.
Faunal remains are sparce and have not been studied
systematically until recently. An exception are some
animal groups as e.g. Insecta and Stegocephala; but
these are not very significant from the stratigraphical
point of view. In addition fishes, crustaceans, amphibians and lamellibranchs have been found. Some lithostratigraphic units are not palaeontologically documented. The stratigraphy of these sediments is based on
marker bands which facilitate the correlation of the

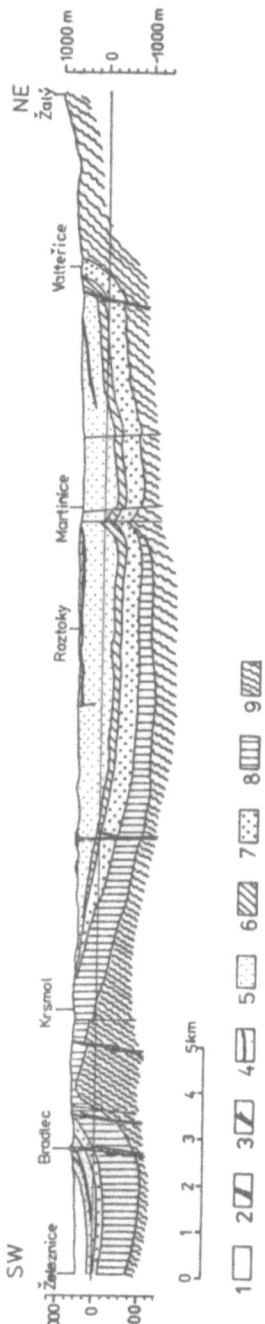

Fig.2. Geological cross-section through the western part of the Krkonoše Piedmont Basin (from SW to NE between Železnice near Jičín and Valteřice near Vrchlabí)

1–Upper Cretaceous (Turonian and Cenomanian); 2–Tertiary basalts; 3–Late Palaeozoic melaphyres; 4–Kalná Horizon of the Libštát Formation (grey and variegated marlstones, limestones, locally bituminous shales) – Permian, lower Autunian; 5– Libštát Formation (brown-red to brown siltstones,mud-stones, sandstones, rarely arkoses; layers of pyroclastics) – Permian,lower Autunian; 6–Rudník Horizon of the Libštát Formation (grey and green-grey siltstones, mudstones, bituminous shales) – Permian, lower Autunian; 7– Semily Member (polymictic conglomerates, sandstones, rarely brown and grey siltstones, mudstones) – Stephanian; 8 – Stupná Member (arkoses,ra-rely conglomerates, red-brown siltstones, mudstones, pyroclastics) – Stephan-ian; 9 – crystalline basement (phyllite, quarzites, limestones, dolomites, altered diabase and diabase pyroclasics, orthogneisses at Žalý Hill.

individual coal districts as well as that of the entire
basin; among these horizons tuffaceous layers are the
most important.

The <u>Stephanian and Autunian</u> have been joined (see
Figs. 3 and 7) because of the absence of a distinct bounda-
ry between the Stephanian and Autunian in some Permo-
Carboniferous basins of the Bohemian Massif.

Let us now have a look at the Carboniferous/Permian
boundary in the Bohemian Massif. There the boundary
sequences are present which are suitable for the study
of the C/P boundary problem. As a decisive factor the
first occurrence of the <u>Callipteris</u> genus representatives are considered. An accurate location of that boundary
into the succession is dependent on completeness of the
sequence and richness of the fossiliferous assemblages,
progress of geological and palaeontological research etc.
The most exact location of the C/P boundary can be done
in the basins, where this boundary coincides with the
occurrence of fossiliferous horizons. As far as lithology is concerned, a lithological change occurs generally
in the time span, which corresponds with the Stephanian C,
or with the Stephanian C/Autunian boundary. In this case
it is necessary to take into account, what principles
are considered; either the boundary is fixed at the base
of a lithological cycle, with the fossiliferous horizons bearing representatives of the genus <u>Callipteris</u>
in the accumulation part of the cycle, or the boundary
is put at the base of a fossiliferous horizon with representatives of callipterids (Boskovice Furrow, Krkonoše
Piedmont Basin).

However, as far as fixing of the C/P boundary is
concerned there are two types of basins in the Bohemian
Massif: 1/basins, where the C/P boundary is sharp and
quite clear. This type of boundary is put at the base
of the sedimentary cycle in which the representatives
of the genus <u>Callipteris</u> are known.These are usually
found in its upper part. This situation appears in the
Intra-Sudetic Basin, and in the Brandov Basin in the
Krušné hory Mts., where the Stephanian (Westphalian)
and Autunian deposits are separated by an outstanding
sedimentation break; in the Intra-Sudetic Basin the
entire Upper Stephanian, in the Brandov Basin the complete Stephanian and Westphalian C and D are not present.
2/ The second type is given by those basins, where the
C/P boundary is not sharply expressed in the geological
character of sediments (lithologically and tectogenetic-
ally), and where the continuous transition can be obser-

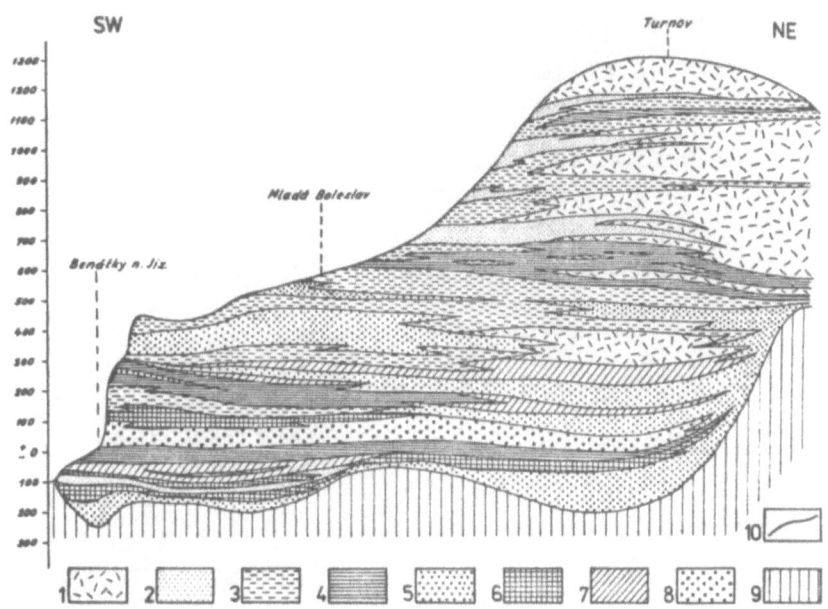

Fig. 3. Lithofacies section through the Mšeno and
         Mnichovo Hradiště basins

        1- volcanic facies; 2- alluvial-deltaic facies
(intermittent streams), 3- intermittent flood plain
facies; 4- persistent and intermittent lake facies
(mudstones, marlstones, siltstones, intercalations of
limestones, silicites and tuffogenous rocks); 5 - pro-
luvial facies (mostly stream bed sediments); 6- alluvial
facies (mostly river bed sediments); 7 - alluvial and
coal bearing peat - swamp facies; 8 - deltaic -lake
and alluvial plain facies; 9 - basement underlying the
Late Palaeozoic; 10 - pre-Cretaceous erosional boundary
of the Permo-Carboniferous.

ved. There the boundary-line between the Stephanian B
and C is rather clearly expressed, while C/P boundary
lies inside the Stephanian-Autunian sedimentary cycle
as far as lithological development is concerned. In this
case determination of the C/P boundary has been done on
the biostratigraphical basis only according to the first
occurrence of callipterids. This means that it has not
been fixed to the base of a lithological cycle. Therefore,
some differences originated, which are to be eliminated

in the future research.

The most complicated situation arose in the Central
Bohemian area.The C/P boundary occurs, according to up-
to-date criteria, inside the Upper Red Formation (the
Líně F.). The greater part of this formation belongs to
the Stephanian (no discoveries of callipterids). On the
basis of the geological correlation with basinal fills
of the Late Palaeozoic of the Sudetic area, it has been
proved, that the top part of the Líně Formation corresponds with deposits in the Sudetic area, which contain
fossiliferous horizons with callipterids, and therefore,
they are affiliated to the sequences of the Autunian age.

In the Krkonoše Piedmont Basin the C/P boundary
is placed at the bottom of the Rudník Horizon (Havlena,
1964), occurring at the base of the Libštát Formation;
the base of the respective sedimentary cycle, however,
lies several tens of metres below the base of the Libštát Formation.

In the Boskovice Furrow the C/P boundary occurs
inside of brownish-grey coal bearing sequence, where
the boundary is located at the bottom of a coal seam
(or at the base of its sedimentary cycle), in which -
-usually in the intercalations, or in its roof- the
representatives of Callipteris genus have been found.
In the Blanice Furrow the C/P boundary appears between
two horizons with thin coals, where the lower horizon
does not contain callipterids, which can be found, however, in the upper one.

The Upper Red Formation is the corresponding interval in Central and western Bohemia, and the Semily Member,
the Libštát Formation, and the Lomnice Member in the
Krkonoše Piedmont Basin. The Chvaleč and Broumov Formations in the Intra-Sudetic Basin and the whole fill of
the Boskovice and Blanice furrows also originated in
this time interval. We assume that sediments of this
span are present in the Orlické hory Mts. Piedmont Basin and also in the little-known territory along the
Lusatian Fault in the vicinity of the towns Děčín and
Česká Lípa. At the same time, the Permo-Carboniferous
deposits of the Železné hory Mts. and the upper part of
the sedimentary fill in the Brandov Basin were formed.
An isopach map of the Stephanian C - Autunian is presented by the Fig. 4.

Saxonian sediments are known from the basins of the
Sudetic region. They are represented by the Trutnov For-

Tab. 1. Stratigraphical review of the Permian (and of the Upper Stephanian and the lower Triassic) basins in the Bohemian Massif.

mation in the Intra-Sudetic Basin and in the Krkonoše
Piedmont Basin; their most complete development is found
in the Trutnov-Náchod Basin. This structure originated
after the Saalian movements in the eastern part of the
Krkonoše Piedmont area between the Krkonoše and the
Orlické hory Mts. Piedmont Basin and rarely occurs in
the Orlické hory Mts. Fig. 5 shows an isopach map of
the Saxonian.

   Thuringian sediments occur only in the Permian
basins of the Sudetic region; they occur in a small
area in the eastern part of the Krkonoše Piedmont Basin
and in the Intra-Sudetic Basin, where they are represented by the Bohuslavice Formation. Fig. 5 shows the
isopach map of the Thuringian. Table 1 presents the
stratigraphical review of the Upper Stephanian and
Autunian basins in the Bohemian Massif.

Palaeogeographical reconstruction

   Palaeogeography is defined as a share of sediments
having different sedimentary environments based on the
composition of the sedimentary fill. The Upper Stephanian and lower Autunian sediments were deposited mostly
under a semi-arid climate, the Saxonian and Thuringian
sediments originated under aridic climatical conditions.
The following types of lithofacies for the units deposited under these climatic conditions have been distinguished: a/ pediplain sediments and alluvial fans or alluvial
apron deposits are suggested for the marginal areas, b/
sediments of flood-plains, ephemeral and perennial lakes,
peats and swamps are distinguished for the central parts
of the sedimentary basins. On the palaeogeographical maps,
the extent of the recent occurrences of sediments, the
presumed area of sedimentation, and the share of sediments
deposited in the individual onvironments are presented.

   Palaeogeography of the Stephanian C - Autunian time
span in the Bohemian Massif is presented in Fig. 7. Palaeogeography of the Saxonian period is given by Fig. 8. Figure 9 shows palaeogeographical conditions of the Thuringian.

Stephanian C - Autunian period

   The Stephanian C - Autunian sediments were deposited
under a predominantly semi-arid climate. This climate
was interrupted by several periods of increased humidity.

Fig. 4. Isopach map of the Stephanian C - Autunian in the Bohemian Massif.

1- thickness in metres, 2- underlying rocks, 3-state frontier, 4- boundaries of the basins.

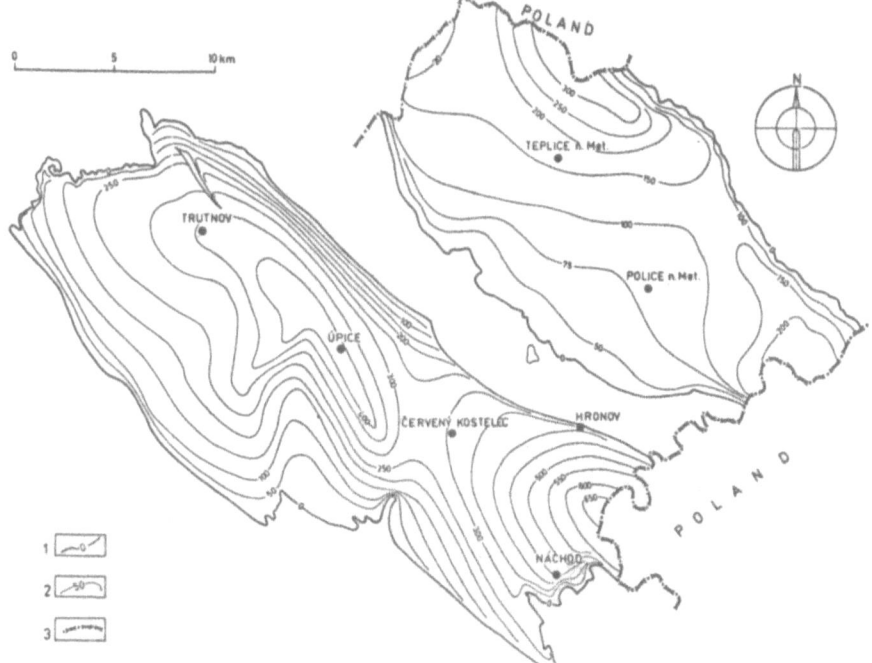

Fig. 5. Isopach map of the Trutnov Formation (Saxonian) in the Bohemian Massif.
1- boundary of the basin , 2-isopach line (thickness in metres), 3-state frontier.

For such climatic conditions the following sequence of
environments is suggested: pediplain, alluvial fans,
alluvial plain and ephemeral lakes. Only a few sediments
can be interpreted as deposits of perennial lakes and
swamps. Evidence was also found for aeolian sediments.
Drifts of silty material and even aeolian dunes probably
existed in the sedimentary area. Proof was found in the
basins of central Bohemia as well as in the Sudetic region. The originally aeolian material became mostly a
component of the water-laid sediments due to erosion and
reworking of the original accumulations.

It is expected that the Late Palaeozoic sedimentation in the Bohemian Massif reached its maximum in
this time span. We do not consider it as proved that the
central part of the Bohemian Massif began to arch at the
end of the Upper Carboniferous as some authors (e.g. Havlena,1964) have suggested. Therefore, no reason exists
to shift the southern boundary of the presumed sedimentary area northwards in comparison with the underlying
Stephanian B. This concept is supported by the fact that
the southern boundary of the Upper Red Formation has an
erosional character in central and western Bohemia. We
explain the increasing thickness of this formation northwards by the variable subsequent erosion. On the other
hand, the relatively great thickness of sediments in
tectonically depressed blocks, together with the absence
of marker horizons for the upper part of the Upper Red
Formation, indicate that the Manětín and Plzeň basins
had relatively high rates of sedimentation. The similar
development of the Upper Red Formation sediments in the
exposures along the southern margin of the Kladno- and
Rakovník basins and of those in the central parts is
additional evidence supporting this concept.

The Boskovice and Blanice furrows originated along
deep radial faults. We assume that an intensive synsedimentary subsidence was characteristic, especially in the
eastern parts of these furrows. Therefore, the eastern
margin was covered by alluvial fan sediments. The clastic material transported along the foot of fans was deposited as sandstones and mudstones in an alluvial and/or
lacustrine environment. The gently sloping western margins were covered by pediment deposits.

Saxonian period

The delimitation of the presumed sedimentary area
is rather difficult. Some data exist for determination

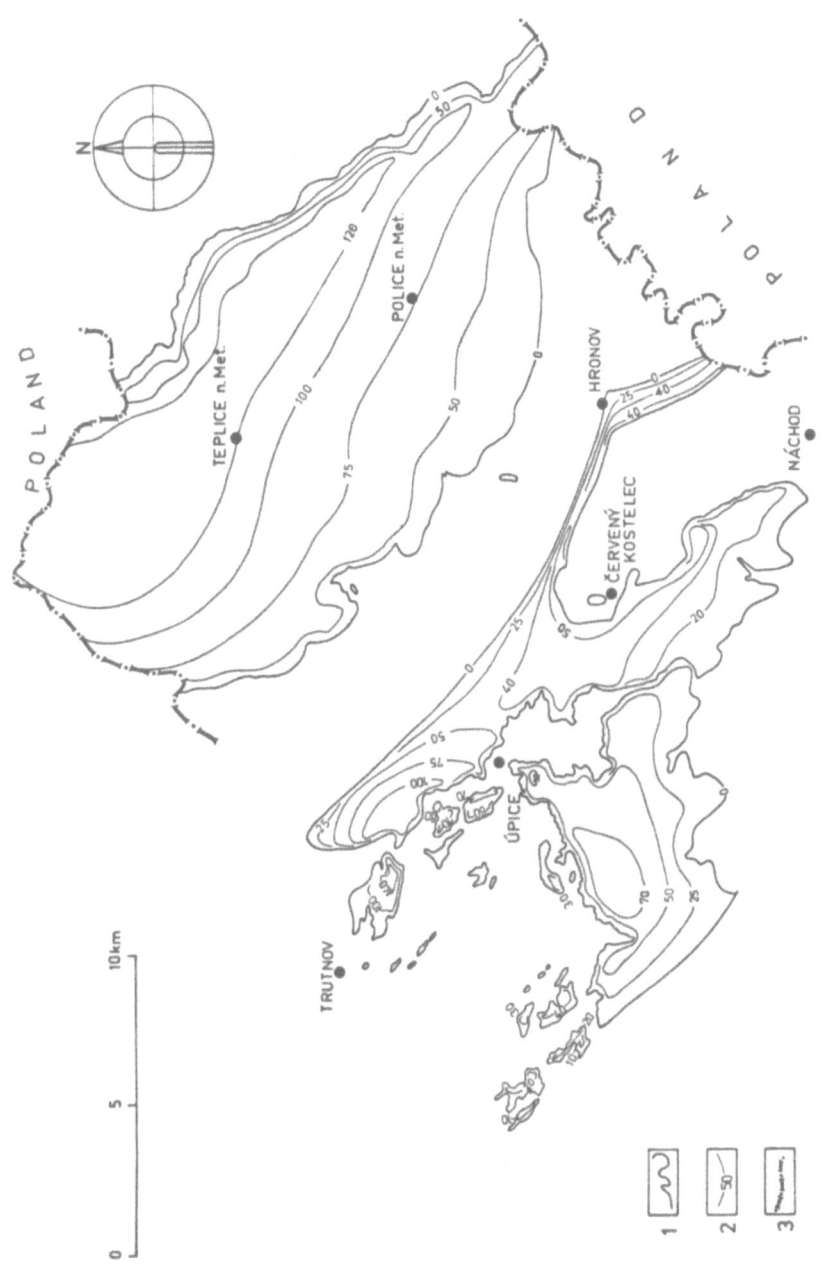

Fig. 6. Isopach map of the Bohuslavice Formation (Thuringian) in the Bohemian Massif. 1-boundary of recent occurrence, 2- isopach line (thickness in metres), 3-state frontier.

of the north-eastern margin (where the sediment composition indicates erosion of rocks outcropping in the
immediate vicinity) but there are no data for the demar-
cation of the basin towards the interior of the Bohemian
Massif. We assume that the peneplanation of the Bohemian
Massif began during Saxonian times. The regions which
had been characterized by substantial subsidence during
Stephanian and Autunian times (i.e. the majority of basins in the central Bohemian region, the furrows, and
the Krkonoše Piedmont Basin) were considerably stabilized
during the Saxonian. Intensive subsidence can be suggested only for the area in which the deposits of this
interval occur at present. The peneplaned surface and
minimal subsidence accompanied by a hot desert climate
were the main reasons for a reduced rate of sedimentation. It is reasonable to expect that the regions having
Early Permian sedimentation became areas of distinctly
intermittent desert sedimentation (drifting dunes, silt
drifts, residual gravels and flood sediments). These sediments did not form large accumulations and therefore
were easily removed by the erosion that followed. We
thus interpret the central part of the Bohemian Massif
as an area of retarded intermittent desert sedimentation.
In view of the low rate of subsidence in central and
western Bohemia, the sedimentary area was reduced in
comparison with the previous unit.

We suggest deposition in alluvial fans for the
Saxonian sediments in eastern Bohemia. Such a facies is
indicated by a large amount of local material and lateral and vertical variations in composition and grain
size. Additionally we suggest flood plain sediments with
ephemeral streams and deposits of temporary lakes (playa
sediments). These types of sediments are mainly formed
in the uppermost part of the Saxonian sequence. We assume
that deluvial, residual and aeolian sediments were formed to a small extent in the central part of the Bohemian
Massif.

Thuringian period

Deposition of the Thuringian sediments was related
to tectonic activity in north-eastern Bohemia which led
to renewed erosion and a fresh supply of clastic material. On the contrary, the central part of the Bohemian
Massif was levelled and tectonically stabilized to a
great extent. The rate of sedimentation was also generally low because of the extremely dry climate. The deposition of typical desert sediments (aeolian sands and silts,

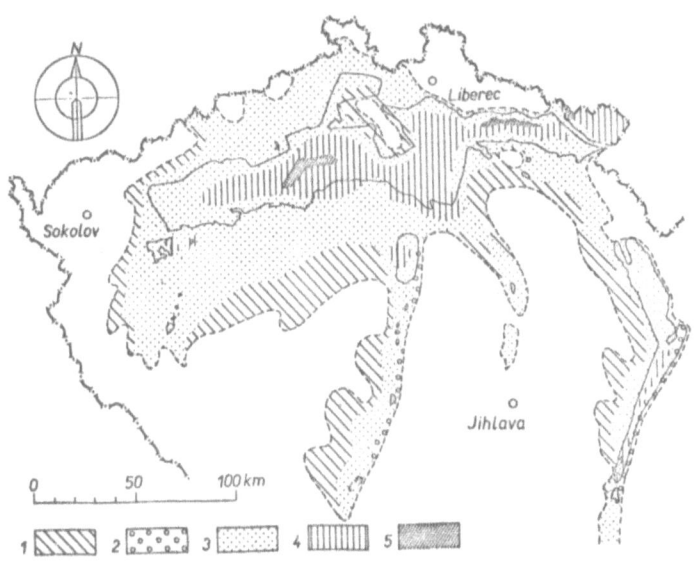

Fig. 7. Palaeogeographical map of the Stephanian C -
Autunian.
1- pediplain, 2- alluvial fans, 3- alluvial plain,
4- alluvial plains and ephemeral lakes, 5- alluvial plain with perennial lakes and swamps.

Fig. 8. Palaeogeographical map of the Saxonian.
1-different environments of desert sedimentation
(desert plains and gentle slopes covered from time
to time by aeolian or water-laid sediments),
2- alluvial apron, 3-alluvial plain with ephemeral lakes.

residual gravels, deposits of ephemeral streams) was intermittent and alternated with periods of predominant erosion (reworking and deflation). That was the reason why the sedimentary cover was rather discontinuous and relatively thin. We presume that during the Thuringian the sedimentary area was reduced in comparison to the underlying Saxonian. This has been observed in the central and eastern parts of the Bohemian Massif. We place the areas of reduced sedimentation only in those regions of the Bohemian Massif where we can presume continuing Saxonian subsidence.

The palaeogeographic interpretation is based on sediments which represent only a small remnant of the original Thuringian sedimentary cover. The homogenous composition of the Bohuslavice Member, mainly consisting of sandstones and arkoses, and having variable admixture of aeolian material (Valín, 1964) shows that these sediments were deposited from ephemeral water courses which simultaneously carried fresh and reworked material from older sediments. We explain the considerable differences in the amount of carbonate cement by differences in the intensity of the desert crust formation. For the central part of the Bohemian Massif we suggest isolated areas of reduced desert deposition. At the present we do not know whether the Bohemian sedimentary area communicated with the North-Sudetic Basin.

TECTONIC HISTORY AND CYCLICITY IN SEDIMENTATION

Basinal fillings have a cyclic character. The development of the sedimentary area was especially influenced by diastrophism. Diastrophic movements which were connected with the final evolution of the Variscan orogeny are the principal causes of the origin of the major sedimentary cycles.

As far as cycles of higher order are concerned, the principal two Permian cycles exist: 1/ Autunian cycle (including Upper Stephanian and the upper Autunian), and 2/ post-Saalian cycle. The last mentioned started after the Saalian phase with Saxonian deposits, continued into the Thuringian and ended in the Lower Triassic.

The most important tectonic movements occurred at the Stephanian B/C, lower/upper Autunian, and Autunian/ Thuringian boundaries. Their manifestations can be observed especially in the marginal parts of the basins.

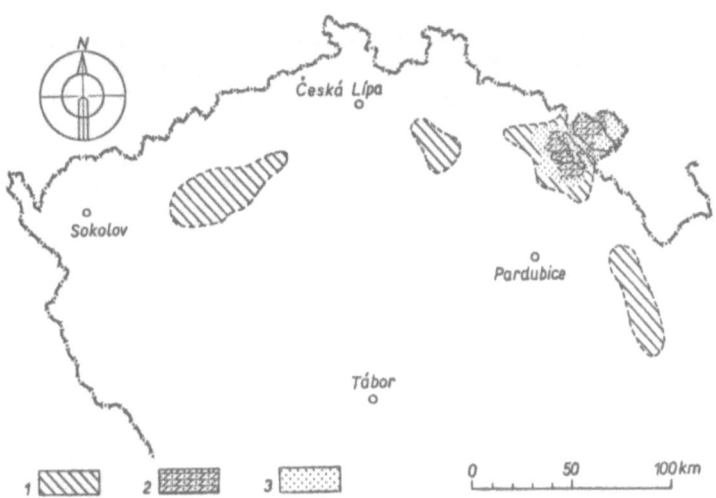

Fig. 9. Palaeogeographical map of the Thuringian.

1- different environments of the desert (desert flats
and gently sloping hills covered from time to time by
aeolian or water-laid sediments), 2- alluvial plain of
ephemeral water courses (carbonate cement common), 3-
alluvial plain of ephemeral streams (without carbonate
cement).

    The variations of subsidence movements in the dif-
ferent basins can be accurately observed e.g. on the
basis of the development of the Stephanian and Autunian
deposits. The intensity of tectonic movements at the
Stephanian B/C boundary was variable; they are generally
manifested in the Central Permo-Carboniferous basins as
a deepening of the sedimentary area without any important
break in sedimentation. The great thickness of the Liné
Formation shows a relatively great subsidence in the
Permo-Carboniferous basins of the Central Bohemian basins during the Intra-Stephanian movements, while subsidence was substantially less intensive in the basins
of the Sudetic area (decreasing from West to East).
Epirogenetic movements having observable symptoms of
breaks in the sedimentation process occur in the Intra-
Sudetic Basin (absence of the Upper Stephanian), and
caused an overlapping of the older units (in the Intra-
Sudetic and the Krkonoše Piedmont basins). During the
Autunian span the inversion of tectonic development
occurred in the Central and Sudetic areas. Sedimentation
continued from the lower Autunian into the overlying

rocks in the basins of the Sudetic area (locally after
a sedimentary break), and simultaneously expressed more
intensive subsidence. This can be observed especially
in the Intra-Sudetic Basin, where sedimentation occurred
during the entire Permian and into the Lower Triassic
with various gaps in sedimentation, especially at the
Autunian/Saxonian boundary, and in various, mostly marginal parts of the sedimentary area between Saxonian
and Thuringian, as well as between Thuringian and Triassic.

Apart from the previously mentioned basins, the
development of Autunian and of Stephanian C can be
followed in the Blanice and Boskovice furrows. The majority of the younger deposits were destroyed during the
subsequent denudation in the Central Bohemian Permo-Carboniferous basins. Only the part, which stratigraphically
corresponds to the Stephanian C, was mostly preserved.

In the Permian of the Sudetic area two major cycles
are distinguished which differ from the cycles in the
Carboniferous deposits. This is observable in the substantially larger extent (both horizontal and vertical)
of the intermittent and perennial lake facies. A greater
stability in the horizontal extents manifested by the
distribution and development of the lithological marker
horizons. As soon as the intra-Stephanian tectogeneti-
cally caused rejuvenation was balanced, sedimentation
took place in a landscape having a peneplanisation
substantially more pronounced than that of the Carboniferous period. The morphological differentiation which
especially affected the development of the volcanodetri-
tic facies in the Autunian sedimentary cycle occurred
inside the basins only during periods of intensive volcanic activity.

The last Late Palaeozoic continental cyclo came into
being as a result of the Saalian phase, which was a very
important tectonic element for the tectonic history of
the Permian in the Bohemian Massif. It changed in an
outstanding form the face of the landscape. The sediments
of that cycle are preserved especially well in the Intra-
Sudetic Basin and in the eastern part of the Krkonoše
Piedmont area. In other places, however, they either had
been mostly destroyed during the denudation process or
there had been no deposition originally. The outstanding
rejuvenation of the relief during the Saalian phase
affected the origin of the fanglomerates at the base of
the post-Saalian filling in some parts of the basins. In
contrast to the preceeding Autunian cycles, the deposits

of perennial lakes are not represented in this cycle. It
is the result of the increasing aridity of the climate.

The continental Late Palaeozoic sedimentation terminated with the sequence of psammites without showing
cyclic structures. These represent rainwashed-sediments
containing reworked aeolian material. They originated
in very flat basins with an oscillating water table,
which affected the carbonate incrustation of these sediments. According to the analogy with the North-Sudetic
Basin in Lower Silesia, these sediments are considered
to be Thuringian in age. They are only preserved in the
Intra-Sudetic Basin and in the eastern part of the Krkonoše Piedmont area. In post-Saalian time the tectonic
activity decreased, and the thickness of sediments as
well.

The structures of the sedimentation basins of the
continental Permo-Carboniferous in the Bohemian Massif
were affected by the principal structural trends: the
Sudetic (NNE-SE), that of the Krušné hory Mts. (NE-SW),
and the Rhinish (NNE-SSW).

The style of the tectonic structure was also influenced by the various types of consolidated basements in
the basins. Fold deformations are most frequent in the
marginal basins of the massif (Intra-Sudetic, and Krkonoše Piedmont basins) than in the basins of the core
area of the Bohemian Massif (the Permo-Carboniferous
basins in Central Bohemia and Blanice and Boskovice furrows).

The tectonic disturbance of the Late Palaeozoic
continental basins in the Bohemian Massif is posthumous
in relation to the Variscan orogeny, and is of a dominantly germanotype character. Only some basins in the Sudetic system are exceptions, especially the Intra-Sudetic Basin, displaying a mediotype structure with fold
structures and overthrusts occurring especially at the
basinal margins.

On the basis of the history of the Permo-Carboniferous continental basins, it can be concluded that
during the Upper Carboniferous and the Permian periods
the Sudetic block appeared more mobile than the Assynthian block in the core area of the Bohemian Massif.

VOLCANIC ACTIVITY

Volcanic activity in the continental Permo-Carboniferous basins expressed itself in cycles. Two principal
culminations can be observed during the Late Palaeozoic
period: 1/ during the Westphalian C substage, and 2/ during the Autunian (see Fig. 10). In the last mentioned
period the Late Palaeozoic volcanic activity reached its
maximum, but it also ended in the same period. The volcanics of the Autunian time span have the largest extent,
especially when the basic ones are concerned.

The volcanics usually are concentrated at the tectonically mobile belts; they occurred mostly at the crossing of the important tectonic lines, and along the outstanding principal fault systems (e.g. Lusatian Fault,
Erzgebirgian Fault etc.). From the regional distribution
of the volcanic products, it has been inferred that the
volcanic centres are associated with zones of intense
Variscan tectogenesis.

The Autunian volcanics are the most frequent in the
Mnichovo Hradiště Basin. They sometimes represent more
than 50 percent of the lower Autunian basinal fill in
the Broumov area (the central part of the Intra-Sudetic
Basin), in addition there are volcanics of upper Autunian age. Further they can be observed underlying the
Bohemian Cretaceous Basin which are of various ages.

As far as the petrological character of the volcanic
products is concerned, the volcanic activity was manifested by alterations of basic to intermediate (the so called
melaphyres) and acidic (palaeorhyolites) volcanics.

In some areas the volcanic activity influenced also
lithofacial development of the upper and even lower Autunian fill. During that time volcanic mountains originated. This substantionally influenced the morphology of
the landscape. In the Autunian period, filling of the
relatively persistent lakes occurred in the Intra-Sudetic
Basin. This is proved by the development of the Libštát
Formation (lower Autunian) and the Broumov Formation
(upper Autunian) in the Intra-Sudetic Basin. The origin
of lakes was often in relation to the products of the
volcanic activity. Depressions among the volcano products were filled by lacustrine and alluvial deposits.
The specific microclimate originated in these depressions.
It appeared during the time when the climate in the other
areas was less suitable for the deposition of lacustrine
sediments. These depressions did not form tectonically,

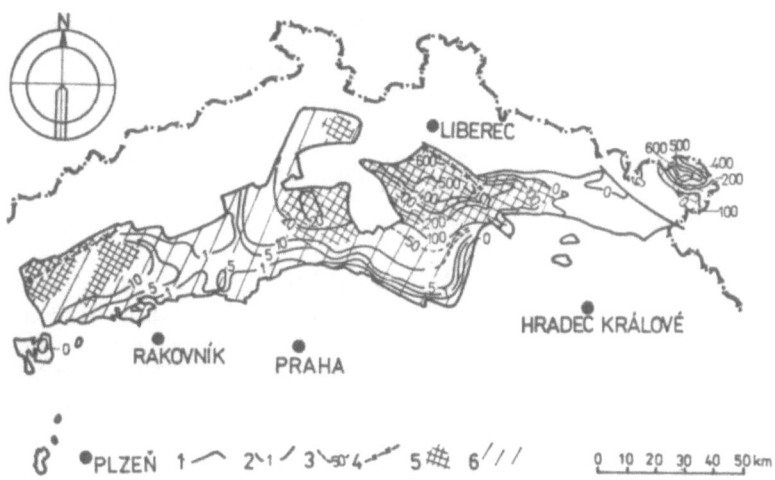

Fig. 10. Volcanic rocks of the Stephanian C - Autunian age.

1- boundaries of the basins, 2-thickness in metres (ascertained), 3- thickness in metres (assumed), 4- centres of
the areal effusions, 5- volcanic rocks of basic to intermediate character, 6- volcanic rocks of acid type (mostly
tuffogenous character).

but rather volcanically and contain sediments of the grey
facies, which are bordered by volcanic mountains. They
appear, for example, in the Intra-Sudetic Basin. The
thickness of volcanic products in the Mnichovo Hradiště
Basin is 800-900 metres, in Broumov Area 300-400 metres.

RATE OF SEDIMENTATION

     Rate of sedimentation measurements during individual
units of the Permo-Carboniferous have been carried out.
The sedimentation rate during the Autunian exceeds three
times the average sedimentation rate of the Carboniferous.
This discovery contradicts the views on the genesis of
sediments. It may be explained by the different intensity
of erosion and denudation during the Carboniferous and
the Permian affected by morphological and especially climatical factors. While the major part of the Permian sediments remained preserved after their deposition, rede-
position and transport of a part of the sediments took
place, however, in the Carboniferous during sedimentation.

In the separate basins the rate of sedimentation
and the evolution of the sediments regressively infer
the intensity of tectonic movements. In the basins
situated on the Sudetic block on the periphery of the
Bohemian Massif (e.g. in the Intra-Sudetic and Krkonoše
Piedmont basins) the rate of sedimentation is higher
than in the basins which are located on the Assynthian
block of the Bohemian Massif (the Permo-Carboniniferous
basins of the Central Bohemian area).

The rate of sedimentation thus indicates that denudation was much more intensive inside the basins during
the Upper Carboniferous than in the lower Permian. Thus
today the sequence of lower Permian strata is much more
complete than the filling of the Upper Carboniferous
basins.

CLIMATICAL CHANGE

In addition to other factors, climatic influences
also affected the variety of the vertical quality of
Late Palaeozoic continental basins. Studies proved the
assumption that the climate periodically oscillated and
changed during the Carboniferous and Permian. An alternation of humid and arid periods existed but, in general,
the aridity of the climate increased gradually and with
oscillations, and, in the Permian, it entirely predominated. It culminated during the late Permian period, in
the Saxonian and, especially, in the Thuringian.

The overwhelming majority of Stephanian sequences
consist of red-beds, and only rarely of phytogenic rocks.
The coal seams came into being in some places during the
Upper Stephanian. They have been evidenced especially in
the Boskovice Furrow, and only slight manifestations of
phytogenic deposits have been found in the Blanice Furrow,
in the Krkonoše Piedmont Basin, and in the Permo-Carboniferous basins of the Central Bohemian area. The period
of formation of grey sediments with caustobioliths was
even more restricted in the Autunian, where the coal-for-
ming activity continued in furrows (esp. the Rosice coals
in the Boskovice Furrow). Small local coal peats accompanied by lacustrine deposits occurred only at the close of
the main sedimentary cycles. The coal seams are often
replaced by sapropel deposits (bituminous pelites). Almost
no traces of the caustobiolith sedimentation have been
recorded in the Saxonian and Thuringian sediments.

The concretions which originated during early diage-

nesis, react sensitively to climatic conditions. The
calcite and dolomite concretions are typical for red-bed
complexes; the major content of dolomite occurs in the
periods for which the greatest aridity is assumed, especially in the Thuringian.

The occurrence of gypsum , in veins and especially
as fillings of joints and intra-bedding planes, is characteristic of the Autunian sequences. The content of illite
in the weathered rocks increases with the aridity of the
climate. Illite is common and the predominant clay mineral of the red-coloured aleuropelitic formations, while
in the grey coloured sequences the content of kaolinite
is appreciably increased.

The climatic conditions may also be inferred from
the mineralogical and textural maturity of the sediments.
During the arid periods, especially with regard to the
irregularity of transport and absolute predominance of
the mechanical type of weathering, structural and mineralogical sediments with low maturity originated. Therefore,
in the Permian, a high proportion of sediments of low
maturity occurs in the lower part of the main sedimentary
cycles. They are considerably restricted in the Westphalian as far as time and space are concerned.

Bimodal sandstones were ascertained in sediments
of Saxonian and Thuringian ages. Valín (1972) explained
their origin in connection with aeolian activity, and he
considered them as a proof for the existence of an arid
climate for the time of their origin.

Chemogenical sediments (limestones, and/or dolomites) are present in the Late Palaeozoic basins. They are
explained as sediments of shallow, intermittent lakes
which originated under an arid climate. They appear in
places as a succession starting in the Westphalian D,
progressing through the entire Stephanian, and their
distribution has been recorded upwards into the Autunian
formations. A study of the chemical composition and of
the extremely mineralized chloride and sulphate ground-
waters from the Late Palaeozoic basins of the Central Bohemian region indicates that the mineralisation was affected by solution of syngenetical salts and their diffusion
into ground waters. In this way it is possible to explain
the lack of evaporites in the Late Palaeozoic basins of
the Bohemian Massif.The mineralogical and petrological
proofs of the climatic changes are discussed in detail
by Skoček, 1974.

REFERENCES

FALKE,H. (1972): Rotliegend Essays on European Lower
     Permian.- Brill (Holland). Leiden.
HAVLENA,V. (1964): Geology of Coal Deposits,2.-Nakl.
     Česk. Akad. věd,437 pp.Praha.
HOLUB,V. (1972): Permian of the Bohemian Massif. (In:
     Falke,H.:Rotliegend Essays on European Lower
     Permian.).- Brill (Holland), p.137-188.Leiden.
HOLUB,V.(editor) (1973): Carboniferous and Permian of
     the Bohemian Massif. - Excursion Guide to the
     S.C.C.S. Field and General Meeting in Czecho-
     slovakia. 155 pp., Geological Survey, Prague.
HOLUB,V. (1976): The Permo-Carboniferous continental ba-
     sins of Bohemia and Moravia.- Symposium on
     Carboniferous Stratigraphy, Proceedings of
     and Contributions to the Field and General
     Meeting of the IUGS S.C.C.S. in Czechoslovakia
     1973. Geological Survey, Prague.
HOLUB,V.-SKOČEK,V. (1968): Problems on Genetic Classifi-
     cation of Coal Basins in the Bohemian Massif.-
     23. Internat. Geol. Congr.,Vol.11,p.105-117.
     Prague.
HOLUB,V.-SKOČEK,V.-TÁSLER,R. (1975): Palaeogeography of
     the Late Palaeozoic in the Bohemian Massif.-
     Palaeogeography-Palaeoclimatology-Palaeoeco-
     logy,18,p.313-332. Amsterdam.
HOLUB,V.-TÁSLER,R. (1972): Late Palaeozoic and Lower
     Triassic underlying the Bohemian Cretaceous
     Basin. (In: Malkovský,M. et al.: Geology of
     the Bohemian Cretaceous Basin and of its ba-
     sement.).- Ústř. úst.Geol., Akademia,p.72-100.
     Prague.
HOLUB,V.-TÁSLER,R. (in press): Filling of the Late Palaeo-
     zoic continental basins in the Bohemian Mas-
     sif as record of their palaeogeographical de-
     velopment.- Compte Rendu of VIII. Congr.Carb.
     Strat. Geol. Moscow.
NĚMEJC,F. (1953): Introduction to the floristic stratig-
     raphy of coal-bearing areas in ČSR. - Nakl.
     Česk. Akad. věd, 173 pp. Praha.
NĚMEJC,F. (1960): Remarks on the dividing line between
     the Stephanian and the lower Permian in Bo-
     hemia and Moravia.- Čas. Mineral. Geol., 5,
     p.452-455. Praha.
NĚMEJC,F. (1966): Stephanian floras in the coal-districts
     of Czechoslovakia in the light of new geological
     investigations. - The Paleobotanist 14,p.70-72.
     Lucknow, India.

PETRASCHECK,W. (1934): Der böhmische Anteil der Mittel-
                sudeten und sein Vorland.- Mitt. Geol. Gesell.
                (Wien),26,pp.136. Wien.
RIEGER,Z. (1968): Phytopalaeontological-stratigraphical
                studies of the Stephanian and Autunian of the
                Krkonoše Piedmont Basin. - Věst.Ústř.Úst.geol.,
                p.449-457. Praha.
SKOČEK,V. (1974): Climate and diatrophism, the princi-
                pal factors controlling Late Palaeozoic sedi-
                mentation in central Bohemia. - Čas. Mineral.
                Geol.,19,p.27-45. Praha.
VALÍN,F. (1972): New evidence of eolian desert sedimen-
                tation in the north-east Bohemia.- Věst.Ústř.
                Úst. Geol., p.141-146. Praha.

LE PERMIEN ET LA PHASE SAALIENNE DANS LE BASSIN DE BRIVE (SW DE LA FRANCE).

R. Feys

Bureau de Recherches Géologiques et Minières, France

RESUME.   Dans le SW de la France, les anciens auteurs attribuaient au Saxonien les terrains rouges des bassins permiens (Lodève, Brive, Saint-Affrique) par opposition àl'Autunien gris sous-jacent. L'ensemble paraissait condordant.

Des études récentes et les levés détaillés effectués pour les nouvelles cartes géologiques au 50.000è ont fait revoir cette conception : la couleur rouge n'est pas un bon critère stratigraphique, et une partie des terrains rouges sont autuniens.

Cette note décrit la stratigraphie du Permien du basin de Brive et montre que dans cette nouvelle conception la discordance saalienne y est reconnaissable, comme d'ailleurs dans les autres bassins permiens du SW de la France: Lodève et Saint-Affrique.

SITUATION GENERALE

Dans le SW du Massif Central, une série de bassins permiens: Brive, Rodez (Decazeville), Saint-Affrique et Lodève se ressemblent beaucoup. La bibliographie a été synthétisée et résumée in Falke (1972).

Tous ces bassins étaient naguère considérés comme débutant par un Autunien parfois concordant sur le Stéphanien, parfois discordant sur cet étage ou sur des terrains plus anciens. Cet Autunien est gris et contient,

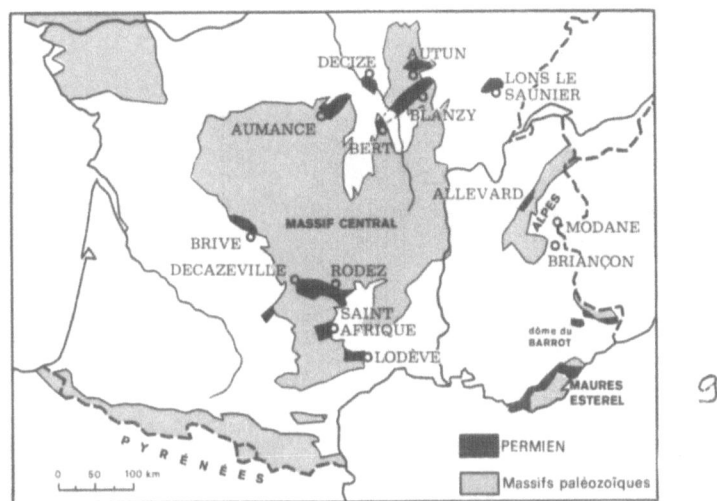

Fig. 1 - Situation générale des bassins permiens du
SW de la France.

vers son sommet, des niveaux calcaires ou calcaréo-
dolomitiques. Le Permien s'y continue par de puissantes
séries rouges (800 à 2.000 m) rapportées au Saxonien,
sans qu'il y ait de limite tranchée entre les deux, par
une zone où alternent des faciès rouges et des faciès
gris. Le volcanisme y parait peu important, sinon totale-
ment absent.

Mais les études récentes, et encore en cours, ont
montré que les anciennes conceptions stratigraphiques
doivent être rectifiées, que la limite entre l'Autunien
et le Saxonien se trouve là où les anciens ne la
cherchaient pas. Ainsi la phase saalienne, si importante
dans le reste de l'Europe, est enfin indentifiée dans
le SW de la France, et cette région ne fait plus exception dans l'édifice de l'Europe hercynienne.

STRATIGRAPHIE DU PERMIEN DU BASSIN DE BRIVE

L'histoire du Permien peut être schématisée comme
l'implantation puis le triomphe du rouge, par transitions ménagées et recurrences. Dès la partie supérieure de l'Autunien et jusqu'à la fin du Permien, les
dépots sont essentiellement rouges, rarement gris, ils

sont formés de détritiques grossiers et de sédiments
très fins : argiles parfois carbonatées, pélites,
silts, qui ont conservé des traces sédimentologiques
particulières : fentes des dessication, ripple-marks,
gouttes de pluie, etc.

Ces séries présentent des variations latérales
rapides qui rendent difficiles les raccordements entre
elles. Ceci s'explique par l'existence d'une zone ou
débouchaient des cours d'eau drainant le socle tout
proche, et dont les différents bras pouvaient divaguer;
il s'est ainsi déposé des sédiments grossiers largement intriques (Lille, 1968).

Les fossiles sont très rares; on n'y a jamais
trouvé que de mauvaises pistes et des perforations de
vers. Certains ont estimé que cette stérilité prouvait
l'existence d'un climat désertique au Permien. Mais
alors, on comprend mal comment dans le bassin permien
de Lodève qui est proche de celui de Brive, avec des
faciès rouges identiques, les traces de vie soient
aussi abondantes? Selon des travaux récents, il s'agis-
sait plutôt d'un climat sub-tropical avec alternance
de saisons sèches et humides. Ce seraient les conditions
oxydantes liées aux faciès rouges qui seraient res-
ponsables de la non conservation des plantes fossiles
- et non l'absence de vie à cette époque; On peut aussi
estimer que cette couleur rouge serait l'héritage du
lessivage des reliefs cristallins; dans cette conception
ce seraient au contraire les sédiments gris ou verdaâtres
qui résulteraient d'une altération par des eaux réductrices, notamment dans les faciès houillers.

Après avoir dominé tous les travaux anciens,
l'oeuvre monumentale - et un peu confuse - de MOURET
(1891) semble avoir découragé tout autre auteur,
jusqu'à ce que ROGER (1968) apporte une salutaire
tentative de clarification et de synthèse. L'ensemble
du bassin est actuellement étudié par la nouvelle
carte géologique détaillée au 50.000 è.

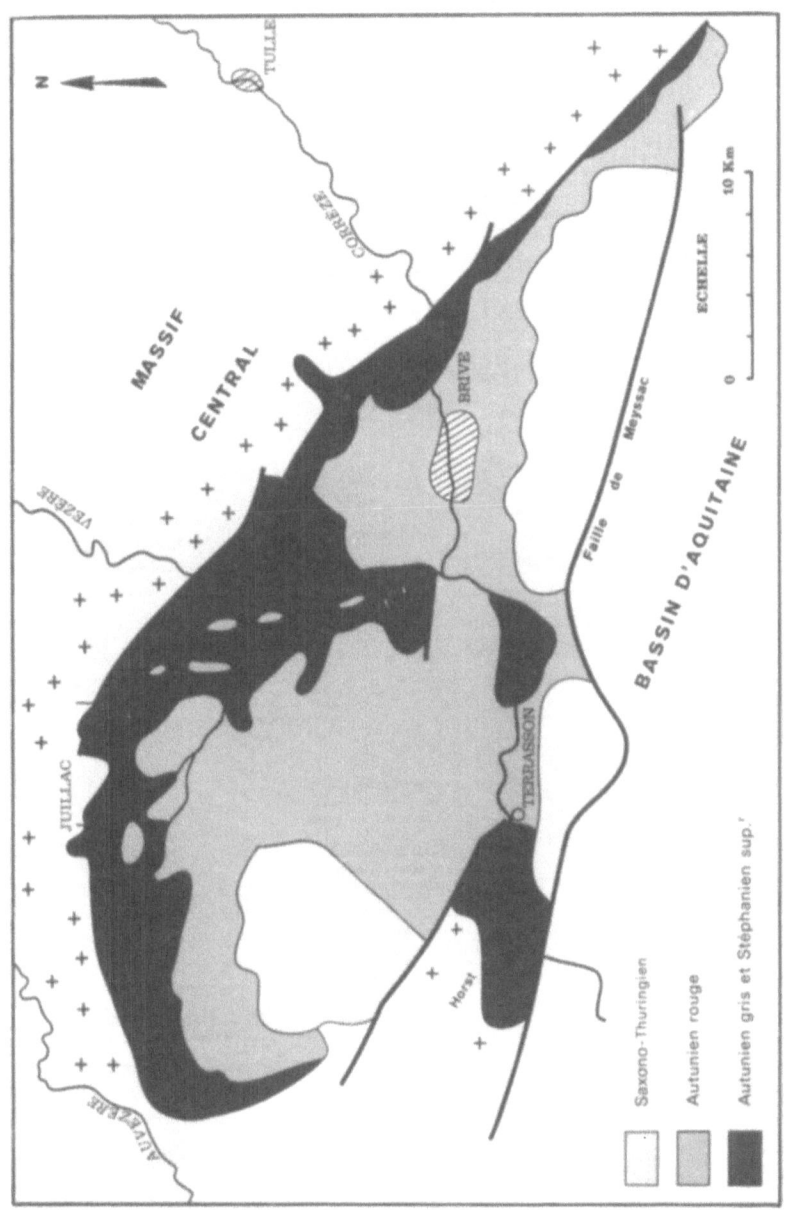

Fig. 2. Esquisse du bassin permien de Brive.

Fig. 3 – Coupe synthétique du Permien de Brive

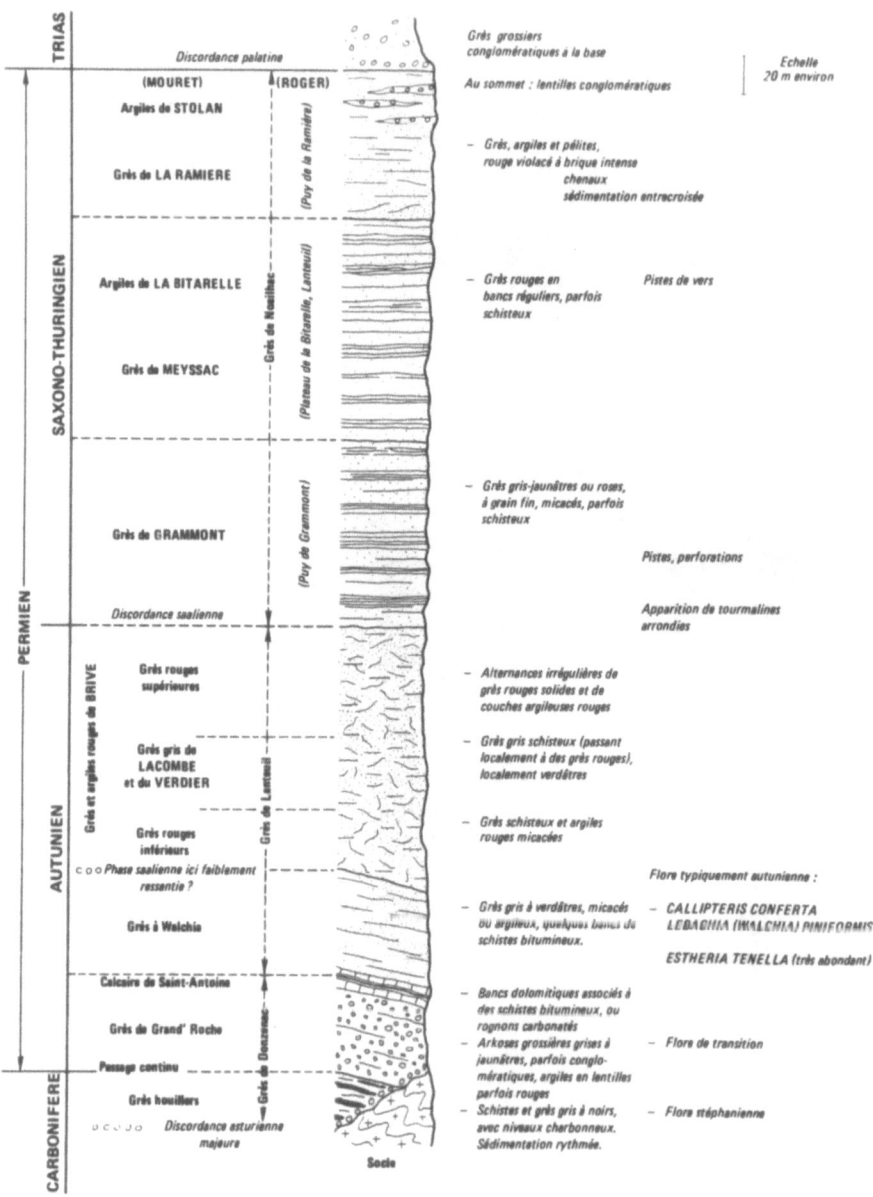

Fig. 3. Coupe synthétique du Permien de Brive.

STEPHANIEN SUPERIEUR

C'est seulement à la fin du Carbonifère, sinon au début du Permien, qu'on observe les premiers dépots sédimentaires non métamorphisés; longtemps après la "phase asturienne" proprement dite, on peut attribuer à des rajustements sous tension l'effondrement de la vaste zone qu'on a souvent appelée, un peu improprement, le "bassin houiller et permien de Brive".

Ces premiers dépots sont, sur les premières pentes ou au pied des falaises du Massif Central, des conglomérats parfois associés à des schistes et grès gris à noirs recélant quelques rares et chiches dépots de charbon. Ces lambeaux houillers sont confinés dans quelques échancrures de la bordure cristallophyllienne. C'est l'indice d'un couvert végétal et de conditions réductrices fugaces.

Les anciens auteurs avaient certainement surestimé l'importance des terrains stéphaniens dans le bassin de Brive. Ne mettant pas en doute la nécessité de Houiller entre le Permien et son substratum, ils avaient attribué au Stéphanien des conglomérats et des arkoses grossières qu'ils avaient cartographiés comme un liseré trop souvent continu le long de la bordure cristallophyllienne, alors que ces arkoses et conglomérats constituent plutot le début du Permien.

Les études paléobotaniques de ZEILLER (1892) et de DOUBINGER (1956) n'excluent pas que le Stéphanien soit entièrement à bannir au profit d'un Autuno-Stéphanian ou d'un Autunien basal très probable.

PERMIEN

Tout le Permien du bassin de Brive est constitué de "formations" qui sont de grandes lentilles largement intriquées, et dont les variations latérales parfois rapides rendent difficiles les raccordements entre elles. Nous conservons autant que possible le nom des formations distinguées par MOURET grace à des différences.pétrographiques parfois difficiles à apprécier, grace aussi à des raisonnements pétrographiques. Il est clair que ces formations ne sont pas toujours de rigoureuses superpositions stratigraphiques mais dans une certaine mesure des extensions géographiques; il en résulte que les contours avec lesquels elles sont cartographiées ne sont parfois que l'enveloppe de cer-

taines particularités faciales. Dans ces conditions on
s'explique les différences d'interprétation d'un auteur
à l'autre.

AUTUNIEN

Grès rouges inférieurs ; grès de Grand'Roche

      L'Autunien peut succéder plus ou moins confusé-
ment aux dépots houillers là ou ceux-ci existent, en
continuité semble-t-il, par des arkoses plus ous moins
conglomératiques, ne contenant plus aucun indice
charbonneux, mais quelques lentilles d'argiles rouges.
Ou bien ce sont des grès rougeâtres, plus ou moins
chargés de conglomérats.

      Ou bien l'Autunien, débutant par une inondation
qui a fait déborder le bassin de Brive, débute par le
dépot, sur des pentes cristallines auparavant exondées,
d'un conglomérat de base analoque à celui d'un
Stéphanien; ainsi ce sont les mêmes conglomérats qui
changent d'étage en jalonnant dans le temps et dans
l'espace l'envahissement des épandages de sables et de
gravillons du bassin de Brive au commencement.

      Dans des lentilles argileuses, la flore contient
encore des espèces stéphaniennes pouvant subsister dans
l'Autunien, mais ne recèle pas encore d'espèces pure-
ment autuniennes, Cette assise ambigüe, "permo-carbonifère", peut aussi bien être attribuée à la partie tout
à fait supérieure du Stéphanien qu'à l'extrème base de
l'Autunien.

      Quoiqu'il en soit, la datation de l'Autunien sera
quelques mètres plus haut hors. de doute avec Estheria
tenella dans le Calcaire de Saint-Antoine et une flore
à Callipteris conferta et Lebachia (Walchia) piniformis
des Grès à Walchia.

Calcaire de Saint-Antoine

      C'est un ensemble formé par l'alternance de bancs
carbonatés, noirs ou gris de fumée, durs, compacts, et
de schistes calcareux ou bitumineux, riches en débris
animaux d'eau douce. Bien développé et visible surtout
autour de Brive où son épaisseur est de 15 à 20 m.
Autrefois cartographié comme le niveau repère du Permien
du bassin de Brive, il est vrai que c'est le meilleur

connu dans ce bassin mais il est malheureusement très
variable latèralement; on constate son amenuisement,
sinon sa disparition dans d'autres secteurs; ou bien
il passe à des rognons calcaires épars dans des argiles
rouges, de sorte qu'on hésite à lui attribuer tel ou
tel des indices carbonatés qu'on observe souvent dans
les Grès à Walchia qui lui succèdent, ou sporadiquement
dans d'autres niveaux. Ce qui est bien facheux pour
un niveau repère.

Quoiqu'il en soit, cet épisode remarquable montre
qu'à cette époque c'était le domaine d'eaux douces
calmes, confinées et réductrices. Après ce court répit,
le démantèlement des reliefs du Massif Central et de
comblement de la zone d'effondrement dite "bassin
permien de Brive", reprennent activement : la sédimentation gréseuse prédomine à nouveau, d'abord avec les
"Grès à Walchia".

Grès à Walchia

Ce sont des grès gris à verdatres, micacés ou
argileux; ainsi nommés à cause de la présence de très
nombreux petits débris charbonneux qui sont souvent de
petites écailles de Walchia. Ceci indique la proximité
de pentes couvertes de Conifères. D'épaisseur très
variable, parfois absents, ces grès gris renferment
les dernières et fugaces manifestations carbonatées et
bitumineuses, sous forme de nodules carbonatés, quelques
dizaines de mètres au dessus du niveau de Saint-Antoine.
Très localement aussi, il leur arrive de présenter un
faciè presque houiller.

Par des intercalations rouges qui se font de plus
en plus fréquentes jusqu'à prendre toute la place, ils
passent progressivement à la formation suivante .

Grès rouges de Brive

C'est le premier envahissement général du bassin
par la couleur rouge : alternances irrégulières de
grès solides, rouges ou clairs, et de couches argileuses, rouges et micacées. D'épaisseur très variable,
cette formation est un faciès qui peut localement
remplacer toutes les formations sous-jacentes. Aucun
fossile n'y a été signalé.

Grès du Verdier

Cette formation distinguée MOURET qui la considé-
rait comme intercalée entre les "Grès rouges supérieurs"
et les "Grès rouges inférieurs" n'est qu'une récurrence
locale des faciès réducteurs de l'Autunien inférieur
("Grès à Walchia"). Sa sédimentologie est identique à
celle des "Gres rouges" dont elle ne diffère que par
la couleur, et auxquels elle passe latéralement par une
intrication moins que centimétrique. Elle existe dans
le SE du bassin, là où les faciès réducteurs ont paru
l'emporter, d'où son intéret paléogéographique. Dans
l'W du bassin, la formation des "Grès de Brignac" se
distinque par la même couleur.

SAXONO - THURINGIEN

Après les alternances très irrégulières des grès
et argiles rouges de Brive, on note un changement dans
la sédimentation, avec l'arrivée de grès plus fermes
et plus réguliers, souvent massifs, avec des intercalations schisteuses plus rares et dont ils sont mieux
séparés. En même temps apparaissent parmi les minéraux
lourds des tourmalines brunes arrondies (ROGER, 1968).
Ces indices dont l'observation n'est pas évidente à
première vue permettent cependant de reconnaitre à ce
niveau une manifestation assourdie de la phase saalienne.

En l'absence de tout fossile, on a interprété ce
changement de style sédimentologique comme l'inaugura-
tion du Saxonien.

Grès de Grammont

Grès gris-jaunatres ou roses, fermes, à grain fin,
à délits micacés, parfois schisteux. Quelques pistes et
perforations.

Grès de Meyssac

Grès rouges en bancs réguliers, parfois schisteux.
Quelques pistes de vers. Ils ressemblent beaucoup aux
Grès de Grammont, et quand ces derniers deviennent plus
rouges, vers l'E du bassin, la distinction des deux
formations est un peu conventionnelle.

Ces deux dernières formations n'ont été identifiées

par MOURET que dans la partie sud-est du bassin. Mais
dans l'W du bassin nous en voyons l'équivalent dans les
"Grès de Louignac".

Grès de la Ramière.

Enfin, après les prédédents grès en bancs épais et
réguliers, reviennent des conditions de sédimentation
anarchiques. La formation de la Ramière est constituée
par des grès, argiles et pélites, rouge violacé à rouge
intense. Sédimentation entrecroisée, nombreux chenaux
ou ravinements intraformationnels. Epais de 60 à 80 m
au maximum, ils semblent raviner légèrement les Grès
de Meyssac. Leux extension est très limitée. Au sommet
ils se chargent de lentilles conglomératiques, en
trainées irrégulières, prémonitoires de la transgression
triasique. C'est enfin l'épogée de la couleur rouge...

La limite supérieure du Saxonien de Brive n'a pas
été précisée. Puisque la sédimentation semble bien
avoir été continue jusqu'à la base du Trias, c'est que
la partie supérieure des "Grès rouges" est contemporaine du Thuringien. C'est pourquoi il paraît objectif
d'attribuer ces "Grès rouges" au Saxono-Thuringien.

L'épaisseur totale du Permien paraît être de l'
ordre de 600 à 1.000 m, peut être davantage.

TRIAS

On y rapporte un ensemble de 50 à 80 m, uniquement
gréseux, qui comprend:
- à la base 15 à 20 m de grès lie de vin sombre, riche
en galets, à sédimentation entrecroisée. Certains grès
micacés violacés rappellent beaucoup le Permien.
- puis 20 à 30 m de grès assez fins, assez compacts,
kaolinisés et blancs.
- à la partie supérieure, 15 à 20 m de grès bariolés,
plus ou moins grossiers, à passées argileuses à
couleurs vives.

Mais ce Trias s'amenuise et disparaît dans la
parte nord-ouest du bassin.

C'est seulement à partir de l'Hettangien supérieur
que s'installent définitivement des dépots marins et
carbonatés.

CONCLUSION

　　Ainsi la base du Saxonien n'est pas indiquée par
l'apparition franche et définitive de la couleur rouge.
Celle-ci arrive dès l'Autunien; elle s'installe par
transitions ménagées et il y a des recurrences de
faciès gris dans le Saxonien.

　　Cette nouvelle consception de la stratigraphie du
Permien de Brive est à rapprocher des observations de
C. GREBER (communication orale) dans le bassin de
Saint-Affrique et de M. HERY reprises par P. VETTER
(1971) dans le bassin de Lodève.

BIBLIOGRAPHIE SOMMAIRE

DAVID, A. (1956): Etude géologique du bassin permien
　　　　de Belmont-sur-Rance (Aveyron) et de ses
　　　　minéralisations uranifères et suprifères.-
　　　　Thèse 3 cycle, Clermont-Ferrand.
DOUBINGER, J. (1956): Contribution à l'étude des flores
　　　　autuno-stéphaniennes.- Mém.Soc.géol.Fr.,
　　　　no 5, 75.
FEYS, R. et GREBER, C. (1972): L'Autunien et le Saxonien
　　　　en France.- in FALKE, H.: Rotliegend. Essays
　　　　on European Lower Permian, Leiden, E.J.Brill.
MOURET, G. (1879): Etude géologique des environs de
　　　　Brive.- Bull.Soc.scient., hist. et archéolo-
　　　　gique de Brive, I.
MOURET, G. (1891): Bassin houiller et permien de Brive.
　　　　Et. Gites minéraux Fr., I. Stratigraphie.
ROGER, P. (1968): Lithostratigraphie et sédimentologie
　　　　des formations détritiques du bassin de Brive.
　　　　Essai de synthèse paléogeographique.- Actes
　　　　Soc. linnéenne de Bordeaux, 105, (B), I.
SABOURDY, G. (1962): Contribution à l'étude pétro-
　　　　graphique et stratigraphique et étude de la
　　　　minéralisation en cuivre du bassin de Brive
　　　　(Corrèze).- Dipl. Et. sup., Clermont-Ferrand.
VETTER, P. (1971): Le Carbonifère supérieur et le
　　　　Permien du Massif Central.- Symposium J.Jung,
　　　　Clermont-Ferrand 1971, p. 169-213.
ZEILLER, R. (1892): Bassin houiller et permien de Brive.-
　　　　Et. Gites mineraux Fr., II. Flore fossile.

LE PERMIEN EN ESPAGNE

C. Virgili, S. Hernando, A. Ramos, et A. Sopeña.

Département de Stratigraphie, Université de Madrid
Département de Géologie Economique. C.S.I.C. Madrid
Espagne.

RESUME. Ces matériaux qui en grande partie étaient jusqu'à main-
tenant attribués au Trias ou au Carbonifère, existent dans maints
endroits de la Péninsule Ibérique. Les affleurements sont presque
toujours nettement discordants au-dessus du socle hercynien et re
couverts aussi en discordance par le Buntsandstein ou par des niveaux plus récents.
      Ils représentent un cycle sédimentaire très spécial et inter
médiaire entre le cycle hercynien et le cycle alpin. La configuration des bassins est conditionnée par la tectonique hercynienne
mais le milieu sédimentaire annonce déjà ce que sera le Secondaire.
      Le Permien se trouve constitué par une série détritique qui
va de l'argile aux gros blocs. La couleur va du rouge au noir et
au gris-vert. L'épaisseur de la série est très variable et elle
atteint jusqu'à 1.500 m. En quelques endroits il y a de la flore
de l'Autunien et du Zechstein mais dans la plupart des coupes une
datation exacte n'est pas possible.

ABSTRACT. The Permian materials; existing in many places along
Spain, have been considered up to now of Triassic or Carboniferous age.
      The outcrops lie always unconformably over an hercynian base
ment and Triassic or younger materials overlap them unconformably.
      The Permian sediments belong to a very special sedimentary
cycle. The sedimentary basins are controled by the hercynian tectonics, but environments are a prelude of the secondary sedimentation.
      The Permian consists of a detrital series red, black or grey
green in colour; thickness is variable, up to 1.500 m.
      Some plant remains of Autunian and Zechstein age have been

H. Falke (ed.), The Continental Permian in Central, West, and South Europe, 91-109. All Rights Reserved.
Copyright © 1976 by D. Reidel Publishing Company, Dordrecht-Holland.

found, but most of the sections cannot be dated.

1. INTRODUCTION

Les données qu'on possède actuellement sur le Permien en Espagne
ne sont pas suffisantes pour donner une image détaillée de la pa-
léogéographie de la Péninsule Ibérique pendant cette période. Il
y a des régions dans lesquelles on n'a pas encore bien différen-
cié le Permien du Buntsandstein ou du Stéphanien et les repères
paléontologiques sont très rares.

Néanmoins, ces dernières années, de nombreuses observations
nouvelles ont été effectuées à la base des séries secondaires
dans différentes régions d'Espagne, spécialement dans la Chaîne
Ibérique, Le Système Central et les Pyrénées. Elles permettent
de compléter les travaux antérieurs et de tenter une vue d'ensemble sur l'existence et les caractéristiques de la sédimentation
permienne.

Il n'est pas encore possible de connaître la configuration
exacte des bassins. C'est pour cette raison que, dans cette étude, les affleurements sont groupés d'après les grandes régions
géographiques espagnoles, c'est-à-dire d'après les structures al-
pines et non pas suivant les aires sédimentaires permiennes qui
sont conditionnées par les structures hercyniennes.

Dans ce travail, nous n'étudions pas le sud de l'Espagne.
Les données sur cette région sont très rares (4) et il est très
probable que de nombreux affleurements attribués au Stéphanien et
au Buntsandstein doivent être rangés dans le Permien.

2. TRAVAUX ANTERIEURS

Même si le Permien est peu connu, il existe des références très
anciennes.

Dans la bibliographie ci-jointe, nous avons retenu seulement
les publications les plus importantes.

Jacquot (15) en 1866 et Perez de Cossío (25), en 1920, parlent de matériaux du centre de l'Espagne qu'ils attribuent au Per
mien.

Dalloni, à partir de 1910, soupçonne l'existence du Permien
dans les Pyrénées et finit par y trouver (5, 6) une riche flore
autunienne. Patac (24) en 1920 trouve aussi dans la région Cantabrique une flore du Rotliegendes.

C'est beaucoup plus tard qu'on reprend les recherches dans
ce domaine. En 1966 Sacher (28) prouve l'âge permien des matériaux
de la chaîne ibérique. Finalement on découvre la flore permienne
(3) (4) (31) au centre et au sud de l'Espagne.

Ces dernières années de nombreux travaux (7, 8, 9, 10, 11,
12, 13, 17, 21, 22, 23, 26, 29, 32, 37, 39) ont apporté de nouvel
les précisions, mais il manque encore une vue d'ensemble et on
continue à attribuer au Buntsandstein ou au Stéphanien une partie

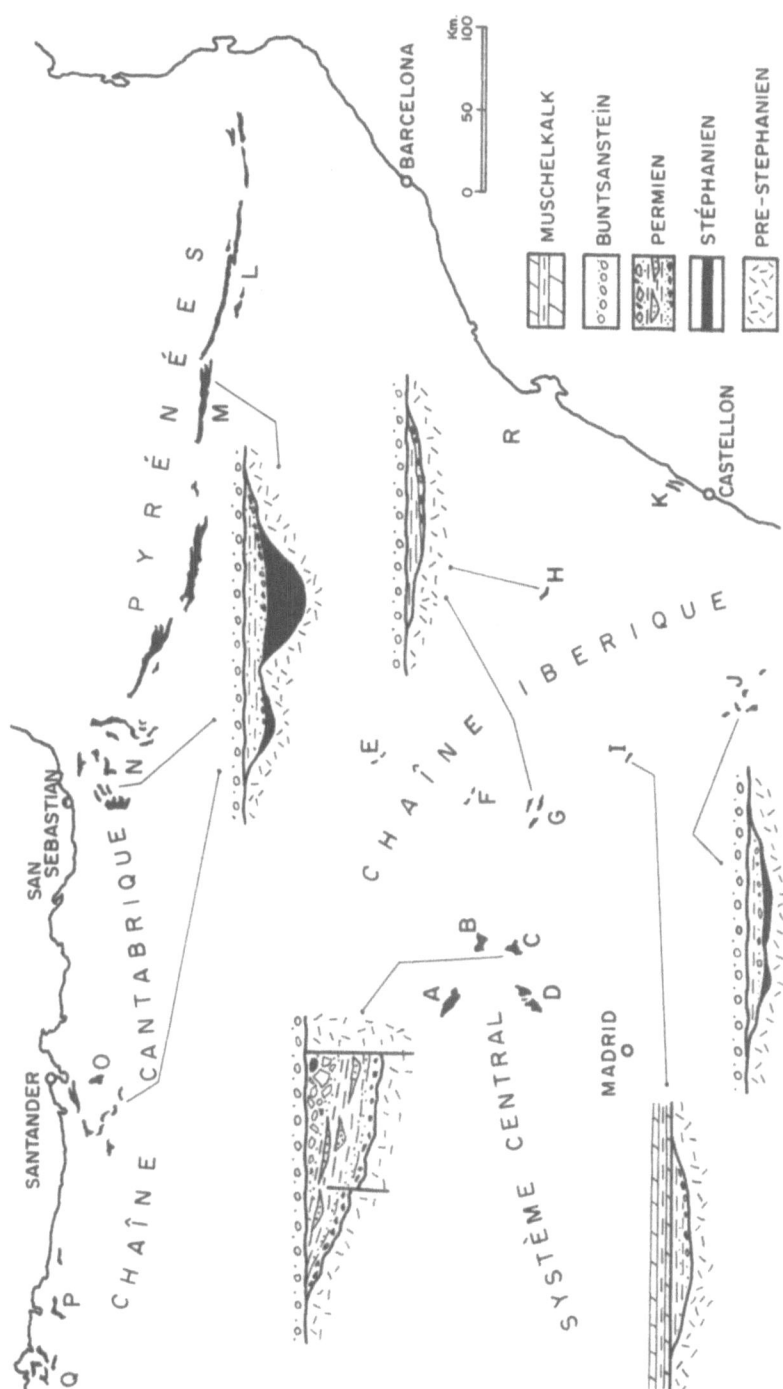

Fig.1.- A: Sierra Pela, B: Atienza, C: Pálmaces, D: Valdesotos-Tamajón, E: Reznos, F: Nuevalos-Monterde, G: El Bosque-Aragoncillo-Molina, H: Montalbán, I: Beteta, J: Landete-Minas de Henarejos, K: Desierto de Las Palmas, L: Seo de Urgel, M: Pont de Suert, N: Valle del Baztán, O: Torrelavega, P: Colunga et Caravia, Q: Oviedo, R: Caspe.

des matériaux qui méritent d'être rangés dans le Permien.

3. OBSERVATIONS DE TERRAIN

Nous avons étudié ou revu l'ensemble des affleurements dont la
situation est représentée sur la figure 1. Ces affleurements sont
groupés suivant quatre régions: les Pyrénées, la Chaîne Cantabrique, le Système Central et la Chaîne Ibérique. Ces deux dernières
sont actuellement les mieux connues et nous donnons la situation
exacte de chaque affleurement. Pour les Pyrénées et la Cahîne Can
tabrique le dessin est moins précis et quelques affleurements fi-
gurés correspondent indifféremment au Permien et au Buntsandstein.

3.1. Les Pyrénées

C'est la première région où il existait des données précises sur
le Permien (5, 6) et plus récemment beaucoup d'études ont été effectuées (10, 11, 14, 16, 20, 21, 22, 23, 35, 36, 39).

Les matériaux permiens forment une ligne d'affleurements qui
borde l'axe pyrénéen de la Méditerranée à l'Atlantique. La sédimentation est localisée dans une série de cuvettes plus ou moins
isolées qui sont en réalité la continuation des bassins stéphaniens.

Dans la partie centrale de ces cuvettes la série permienne
est plus complète et épaisse et repose en concordance sur le Sté
phanien; elle s'amincit progressivement vers les bordures et une
discordance angulaire apparaît à sa base. Ainsi le Permien a une
épaisseur très variable, pouvant atteindre 2.000 m. (Seo d'Urgell,
fig. 1.L) ou rester à quelques dizaines de mètres. Il apparaît
sous forme de lentilles qui remplissent les irrégularités du subs
tratum primaire, avant le dépôt du Trias qui le recouvre en nette
discordance et repose même en transgression sur le Cambro-Ordovi-
cien.

Les coupes les plus connues (5, 6, 11, 20, 21, 23, 35) ayant
fourni des fossiles qui permettent une datation, s'étendent près
de Pont de Suert, dans les vallées du Noguera Pallaresa et du Noguera Ribagorzana. Les études effectués le long des Pyrénées (10,
14, 16, 22, 39) démontrent que les conclusions obtenues dans cette aire sont valables le long de toute la chaîne.

Près de Pont de Suert une série permienne de 600 m. d'épaisseur (35) se trouve en concordance sur le Stéphanien des mines de
charbon de Malpas. Elle comporte à la base un conglomérat de galets peu arrondis et hétérométriques formés de dolomie, schistes
paléozoïques et roches éruptives. (fig.2)

Au-dessus de ce conglomérat se trouve un grès à ciment calcaire et argileux comportant de rares grains de quartz et de felds
paths et d'abondants fragments de roches carbonatées de schistes
et de roches éruptives. Des lits de calcaire s'y intercalent ain-

PYRENEES

PONT DE SUERT (M)

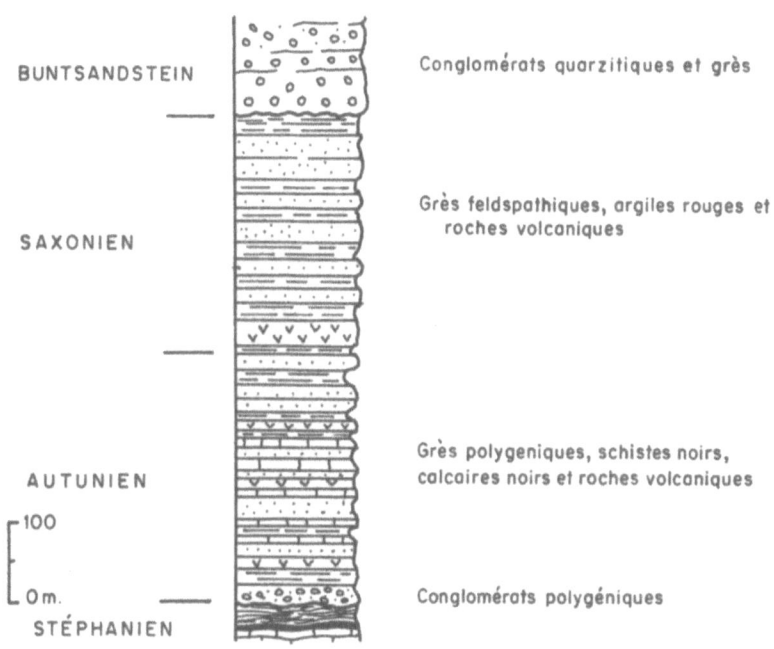

BUNTSANDSTEIN — Conglomérats quarzitiques et grès

SAXONIEN — Grès feldspathiques, argiles rouges et roches volcaniques

AUTUNIEN — Grès polygeniques, schistes noirs, calcaires noirs et roches volcaniques

100

0m. — Conglomérats polygéniques

STÉPHANIEN

COUPE GEOLOGIQUE DU PERMIEN DE PERANERA

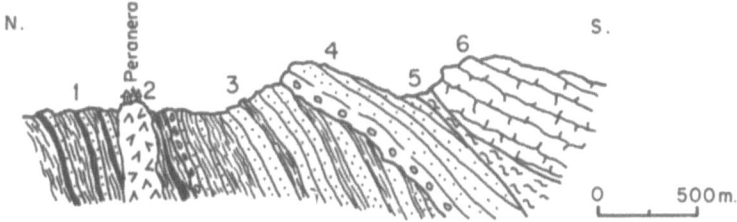

N.　Peranera　1　2　3　4　5　6　S.

0　　500m.

Fig. 2

si que d'abondantes roches volcaniques et des schistes noirâtres
à particules charbonneuses. Ces mêmes schistes noirs dans d'autres endroits très proches (5, 6) contiennent une flore qui dé-
termine l'âge autunien de ces formations (6, 14, 20):

> <u>Annularia stelate</u>. Schl.
> <u>A. spicata</u>. Gut.
> <u>Calipteris conferta</u>. Ster.
> <u>Odontopteris Duponti</u>. Zeil.
> <u>Psygmophyllum expansum</u>. Zal.
> <u>Psy. mongolicum</u>. Zal.
> <u>Walchia filiciformis</u>. Schot.

La partie supérieure de la série permienne présente par contre un facies saxonien avec d'abondantes argiles rouges. Même si
l'aspect général fait penser au Buntsandstein, beaucoup de caractères sont différents.

Dans le Permien les matériaux ferrugineux sont toujours détritiques, les grès interstratifiés sont mal classés et très felds
pathiques, les micas sont rares et le ciment très abondant est ar
gileux et calcaire. Les carbonates y forment des nodules d'origine pédogénétique et les matériaux volcaniques sont fréquents.

Le Permien supérieur "rouge" est plus important et mieux développé que le Permien inférieur "noir" et souvent il est seul re
présenté.

Le Buntsandstein discordant et transgressif recouvre le Permien en débutant par un conglomérat quartzitique et quartzeux
constant et caractéristique qui déborde largement les bassins per
miens.

Nagtegaal, Mey et d'autres (10, 21, 23, 39) ont décrit, dans
cette région de Pont de Suert, la "Formation Peranera" qui semble
correspondre au Permien supérieur "rouge". Pour eux le Permien in
férieur "noir" qui à Peranera n'est pas fossilifère resterait in-
clus dans la "Formation Malpas" avec le Stéphanien.

Entre le Permien et le Trias il y a eu probablement une interruption sédimentaire et une érosion très importante. Pour cette raison, il faut employer avec prudence le terme "supérieur"
pour le Permien rouge de type saxonien.

Comme nous l'avons déjà dit, cette coupe est valable, avec
ses épaisseurs différentes tout le long des Pyrénées sauf l'extrémité occidentale. Dans le Pays basque (16, 22) les niveaux inférieurs du Permien sont plus carbonatés, quelquefois ce sont de
vrais calcaires plus ou moins brechiques avec des intercalations
détritiques et des roches volcaniques.

Mais nous ne pouvons pas assurer que ces séries carbonatées
correspondent aux schistes noirs des Pyrénées centrales qui sont
d'âge autunien. Par contre les niveaux plus élevés du Permien
sont des grès et des argiles rouges très développés, semblables à
la formation peranera et très probablement de même signification.

Les dispositions des affleurements des bassins permiens présentent des directions différentes dans les Pyrénées centrales et orientales mais c'est toujours en relation avec la tectonique hercynienne. On observe plus à l'ouest, aux alentours de San Sébastian, un phénomène généralisé dans le reste de la Péninsule Ibérique: quand le socle paléozoïque est recouvert par le Permien il n'est pas altéré. Quand ce socle est recouvert par le Buntsandstein transgressif, il est altéré et rubéfié sur une ou plusieurs dizaines de mètres d'épaisseur.

Nous reviendrons, à la fin de cette étude, sur ce problème qui peut être étudié dans de meilleures conditions d'affleurements dans le Centre de l'Espagne.

3.2. La Chaîne Cantabrique

Nous possédons moins d'observations sur cette région (9, 14, 24, 19).

Au sud de Torrelavega, en bordure du bouclier paléozoïque cantabrique (fig. 1, 0) se trouve une série de petits bassins permiens comparables aux bassins pyrénéens. Ils reposent sur le Stéphanien ou sur le paléozoïque plus ancien. L'épaisseur est variable mais ne dépasse pas les 300 m. La série est analogue à celle que nous avons décrite près de Pont de Suert (fig. 1, M) avec une nette prédominance de grès et d'argiles rouges. Elle est recouverte par le Trias grâce à une discordance plus ou moins importante.

A l'Ouest du bouclier cantabrique, se trouvent quelques affleurements isolés, difficiles à rattacher à ceux du reste de la Péninsule Ibérique. Les plus orientaux (fig. 1, P) sont ceux de Colunga et de Caravia. Des affleurements de caractère analogue s'étendent plus à l'ouest, jusqu'à proximité d'Oviedo (fig. 1, 0) (fig. 1, 14, 19, 24).

Les séries sont très différentes de celles qui existent de l'autre côté du bouclier paléozoïque cantabrique (fig. 1, 0). Par contre elles ont une certaine ressemblance avec celles des Pyrénées occidentales (fig. 1, N). Les séries de base sont très carbonatées et les roches volcaniques abondantes.

La série que nous allons décrire correspond à la région de Colunga et de Caravia (fig. 1, P) où les affleurements sont meilleurs et où il y a, de plus, des exploitations minières de fluorine.

Le Permien est tantôt concordant sur le Stéphanien, tantôt discordant sur les formations plus anciennes.

Le niveau de base d'une épaisseur approximative de 100 m. est bréchique et carbonaté. C'est une formation hétérogène composée de marnes noirâtres, de grès gris ou rougeâtres, et en prédominance d'une brèche calcaro-dolomitique foncée très compacte.

Le niveau qui le surmonte et lui fait suite graduellement est formé de roches volcaniques presque exclusivement. Son épaisseur est variable.

La partie supérieure du Permien est surtout détritique et malgré des passages de roches volcaniques et carbonatées la couleur dominante est le rouge et l'épaisseur varie de 40 à 70 m.

Quelquefois il n'existe que le niveau détritique supérieur reposant sur du Paléozoïque moyen ou inférieur.

Des discordances internes se présentent dans la série permienne et le tout est recouvert par le Trias, aussi en discordance.

Les bassins sédimentaires sont conditionnés par la tectonique hercynienne, toujours très compliquée et souvent les accidents ont continué après la sédimentation du Permien.

Patac (24) a trouvé une flore (Walchia piniformis, W. hypnoides, Calipteris conferta) qui lui permet d'attribuer à l'Autunien les niveaux de base du Permien des Asturies et au Saxonien (Pinites permiensis Ren. et Antholithus permiensis Ren.) les niveaux supérieurs.

3.3. Le Système Central

Le Permien de cette région était peu connu il y a encore deux ou trois ans, et sauf de rares exceptions les matériaux étaient attribués au Trias.

C'est donc sur nos travaux personnels que nous nous appuyons pour donner la description suivante (12, 13, 26, 31, 37).

Le Permien s'est déposé dans des bassins isolés, conditionnés par la tectonique tardihercynienne. Il est nettement discordant sur le Paléozoïque plus ancien (fig. 1, A, B, C, D) mais la base n'est pas toujours visible. Les affleurements très épais dépassent souvent les 1.000 m.

Dans la série de Palmaces (fig. 1, C) et la série de Sierra Pela (fig. 1, A) il y a trois niveaux, qui changent sans délimitation nette. (fig. 3).

Le niveau inférieur est formé de conglomérats, de tufs volcaniques et d'argiles.

Les argiles prédominent dans le niveau moyen avec de fines intercalations de grès.

Le niveau supérieur est formé de grès et de conglomérats à gros blocs.

Les conglomérats sont toujours formés de galets anguleux et hétérométriques, de nature variée: quartzites, schistes, blocs cristallins. Des matériaux volcaniques plus abondants s'y ajoutent dans les niveaux inférieurs. Tous ces galets n'ont subi aucune altération.

Les argiles et les grès sont toujours rouge-violacé avec quelques bandes et taches verdâtres. Il y a des nodules calcai-

SYSTEME CENTRAL

ATIENZA (B)   PÁLMACES (C)

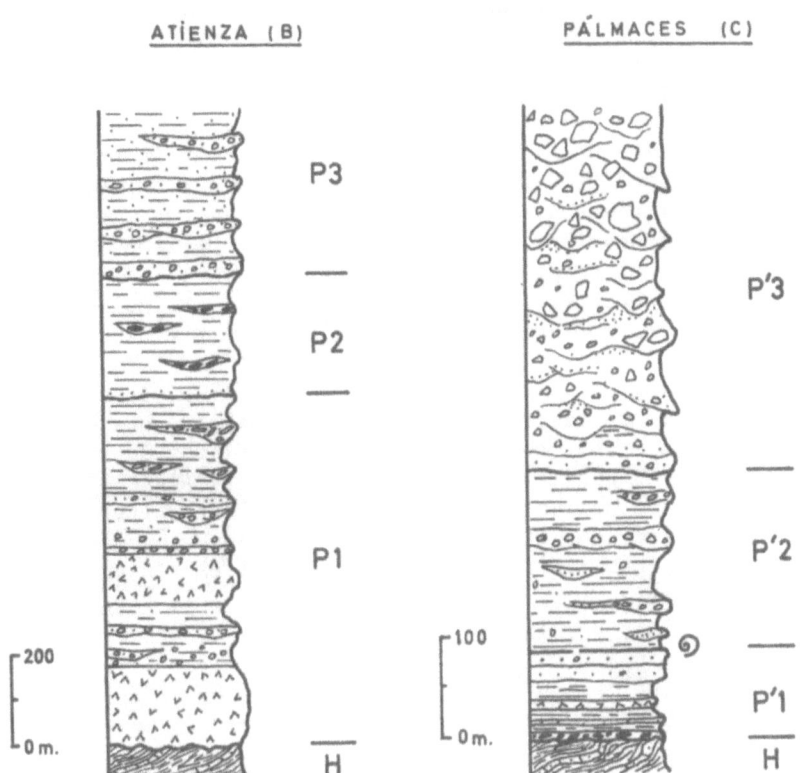

COUPE GEOLOGIQUE DU PERMIEN DU PÁLMACES

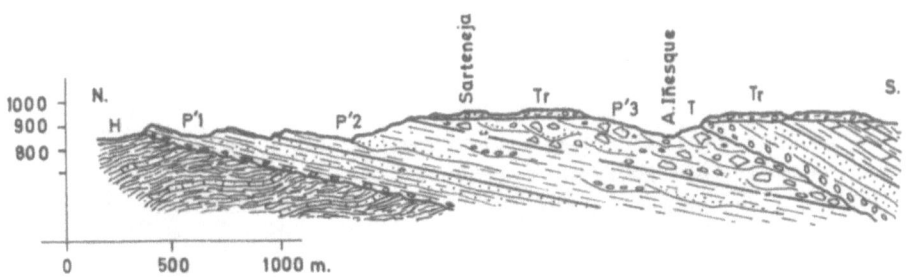

Fig. 3 - Pre-Stephanien (H). Permien : brêches, roches volcaniques, grès et argiles (P1, P'1); argiles avec grès et conglomérats (P2, P'2); conglomérats et grès (P3, P'3). Triasique (T). Tertiaire (Tr). 𝒢 : Niveau à Estherias.

res d'origine pédogénétique et des galets éolisés quelquefois
abondants à la partie supérieure de la série.

La stratification est irrégulière, les corps sédimentaires
lenticulaires et les paléocanaux fréquents et profonds.

Le socle, quand il est visible, est du Paléozoïque préstē
phanien. Le tout est recouvert par le Buntsandstein avec une
discordance angulaire atteignant parfois 45°.

Dans la série d'Atienza (fig. 1, B) certaines parties sont
comparables à celles de Sierra Pela et Palmaces mais la partie
inférieure est constituée par deux importantes coulées rhyolitiques interstratifiées avec des argiles, grès et conglomerats.

La partie moyenne est formée d'argiles avec de fines inter
calations de grès et conglomérats.

La partie supérieure comporte une alternance irrégulière
de grès, conglomérats et argiles. L'ensemble mesure plus de
1.200 m. (fig. 3).

Dans la région de Valdesotos-Tamajon (fig. 1, D) le problè
me est un peu différent: les niveaux de base, discordants sur
le Paléozoïque inférieur sont constitués de matériaux détritiques grossiers et d'argiles noires formant plusieurs cyclothè-
mes. La présence de pollens, de spores et macroreste végétaux
nous permet (31) de leur attribuer un âge autunien.

Ce gisement fossilifère contient, entre autres restes,
Callipteris conferta Ster, Callipteris raymondi Zei, verruco-
sisporides Kaipingiensis Ign. Spinosporides pinosus Alp., Vittatina costabilis Wil.

Ces formations étaient attribuées autrefois au Stéphanien,
et constituent actuellement l'unique gisement autunien du centre de la Péninsule Ibérique.

Il est intéressant de noter que cette série avec la flore
qu'elle comporte est semblable à celle que Broutin (4) a trouvé
dans l'extrémité méridionale de la meseta espagnole. La flore
a aussi des analogies avec la flore autunienne des Pyrénées
(5, 6) et de même que celle-ci, dans son contexte euroaméricain,
elle a aussi des affinités avec la flore asiatique.

Pour le moment nous ne pouvons pas encore établir une corrélation claire entre ces niveaux autuniens et les niveaux de
base des autres affleurements (fig. 1, A, B, C).

Un fait général à toute la région et qui peut s'appliquer
aussi à la Chaîne Ibérique est que le socle paléozoïque n'est
jamais altéré lorsqu'il est recouvert par le Permien. Au contraire lorsqu'il est recouvert par le Buntsandstein, parce que
le Permien manque, il est altéré et rubéfié sur plusieurs mètres d'épaisseur.

3.4. La Chaîne Ibérique

C'est le domaine le plus vaste, riche en affleurements (fig. 1,

CHAÎNE IBERIQUE

MOLINA – ARAGONCILLO – VENTOSA

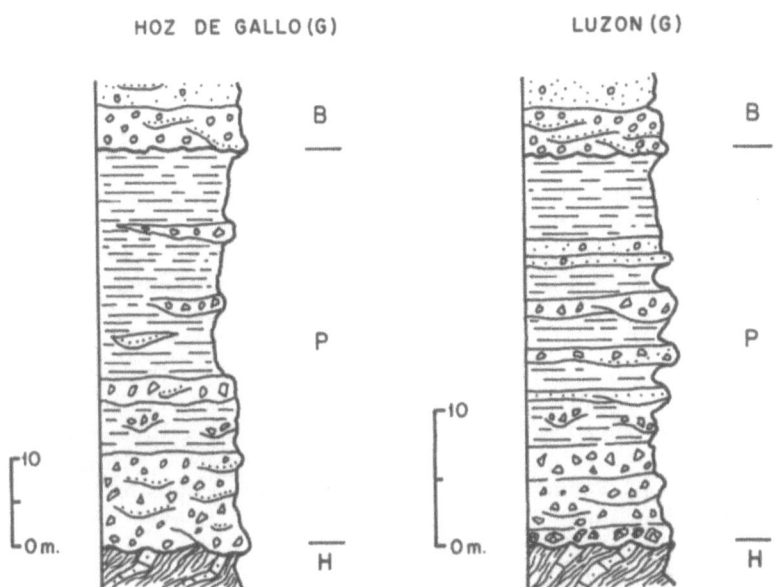

Paléozoïque inférieur(H). Permien: brêches, argiles et grès (P)
Triasique: conglomérats et grès du Buntsandstein (B)

COUPE GEOLOGIQUE DE LA GORGE DU "GALLO"

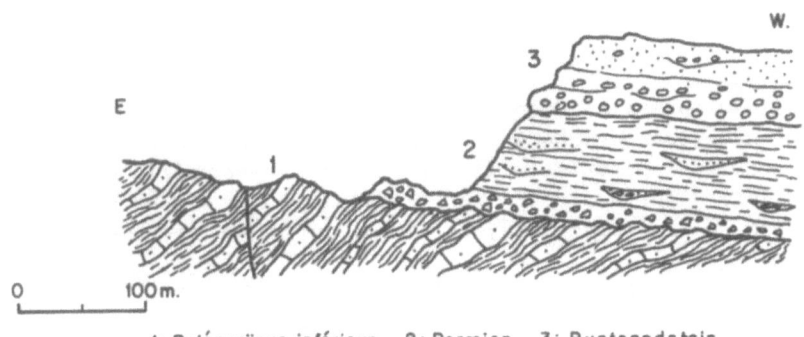

1: Paléozoïque inférieur   2: Permien   3: Buntsandstein

Fig. 4

E, F, G, H, I, J, K) et le Permien s'y trouve aussi en profon-
deur, recouvert par le secondaire ou le tertiaire (30).

Déja Jacquot (15) et Riba et Rios (27) parlent de matériaux qu'ils attribuent au Permien, mais c'est Sacher (28) en
1966 qui démontre son existence dans la région de Aragoncillo-
Molina de Aragón (fig. 1, G.). Il trouve des matériaux discordants sur le Stéphanien, eux-mêmes recouverts en discordance
par le Buntsandstein.

Boulard et Villard (3) apportent les preuves paléontologiques de l'existence du Permien. Ils trouvent près de Landete
(fig. 1. J) une flore attribuable au Zechstein. Ils citent, entre autres, Lueckisporites virkkiae Pot., Taeniaesporites al-
bertae Jan., T. Noviatilensis Lesch., Ingasporites delasancei
Les., Vesicasspora ovata Hart.

Ces dernières années de nouvelles observations ont été fournies par les travaux de Villena (34) Marfil et Perez Gonzalez
(17) Marin (18) Gabaldon et Peña (8) Talens et Melendez (32)
Arche, Hernando, Ramos, Sopeña et Virgili (2, 37).

Au long de la Chaîne Ibérique l'épaisseur des séries varie
de moins de 10 m. à 100 m. a peine. Des matériaux détritiques
fins y prédominent avec quelques niveaux bréchiques à la base.

Le nom de "Formation Montesoro" donné au Permien d'Aragoncillo (fig. 1. G) par Sacher (28) peut être utilisé dans toute
la région, même si on peut discuter si la coupe de Montesoro
est la meilleure pour définir la formation, nous pensons qu'il
ne faut pas multiplier les noms des unités lithostratigraphi-
ques, comme ceci vient d'être récemment fait. Ainsi Marfil et
Perez Gonzalez (17) ont proposé le nom de "Formation Bosque"
pour ces mêmes séries dans une coupe voisine (fig, 1, G) et
Marin (18) a proposé le nom de "Formation Feliciana" pour un
affleurement (fig. 1, H) proche et de lithologie analogue.

La description de Sacher (28) est toujours valable pour
la plupart des affleurements. Il y a deux niveaux bien diffé-
renciés malgré un certain passage latéral de l'un à l'autre.

Le niveau inférieur est constitué par des brèches polygéniques faites de gros blocs de schistes, de roches cristallines à peine altérées dans un ciment ferrugineux très important. (fig.4)

Le niveau supérieur comprend des argiles et des grès rouge-violacé avec quelques taches verdatres, des nodules calcaires et des galets parmi lesquels des quartz et quartzites éolisés. Ce niveau supérieur où l'argile prédomine sur les grès,
est toujours le plus développé; le niveau inférieur peut même
ne pas exister.

L'affleurement de Landete (fig. 1, J) est le plus méridio-
nal que nous ayons étudié. Ici la série est un peu différente.
Elle est plus détritique et plus épaisse (250 m.) Les argiles et

les grès rouges comportent dans sa partie plus élévee des niveaux noirâtres avec un ensemble de spores (3) qu'on a trouvé
associé en Tunisie, à une faune marine du Zechstein.

La série de Landete ressemble beaucoup à l'affleurement per
mien oriental du Desierto de Las Palmas (fig. 1, K) et lors de
sondages (30) une flore semblable à celle de Landete a été trouvée près de Caspe (fig. 1, R).

Le socle de ces séries est tantôt le Primaire pré-stépha-
nien discordant, tantôt le Stéphanien plus ou moins discordant
aussi. Dans ces aires où il y a une certaine continuité sédimentaire l'épaisseur du Permien est plus importante. Il est recouvert par le Buntsandstein légèrement discordant et largement
transgressif.

4. CONCLUSIONS

Nous avons préféré faire un exposé des observations de terrain
plutôt que donner des conclusions prématurées. Les données que
nous possédons nous permettent toutefois de mettre en évidence
quelques résultats.

4.1. Chronostratigraphie

L'existence de l'Autunien est prouvée (3, 5, 6, 24, 31) dans
des bassins du nord, du centre et du sud de l'Espagne (fig. 1,
M, P, D, et Seville) mais pas dans la partie orientale.

Dans la plupart des affleurements la sédimentation débute
par des séries rouges de faciès saxonien. Flore attribuable au
Zechstein est présente quelquesfois en surface (fig. 1, J, R) au
en sondages (30), dans la partie orientale de la Péninsule.

Nous pouvons donc affirmer que la série permienne est complète dans la partie profonde de quelques bassins. Mais souvent
les niveaux inférieurs font défaut et il y a une interruption a
la fin de la sédimentation permienne et le début du Buntsandstein.

4.2. Le passage du Permien au Buntsandstein.

Une discordance angulaire et cartographique sépare le Permien
du Buntsandstein. Dans la Chaîne Ibérique la discordance est
plus difficile à voir car elle s'affaiblit.

Les horizons de base du Buntsandstein présentent souvent
le caractère des dépôts remaniés du Permien. En 'effect, des galets proviennent des conglomérats permiens sous-jacents mais
sont toujours très altérés. Parfois, comme à Sierra Pela (fig.
1, A), il y a même des galets de grès permiens.

Nous n'allons pas insister sur les différences de facies
du Permien et du Buntsandstein car elles sont tres nettes. Les
formations rouge violacé du Permien contiennent des galets non

altérés de schistes et de roches cristallines, et dans n'impor-
te quelle lithologie. La présence de galets éolisés a patine
vert-bleu est caractéristique; il s'en trouve aussi dans le
Buntsandstein mais leur surface est toujours dégradée.

Les niveaux de conglomérats du Buntsandstein contiennent
des galets de quartz arrondis et bien classés et alternent avec
des grès quartzitiques ou feldspathiques de couleur rose à stra
tification paralléle ou subparalléle laquelle contraste avec la
stratification lenticulaire du Permien.

Une autre donée utile pour l'interprétation paléographi-
que est que les sédiments rouges comportent toujours beaucoup
d'illite, une petite proportion de Kaolinite (0 a 15%) et des
interstratifiés (10 a 14 μ). La teneur en kaolinite augmente
brusquement au niveau du passage au Buntsandstein. Les niveaux
inférieurs du Buntsandstein sont spécialement riches en interstratifiés (10 - 14μ).

4.3. L'altération du soubassement de la couverture permo-tria-
     sique.

Quand le Permien n'existe pas et que le Buntsandstein recouvre
directement le socle paléozoïque, celui-ci est toujours altéré et rubéfié sur une ou plussieurs dizaines de mètres d'épais
seur.

A. Alcorlo (fig.1, D) le socle est composé de gneiss et
son altération est facile à suivre.

Entre le Buntsandstein et le socle il y a un niveau blanc-
verdatre trés argileux (illite et interstratifiés gonflants)
de 50 cm. Au-dessous se trouve un niveau bréchique avec des
fragments du socle très altérés et beaucoup d'oxyde de fer (géo
tite). Ce niveau a une couleur violette. Il est délimité par
rapport à l'horizon supérieur et passe insensiblement à l'horizon inférieur. Environ à 2 m. de la base du conglomérat triasique se trouve le gneiss altéré mais à texture conservée. En la-
me mince, les plagioclases apparaissent presque complètement
dissouts.

L'analyse minéralogique de la fraction inférieure à 2 μ
montre la présence d'une grande quantité de kaolinite (50%) et
la disparition progressive des interstratifiés gonflants. L'
altération va en s'atténuant jusqu'à 10 à 12 m. de la surface
de base du Buntsandstein.

Des profils semblables peuvent être observés en d'autres
endroits de la Péninsule.

Le socle hercynien présente toujours une altération rubéfiante de 10 à 40 m. Les coupes étudiées démontrent que ces
profils d'altération présentent une kaolinisation très ménagée
sur les 4 à 10 m. supérieurs. Les roches mères paléozoïques ne
contiennent pas de kaolinite (38).

Quand le socle paléozoïque est recouvert par les formations détritiques du Permien, il ne présente jamais les rubé-
factions et altérations décrites ci-dessus. Il est évident que
la couverture permienne a protégé le socle de ces altérations.
Toutefois dans le cas de couvertures permiennes minces (2 à 10
m.) comme à Nuevalos et Monterde (fig. 1, F) l'altération exis
te aussi.

Quand des altérations se développent à la base d'une formation gréseuse, un point délicat est à déterminer: si l'altération est antérieure à la transgression gréseuse il s'agit
d'une altération climatique; si elle est postérieure, due à
l'influence des nappes circulant à la base des gres, il s'agit
d'une altération posthume.

Nous sommes en mesure (38) de conclure qu'il existe presque toujours une altération posthume à la limite inférieur des
grès triasiques. Elle est blanchissante et engendre des interstratifiés.

De plus, elle s'est souvent superposée à une altération
climatique, rubéfiante et kaolinisante, déjà présente. Cette
altération climatique est postérieure à l'époque du dépôt de
la base des couches permiennes et antérieure à la couverture
du Buntsandstein.

Ces altérations sont donc permiennes et nous en voyons
les traces accumulées sous forme de séries rutilantes dans
les bassins sédimentaires. Elles se sont prolongées pendant
la sédimentation permienne sur les socles émergés. La transgression triasique les a "surprises" en les conservant dans
beaucoup d'endroits.

Enfin ces altérations rubéfiantes et kaolinisantes ont
continué à se produire sur les séries permiennes déposées,
pendant la période de répit qui a précédé l'arrivée des grès
du Buntsandstein. Et ceci explique la présence de kaolinite
dans la zone de passage.

4.4. Paléogéographie

Le Permien s'est déposé dans des bassins continentaux isolés,
au relief très énergique du aux dernieres manifestations de la
tectonique hercynienne.

Il est possible qu'une influence marine se soit manifes-
tée dans certaines aires du NW de l'Espagne (fig. 1, N, P, Q).

Au Buntsandstein correspond un changement total de la géo
métrie des bassins sédimentaires.

Les directions de courants, mesurées dans le Permien,
sont variables et contrôlées par des facteurs locaux et la
forme des corps sédimentaires démontre une sédimentation de
tipe alluvial-fan. Dans le Buntsandstein les directions des

courants sont plus constantes et la sédimentation plus calme
(2). Ainsi le Buntsandstein représente la base de la transgres
sion secondaire qui recouvre d'une façon générale et presque
uniforme la partie centrale et orientale de la Péninsule Ibérique.

Les dépots du Buntsandstein sont en grande partie des matériaux du Permien remaniés, l'érosion ante-triasique étant
très importante. C'est pour cette raison et aussi par l'ac-
tion de la tectonique alpine que le tracé des affleurements
permiens est très différent du tracé des bassins dans lesquels
les matériaux se sont déposés.

Même si la dyagénèse a sans doute été importante, la petrographie des sédiments et des altérations donne des indications sur la tectonique et les climats qui ont permis l'accumulation des grès permiens (38). Les galets polycristallins
sont frais, les feldspath alcalins sont conservés, les plagioclases et les ferro-magnésiens ont disparu, les quartz sont
ferruginisés mais non cariés, l'illite est dominante, la kaolinite est présente, la gibbsite est absente.

Ces caractères n'ont rien à voir avec ceux des altérations des pays ferrallitiques ou latéritiques. Ils indiquent
au contraire des altérations ménagées de climats intermédiai-
res chauds alternativement humides et secs, tels que ceux qui
président à la naissance de sols fersiallitiques.

Il est possible que sur une pénéplaine stable ces climats
soient parvenus à développer des hydrolyses et des kaolinisa-
tions plus puissantes.

Mais ces altérations se sont développées sur des socles
animés d'une tectonique positive vivante.

Avant que l'altération n'ait développé tous ses effets,
ses produits sont entraînés dans les bassins et fossés voi-
sins.

REFERENCES

1. A.Almela y J.M.Rios, Investigación del hullero bajo
    los terrenos mesozoicos de la costa cantábrica. Em.Nac.
    Adaro Inv.Min., 1962.

2. A.Arche, S.Hernando, A.Ramos, A.Sopeña and C.Virgili,
    Distribution between environments in the Permian and Basal
    Triassic Red Beds in Central Spain. IXe Congrés Internatio

nal de Sédimentologie, Nice, 1975.

3. Ch. Boulouard et P. Viallard, Identification du Permien dans la Chaîne Ibérique, C.R.Acad. Sc. Paris, (273), 1971.

4. J. Broutin, Découverte de l'Autunien dans le bassins de Guadalcanal (Nord de la Province de Seville, Espagne du Sud), C.R.Acad. Sc. Paris, (278), 1974.

5. M. Dalloni, Etude géologique des Pyrénées catalanes, Ann. Fac. Sc. Marseille, (26), 1930.

6. M. Dalloni, Sur les dépôts permiens des Pyrénées a flore de l'Angaride, C.R.Seances de l'Academie des Sciences, (206), 1938.

7. R. Desparmet, H. Monrose und U. Schmitz, Zur Altersstellung der Eruptiv- Gesteine und Tuffite in Nordteil der Westlichen Iberichen Ketten (NE.-Spanien), Munster Forsch. Geol. Palaont., (24), 1972.

8. V. Gabaldon y J.A. de la Peña Blasco, Estudio petrológico del Carbonífero, Pérmico (?) y Triásico inferior del NW. de Molina de Aragón (Guadalajara), Est. Geológicos, (XXIX), 1973.

9. M. Gervilla y A. García-Laygorri, Sobre la datación del Permo Estefaniense en el extremo Noroccidental de la cuenca Carbonífera central Asturiana, E.N.A.D.I.N.S.A., Comunicación oral, 1975.

10. J.J.A. Hartevelt, Geology of the Upper Segre and Valira valleys, Central Pyrenées, Andorra, Spain, Leidse Geol. Mededelingen, (45), 1970.

11. J. Hartevelt et P. Roger, Quelques aspects de la topographie permo-triassique dans le Haut-Segre et la Haute-Pallaresa (Lérida, Espagne). Interpretation structurales et paléogéographiques. C.R. somm. S.G.F. (6), 1968.

12. S.Hernando. El Pérmico de la región Atienza-Somolinos (Provincia de Guadalajara), Bol. Geol. y Min. (LXXXIV), 1973

13. S. Hernando, El Pérmico y Triásico de la región Ayllón-Atienza (Provincias de Segovia, Soria y Guadalajara), Tesis Doctoral. Dpto. Estratigrafía y Geología Histórica. Universidad Complutense de Madrid, (En prensa), 1975.

14. I.G.M.E., Mapa Geológico de España a 1:50.000. y Mapa Geológico de España a 1:200.000. Síntesis de la cartografia existente.

15. E. Jacquot, Sur la composition et sur l'âge des assies que dans la Péninsule Ibérique séparent la formation carbonifere des dépots jurassiques, Bull. Soc. Geol. Fr.(24) 1866.

16. P. Lamare, Sur l'éxistence du Permien dans les Pyrénées basques. C.R.Soc. Geol. France, (31), 1931.

17. R. Marfil y A. Pérez-Gonzalez, Estudio de las series rojas pérmicas en el Sector Nor-Occidental de la Cordillera Ibérica (Región de El Bosque, Alto Tajuña), Est. Geológicos, (XXIX), 1973.

18. P. Marin, Estratigraphie et évolution paléogéographique post-Hercynienne de la Cahîne Celtibérique orientale anse confins de l'Aragón et du Haut Maestrazgo (Provinces de Teruel et Castellón de la Plana, Espagne), Tesis Doctoral. Universidad Claude-Bernard - Lyon, 1974, (Inédito).

19. B. Melendez, Nota previa sobre los terrenos Pérmicos de Colunga y Caravias. Bol. R. Soc. Esp. Hist. Nat. (48), (2), 1950.

20. J. Menendez Amor, Algunas plantas fósiles permianas de la provincia de Lérida, Notas y Comuns. Inst. Geol. y Min. de Esp., (28), 1952.

21. P.H.W. Mey, Geology of the Upper Ribagorzana and tor Valleys, Central Pyrenees, Spain Sheet 8, 1:50.000, Leidse Geol. Mededeling., (41), 1968.

22. D. Muller, Perm und Trias im Valle de Baztán (Spanische West pyrenaen), Dissert., Fak. Natur. Geist. wiss. T.U. Chausthal. 1969.

23 P.J.C. Nagtegaal, Sedimentology, Paleoclimatology, and Diagenesis of post-Hercinian continental deposits in the South-Central Pyrénées, Spain, Leidse Geol. Mededelingen (42), 1969.

24. I. Patac, La formación Uraliense Asturiana. Estudios de Cuencas carboníferas, Comp. Asturiana de Artes Gráficas, S.A. Gijón, 1920.

25. L. Pérez de Cossio, El terreno carbonífero de Tamajón, Retiendas y Valdesotos, en la provincia de Guadalajara, Bol. Comp. Map. Geol. Esp. (6), 1920.

26. A. Ramos y A. Sopeña, Estratigrafía del Pérmico y Triásico en el sector Tamajón-Pálmaces de Jadraque (Provincia de Guadalajara), Estudios Geológicos, (En prensa).

27. O. Riba y J.M. Rios, Observation sur la structure du secteur sudouest de la chaîne Ibérique (Espagne), Livre Mém. Prof. P. Fallot. Mem. Hors série Soc. Geol. Fr. (1), 1960-62.

28. L. Sacher, Stratigraphie und Tektonik der NorWeslichen Hesperischen Ketten bei Molina de Aragón (Spanien), Teil I. Stratigraphie (Palaozoikum), N. J. Geol. Palaont. (124), (2 T.S.), 1966.

29. E. Soers, Stratigraphie et Géologie structurale de la partie oriental de la Sierra de Guadarrama, Studia Geologica IV. Universidad de Salamanca, 1972.

30. R. Soler, Datos sobre el Pérmico de Castellón de la Plana y zonas próximas, Auxini-Hispanoil, Comunicacion oral, 1975.

31. A. Sopeña, J. Doubinger y C. Virgili, El Pérmico inferior
     de Tamajón, Retiendas, Valdesotos y Tortuero, (Borde S.
     del Sistema Central), Tecniterrae (1), 1974.
32. J. Talens y F. Melendez, El anticlinorio de Cueva del Hie-
     rro. El Pérmico del Barranco de la Hoz, Este de Masegosa.
     (Serrania de Cuenca), Est. Geologicos, (XXVIII), 1972.
33. P. Viallard, Recherches sur le cycle alpin dans la Chaîne
     Ibérique sud-occidentale. Tesis doctoral, Trav. du Lab.
     de Géol. Meditérranée. C.N.R.S. Université Paul Sabatier,
     Toulouse, 1973.
34. F. Villena, Estudio geológico de un sector de la Cordille-
     ra Ibérica comprendido entre Molina de Aragón y Monreal
     del Campo. Tesis doctoral. Universidad de Granada. Inédi-
     to.
35. C. Virgili, The sedimentation of the permotriassic rocks
     in the Noguera Ribagorzana valley (Pyrenees - Spain). In-
     tern. Assoc. of Sedimentol. Copenhagen, 1960-61.
36. C. Virgili, Le trias du Nord-Est de l'Espagne. Ext. Liv.
     Mem. du Prof. P. Fallot. Soc. Geol. France (I), 1960-62.
37. C. Virgili, S. Hernando, A. Ramos y A. Sopeña, La sedimen-
     tation permienne au centre de l'Espagne. C.R. somm. S. G.
     F. (XV), (5-6), 1973.
38. C. Virgili, H. Paquet et G. Millot, Alterations du soubas-
     sement de la couverture permo-triasique en Espagne. Bull.
     Groupe Franc. Argiles, (26), 1975.
39. J.H.N. Wennekers, The Geology of the Esera valley and the
     Lys Caillanas Massif. Central Pyrénées, Spain, France.
     Leidse Geol. Mededelingen, (41), 1968.

THE PERMIAN/TRIASSIC BOUNDARY IN THE CONTINENTAL AREA OF MIDDLE EUROPE

Egon Backhaus

Geologisch-Paläontologisches Institut,
TH Darmstadt, D 61 Darmstadt, Germany FR

ABSTRACT. The boundary stratotyp in the South German branch of the UpperPermian/Triassic Basin is described and will be compared with the North German Basin. It is tried to attach this boundary to the worldwide correlated marin biostratigraphic units as well as to scale of palaeomagnetism.

ZUSAMMENFASSUNG. Die fazielle Ausbildung der Gesteine im Grenzbereich Zechstein/Buntsandstein des süddeutschen Teilbeckens wird beschrieben! Die Grenzziehung wird mit der im norddeutschen Becken üblichen verglichen. Es wird versucht, diese Grenze mit der weltweit gültigen biostratigraphischen Perm/Trias Grenze und der paläomagnetischen Skala zu korrelieren.

When F.von ALBERTI established the Trias System in 1834, he was unable to give a stratotype. It was not possible then to define the boundary. We are not better off in the south german area today, because the continental sediments of this area are rather unfossiliferous and full of gaps. The international biostratigraphic standard-scale cannot be applied. The sediments beneath and above the boundary are terrestrial deposits - red coloured silts, sands and arkosic sandstones and conglomerates.
The questions are:
1. Which is the local facies in the boundary zone and how to substantiate and to proof it.
2. How is it to correlate with the North German Basin.
3. Is it possible to correlate the Zechstein/Bunter

boundary with the international standard scale
(Permian/Triassic).

In the North German Basin the red coloured Rotliegend
and Bunter layers are separated by saliniferous and
carbonatic Zechstein sediments. The "Grenzanhydrit"
A 4 r forms the top of the Zechstein. The Bunter
begins with the redbrown coloured sandy-clayish silt
facies of the "Bröckelschiefer".
Thin dolomitic and anhydritic intercalations suggest
a small connection with the facies of the Upper Zechstein sediments. It was often discussed whether to
add these red pelites to the permian stage or not.

Fig. 1. Simplified geological map of the Southwest
Germany area, where Permian/Triassic layers overlap
crystalline (+ +). B=Basel,S=Straßburg,K=Karlsruhe,
H=Heidelberg,M=Mainz,F=Frankfurt.
Abb.1. Vereinfachte geologische Karte des südwest-
deutschen Raumes in dem Perm/Trias Schichten auf
Kristallin (+ +) übergreifen.

But the sediments become increasingly sandy to the
south. Therefore these sediments were accounted to
the Bunter. DIEDERICH thought a sandlayer on the
bottom to be the base surface of the Bunter in the
southern basin. It was a compromise to take the Brök-
kelschiefer for an intercalated member between Zechstein and Bunter. This resembles to the old opinion
that this part was the whole Lower Bunter in southern
Germany.

In the course of the boundary from north to south it
becomes more and more difficult because the sandy
components increase. The three sand beds separated by
DIEDERICH in the Werra-region within the Bröckelschiefer increase in number and thickness to the Odenwald
region. Most of these sand beds are lenses and their
durability as horizons becomes doubtful.

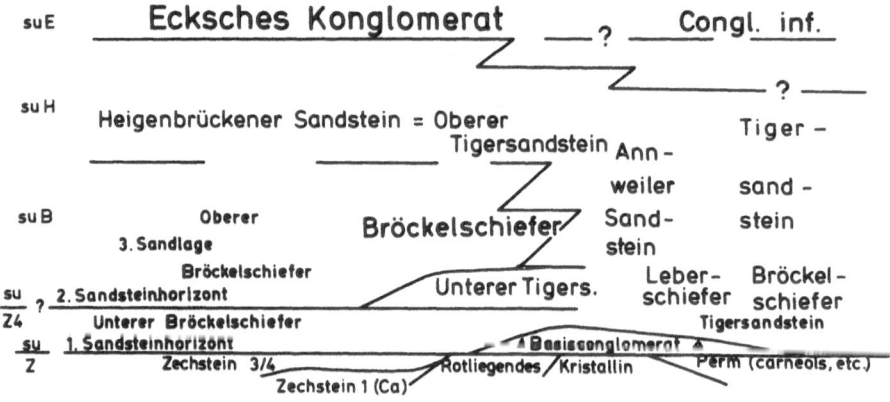

Fig. 2. Different terms of the basal layers of the
Lower Bunter in Southwest Germany subjected to facies.

Abb. 2. Unterschiedliche Benennungen der basalen
Schichten des südwestdeutschen Unteren Buntsandsteins
in Abhängigkeit von der Fazies.

Another insecurity results of the boundary itself. In
the Thuringian area GRUMBT takes the 2. sand bed in
DIEDERICH's scale to be the base of the Bunter, therefore the Lower Bröckelschiefer has to be accounted to
the Upper Zechstein (s.fig.2). In the regions of Heidelberg and of the Palatinate it is doubtful which of
the sand lenses adequates to the first and to second
horizon.
In places like Spessart and Odenwald f.e. the Bunter
lies upon crystalline directly. A basal conglomerate
is developed of local material sometimes. On top of
it lies the Bröckelschiefer facies. The intercalated
sand beds are often bleached, spots of one cm diameter
filled by Fe-Mn yielded the name "Tigersandstein" to
these banks, particularly to Lower T.The subsequent
fine-grained sandstone is called "Oberer (upper) Tigersandstein". In the lateral continuation to the north
(Spessart area) the sandstone is named "Heigenbrücke-
ner Sandstein". In the Palatinate both facies-types
are represented by the fine to medium grained sandstone called "Annweiler Sandstein".
In the Black Forest the same different facies which
we know of the Odenwald lies beneath the "Ecksches
Konglomerat". The results of LEIBER show, that the
basal conglomerate, the Bröckelschiefer facies and
the Tigersandstein are not only a vertical succession
of beds but on the contrary a regional differentiated
facies in the time.
In the Vosges area the research by HOLLINGER 1970 and
KONRAD 1971 yielded the same spotted sandstone of the
type "Tigersandstein" beneath the conglomerat inférieur (= Ecksches Konglomerat), these beds are named
Senones-Schichten by HOLLINGER.
These analogous successions of beds are spreaded on
both sides of the Rhine till the line Bruyères -
Ribeauville - Schramberg to the south and get the
greatest thickness and the most considerable differentiation in the region of the old permian troughs of
St Dié, Weiler - Offenburg and Schramberg.
The superposing Ecksches Konglomerat overlaps the
same differentiated stratigraphic sequence in the
eastern Palatinate in the permian bay of Albersweiler
- Annweiler. There the beneath the Annweiler Sandstein
situated "Leberschiefer" is not only equal to the
Bröckelschiefer-which may be accounted here to the
Zechstein - but after the research of HENTSCHEL it
may be its lateral gradiation of facies. This facies
difference can be recognized in a permian subsidence
trough too.

The "Staufer Konglomerat" in the Palatinate is controversial in the geological age determination, but it
seems quite sure to be older than the "Ecksches
Konglomerat". It is considered to be the marginal
facies of the "Oberen Sandstein (=Zechstein 1 after
HENTSCHEL) or even the Annweiler Sandstein. Under the
"Annweiler Sandstein" respectively the "Leberschiefer"
follows in the Palatinate a dolomitic sandstone or
shaly clay with dolomitic nodules, which contain
pelecypods and gastropods (HENTSCHEL) described by
BACKHAUS (1961) of the Zechstein dolomites of the
Odenwald area. - The carbonate facies of the marine
Zechstein crops out in the Odenwald and Spessart areas
with dolomites of Ca 1 and Ca 3 (Spessart only). These
dolomites are overlapping thin Rotliegend or crystalline directly. In this case there is no difficulty to
separate Rotliegend and Bunter. -
The sequences of Bunter are underlain in the Vosges
and the Black Forest by sedimentary complexes which
include one or more horizons with carneols and dolomites. These carneols and dolomites are developed in
beds, streaks or nodules. The main range and thickness is given in the areas of the former Rotliegend
troughs. In the Vosges HOLLINGER named this formation
St.Dié-Schichten, they are underlain by the so called
Champeney-Schichten. He puts them both together into
the Thuringien and compiles them with the "Rötel-
schiefer" of the Baden-Baden trough. We know the same
carneols in the"Rotliegend facies of Zechstein" sediments in the northern margin of the Odenwald (BACK-
HAUS,1965). Towards the Kraichgau Basin, where the
Zechstein is particulary represented by carbonatic
facies, the carneols are lost. The same abundance is
found towards margin of the Odenwald-swell.
The today-still-recognizable distribution of the Zechstein sea in the south is well known(s.fig.1). The
appoarence of the carneol-dolomite horizons in the
surroundings of the basin was shown manifold by FALKE
(1971, 1972). The distribution goes further to the
south than the "Tigersandstein". The carneol-dolomitic
-horizons are limited at the top by the Bunter. The
permian internal classification is to be discussed.
In the trough of the Black Forest (SCHNEIDER, STELL-
RECHT) and the Vosges (HOLLINGER) the permian stratigraphic successions are divided and spliced so detailed into the overcame stratigraphic division of the
Autun and Saxon, that the carneols appear always above
the top of the Rotliegend in the Zechstein range. In
the northern Odenwald they exist in lateral distribution of the "Zechstein in Rotliegend-facies".

Since the dolomites and carneols in sediments of Rotliegend facies look so much like those of the Middle/
Upper Bunter stage, it seems possible to get an as-
sertion of their stratigraphic range by a comparision
of genesis. In this we can refer to the research of
LANGBEIN, who gave practical proof of the description
and interpretation by BACKHAUS (1968).
The environment in the carneoliferous basal Upper
Bunter of Thuringia suggests a marginal marine situation in a milieu with high salinity. Carbonatecon-
cretions, anhydrite accumulation, poor in feldspare,
heavy mineral placers and grain-size distribution
belong to it- At a low water level and a high salinity
calcareous algae wake up. In near-shore flats the
crusts of algae burst and are redeposited by over-
flooding. Relic structures cased by early diagenetic
silification of algae carbonates can be recognized,
as they are described by WOLF & CONOLLY from the Upper
Devonian of New South Wales.
The origin of the silica may be the weathering of feldspar and first of all in this case the surface evaporation in arid areas. The silica for chalcedon genesis is be derivated from the saturated pore water
(ground water). Chalcedon coagulates after incidence
on carbonate.
The deposition of carneols in the Bunter seems to be
connected with a certain stratigraphical position.
There must have been conditions of critical influence,
which was not general during the Bunter. During Upper
Bunter it was the area of ingression of the Röt sea,
which was connected to a high concentration of salinity (anhydrit, halite). The pore waters of the underlying beds, highly saturated of SiO_2, are mixed with
the also high salinar concentrated waters of the
ingressiv Röt sea and the solubility of SiO_2 decreased. $CaCO_3$ will become more soluble and SiO_2 will be
precipitated. The environment was reducing, so it
could bleach the sandstone beneath.
In the marginal environment of the Zechstein sea the
author in 1965 mentioned structures of marine algae
and stromatolithes in carbonatic beds intercalated to
arkosic sandstones, which could be classed into the
Zechstein by their palaeogeographic situation. Similarly there are carneols in the margin of the Zechstein
sea (see Black Forest and Vosges area too).
The vicious circle is apparent. The circumstances of
carneol development in the Upper Bunter in facies
and paleogeography are nearly the same as in the
Upper Permian. The gradual encroachment of the Röt
sea to the south in time is well known to us.

The same mechanism is described by the author of the
uppermost Rotliegend in North Germany before Zech-
stein-transgression sensu strictu begins (BACKHAUS,
1964). The transgression and regression of the Zechstein sea in the South German area further than the
distribution of Zechstein dolomite today can be explained with the process above.
The origin of the carneoliferous facies in the period
of Zechstein(Upper Permian) has after all the largest
probability. Its intercalation in sediments of Rotliegend facies is no argument against this (s. BACK-
HAUS 1965, FALKE 1969).
To which formation of the Zechstein range the carneols
belong remains undecided. Due to the distribution of
the Z 1-dolomite even further south than Stuttgart,
it is probable that they belong into this period. But
the gap in Zechstein-2 allows only the period of Z 3,
also since the carbonatic facies can be identified in
the Spessart. The transition facies from Zechstein to
Bunter is a red pelite.
Whether the boundary between Zechstein and Bunter in
comparison with the north german basin is correct and
whether the consens about the boundary there harmoniz-
es with the Permian/Triassic boundary in the worldwide marine facies has to be tested.

For the chronological classification we dispose of
the biochronology and we can use the palaeomagnetism.
In a not yet published paper, DACHROTH (1074) classi-
fies all hitherto discussed layers of Lower Bunter by
palaeomagnetic measurements into Thuringien (Zechstein).
All the different facies of the Lower Bunter beneath
the "Ecksches Konglomerat" are considered to be a
marginal facies of the germanic Zechstein. - Since we
know formerly the carbonate facies of Z 1 and Z 3 in
the south, it can be only the facies of the period
of Z 4 and Z 5. - The details cannot be discussed here,
but the period of mixed polarisations is replaced by
a period of normal magnetism on the bottom of the Ecksches Konglomerat. The Bunter would start with it and
therefore with a real "Sohlbank"-cycle, as in former
times the Middle Bunter in South Germany was supposed
to do - This new beginning with a conglomerate would
bring a very interesting parallelism to the Verrucano
typico in the mountains of Pisa. -
In comparison with the international palaeomagnetic-
standard-scale, DACHROTH has to put the Permian/Trias-
sic boundary (Illawarra Reversal is the mark) into the
Saxonien (Ober Rotliegendes). - Upper Saxonien and
Thuringien would become triassic. -

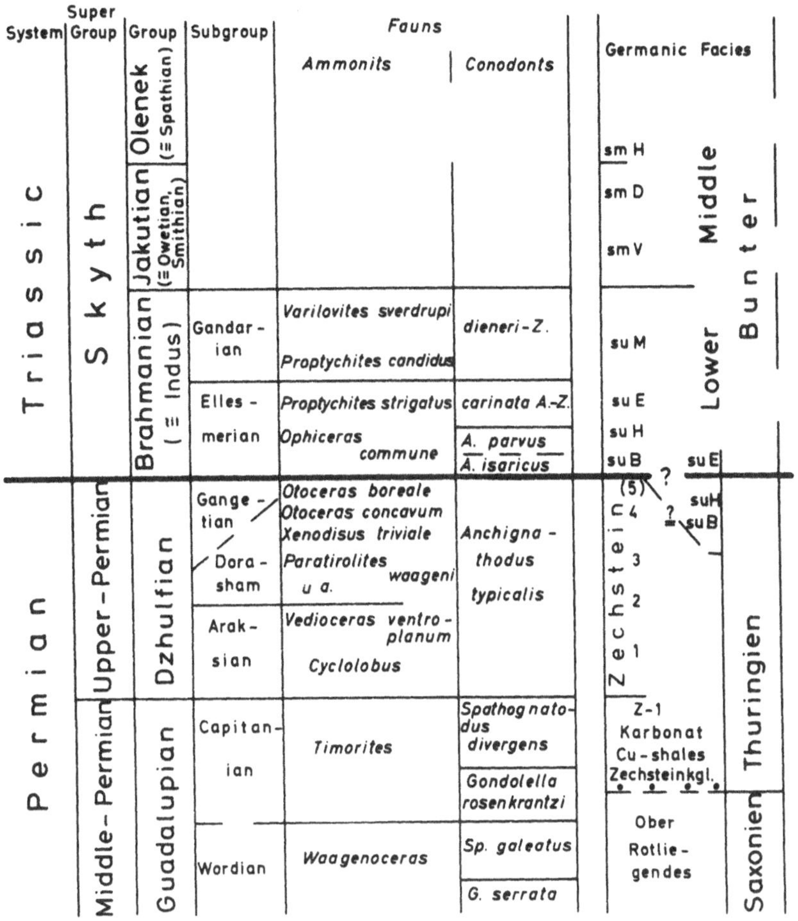

System	Super Group	Group	Subgroup	Ammonits	Conodonts	Germanic Facies
Triassic	Skyth	Olenek (≡ Spathian)				sm H / sm D / sm V — Middle Bunter
		Jakutian (≡ Owetian, Smithian)				
		Brahmanian (≡ Indus)	Gandar-ian	Varilovites sverdrupi / Proptychites candidus	dieneri-Z.	su M — Lower Bunter
			Elles-merian	Proptychites strigatus / Ophiceras commune	carinata A.-Z. / A. parvus / A. isaricus	su E / su H / su B ? suE / suH / suB
Permian	Upper-Permian	Dzhulfian	Gange-tian / Dora-sham	Otoceras boreale / Otoceras concavum / Xenodisus triviale / Paratirolites waageni u a.	Anchigna-thodus typicalis	Zechstein (5) 4 ? suH / suB — Thuringien
			Arak-sian	Vedioceras ventro-planum / Cyclolobus		Z 3 / Z 2 / Z 1
	Middle-Permian	Guadalupian	Capitan-ian	Timorites	Spathognatodus divergens / Gondolella rosenkrantzi	Z-1 Karbonat Cu-shales Zechsteinkgl. — Saxonien
			Wordian	Waagenoceras	Sp. galeatus / G. serrata	Ober Rotlie-gendes

Fauns

Fig. 3. Classification of Permian/Triassic boundary-beds, the marine standard scale (after KOZUR) compared with the Germanic Facies.

Abb. 3. Gliederung der Perm/Trias Grenzschichten. Vergleich der germanischen Fazies mit der Standard-Skala im hochmarinen Bereich nach KOZUR.

But in the phanerozoic the biochronological events are our scale. If it is not possible by the paucity of fossils in the continental sediments, yet we have to try this step in the northern part of the Germanic Basin however.

TRÜMPY compared in 1960 the East Greenland conditions
with our Zechstein/Bunter boundary. The Z./B.-boundary
would range higher than the faunistic boundary (Cyclo-
lobus/Otoceras) according to his hypothetical thoughts.
After an examination of the validity of the worldwide
standard for the triassic period, the boundary was
newly defined by KOZUR (1974 b). He pointed out (s.
fig.3) that the zone with Otoceras woodwardi (=bore-
ale) is of permian (gangetian) age. The triassic system
begins with Ophiceras commune. - That is why the Lower
Ophiceras layers in Greenland remain of permian age
like the benthonic faunes.
The newly defined boundary is characterized by an elimination not only of the index fossils but also of
rugose corals, permian brachiopods, bryozoes and crinoids. After the elimination of the Upper Permian
conodont-association with Anchignathodus typicalis,
the new triassic association with A.isarcicus and A.
parvus manifests. Just as well the coming of the
pelecypod Clareie clareia and C.stachei is marked.
These pelecypods are index fossils in marin beds
without ammonites (South Tyrol, Balaton).

With conodont-assamblage-zones it is possible to con-
nect the Zechstein of the Germanic Basin with the international biochronological standard scale (KOZUR,
1974 a). The conodont Spathognatodus divergens described by BENDER & STOPPEL in the Ca 1 (Zechstein
carbonate) can be found in the American Upper Capi-
tanian. Therefore the Zechstein carbonates belong into the Guadalupian group (Middle Permian). The higher
Zechstein is situated in the Upper Permian(Dzhulfian).
The other faunas of our Zechstein correspond to the
period from Wordian to Lower Dzhulfian.

As new investigations of the conodonts of our Zechstein
will stabilize the agreement between our Zechstein
and the international biostratigraphic standard scale,
there is hope to achieve biochronological correlation-
ship in our continental environment. It will be necessary to utilize the conchostracies, because in the
germanic Lower Triassic species of the genus Estheri-
ella are found which are missed in the Permian. The
forms of the Russian platform are well known. Charo-
phytae, spores and vertrebrates seem to be useful
in biostratigraphy to our sediments. Whether the subdivision of $L_o$wer Triassic on the Russion platform
can be transposed to the Germanic Triassic remains
undecided. But if we can subdivide the Lower Triassic
group into three subgroups by fossils, it will be an
success. It is possible that the footprints of Chiro-

therium and other reptiles can be useful for stratigraphy.

If it is still necessary to clear up some problems
of the Permian/Triassic boundary in worldwide marine
standard classification, it is without doubt necessary
in the continental environment, but in both cases it
seems possible that an answer can be found.

REFERENCES

ALBERTI,F.v.: Beitrag zu einer Monographie des Bunten
   Sandsteins,Muschelkalks und Keupers und die Verbin-
   dung dieser Gebilde zu einer Formation. -366 p.,
   2 Tab., Stuttgart 1834 (Cotta).
BACKHAUS,E.: Das fossilführende Zechsteinvorkommen von
   Forstel-Hummetroth(Nordodenwald) und Bemerkungen zur
   südwestdeutschen Zechsteingliederung. - Notizbl.hess.
   L.-Amt Bodenforsch.,89, 187-2o2, 4 fig.,Wiesbaden 1961
BACKHAUS,E.: Die Lamellibranchiaten aus den roten per-
   mischen Schichten von Lieth bei Elmshorn(Holstein).-
   Geol.Jb.,82, 1o3-13o, 3 fig., 1 Tab.,Hannover 1964.
BACKHAUS,E.: Die randliche "Rotliegend"-Fazies und die
   Paläogeographie des Zechsteins im Bereich des nörd-
   lichen Odenwaldes.- Notizbl.hess.L.-Amt Bodenforsch.,
   93, 112-14o, 7 fig.,Tab.1o, Wiesbaden 1965.
BACKHAUS,E.: Fazies,Stratigraphie und Paläogeographie
   der Solling-Folge (Oberer Buntsandstein) zwischen
   Odenwald - Rhön und Thüringer Wald.- Oberrh.geol.
   Abh.,17,1, 1-164, 16 fig., 14 pl.,4 Tab.,Karlsruhe
   1968.
BENDER,H. & STOPPEL,D.: Perm Conodonten - Geol.Jb.,
   331-364, 1 fig., 3 pl., 1 Tab., Hannover 1965.
DACHROTH,W.: Gesteinsmagnetische Marken im Perm Mitteleuropas. - Unveröff. Habilitationsschr. Universität
   Heidelberg, 98 p., 19 fig., 3 pl., Heidelberg 1974
DIEDERICH,G.: Die Grenze Zechstein/Buntsandstein in
   der südlichen Randfazies.- Notizbl.hess.L.-Amt Boden-
   forsch., 98, 81-92, 2 fig., 1 Tab., Wiesbaden 197o
FALKE,H.: Das vermutete Perm der Bohrung Lichtenau
   bei Rothenbuch (Buntsandsteinspessart).- Notizbl.
   hess.L.-Amt Bodenforsch.,97, 117-129, 6 fig.,
   Wiesbaden 1969
FALKE,H.: Zur Paläogeographie des kontinentalen Perms
   in Süddeutschland. - Abh.h ess.L.-Amt Bodenforsch.,
   6o, 223-234, 4 pl., Wiesbaden 1971.
FALKE,H.: The continental Permian in North- and South
   Germany.- in FALKE,H.: Rotliegend. Essays an European
   Lower Permian. - Inter.Sedim.Petrogr.Ser.,15,43-113,
   11 fig.,Leiden 1972

GRUMBT,E.: Sedimentgefüge im Buntsandstein Südwest-
    und Südthüringens. - Schriftenr. Geol.Wissensch.,1,
    2o5 p, 8o fig., 18 Tab., 15 pl., Berlin 1974.
HENTSCHEL,H.E.: Die permischen Ablagerungen im öst-
    lichen Pfälzer Wald (Haardt) zwischen Neustadt-
    Lambrecht und Klingenmünster-Silz.- Notizbl.hess.
    L.-Amt Bodenforsch., 91, 143-176, 11 fig.,1 Tab.,
    1 pl., Wiesbaden 1963
HOLLINGER,J.: Beitrag zur Gliederung des Deckgebirges
    der Nordvogesen. - Z.deutsch.geol.Ges.,Jg.1969,121,
    79-91, 3 fig.,1 Tab., Hannover 197o.
KONRAD,H.-J.: Über die Verbreitung von Unterem Bunt-
    sandstein in den Nordvogesen.- Abh.hess.L.-Amt
    Bodenforsch., 6o, 258-262, Wiesbaden 1971
KOZUR,H.: Zur Altersstellung des Zechsteinkalkes(Ca 1)
    innerhalb der "tethyalen" Permgliederung. - Freiber-
    ger Forschungsh., C 298, 45-5o, 1 fig.,Laipzig 1974.
KOZUR,H.: Probleme der Triasgliederung der germanischen
    und tethyalen Trias, Teil I: Abgrenzung und Gliede-
    rung der Trias.- Freiberger Forschungsh., C 298,
    139-197, Leipzig 1974
LANGBEIN,R.: Zur Petrologie der Karneole des thürin-
    gischen Chirotheriensandsteins (Solling-Folge).-
    Chemie der Erde, 33, 3o1-325, 7 fig,,Jena 1974.
LEIBER,J.: Zur Gliederung des "Tigersandsteins" (Un-
    terer Buntsandstein,Trias) im Schwarzwald. - N.Jb.
    Geol.Paläontol.,Mh..Jg.1971, 461-467, 2 fig,1 Tab.,
    Stuttgart 1971
SCHNEIDER,H.: Sedimentation und Tektonik im Jungpaläo-
    zoikum der Baden-Badener Senke(Stephan,Rotliegendes,
    Zechstein). Unveröffentl. Inaug.-Dissert. Univers.
    Heidelberg, Heidelberg 1969.
STELLRECHT,R.: Über das Oberrotliegende im Gebiet der
    Schopfheimer Bucht. - Jber.Mitt.oberrh.geol.V.,N.F.
    45, 1-11, 2 fig., Stuttgart 1963.
TRÜMPY,R.: Über die Perm-Trias-Grenze in Ostgrönland
    und über die Problematik stratigraphischer Grenzen.
    - Geol. Rdsch., 49, 97-1o3, Stuttgart 196o.
WOLF,K.H. & CONOLLY,J.R.: Petrogenesis and paleoen-
    vironment of limestone lenses in Upper Devonian
    red beds of New South Wales. - Palaeogeogr.,Palaeo-
    clim.,Palaeooecol., 1, 69-111, fig.6, pl.1-18,1 Tab.
    Amsterdam 1965.

SUMMARY   by H. Falke

The papers on the Northern or Subvariscan province
(Smith, van Adrichem Boogaert) published in the proceedings confirm the results received up to now, namely
that the upper sequence of the Lower Permian has been
deposited under a semiarid to arid climate. Palaeogeo-
graphically this sequence represents an alluvial plain
basinwards of the coast with sheet floods and an intermittent dune zone which is followed by a sebka-line zone
and finally by salt lakes. With appearance of a coarse
marginal facies this landscape continues into East to
Southeast direction into the northern part of GDR. At
the East margin of the Harz it is connected with the
Central or Variscan province. For that region unfortunately a paper from the GDR was not available. Nevertheless, the lower sequence of the Lower Permian in
this basin is principally composed of volcanic rocks
of acid to basic character.   These volcanics obviously extend up to the northwestern part of the basin
located close to Scotland. To what extent they are
connected with those in the Oslograben which demon-
strates remarkable fault tectonics (Oftedahl) is not
known as there is the Fünenridge between both areas.
In this whole NW-SE extending area it is difficult to
draw the boundary between the Lower Permian and the
Stephanian. The "Saalian" unconformity at the top of the
volcanic rocks has not yet been proved completely. The
boundary between Upper Rotliegend and Zechstein, however, is defined by use of the copper shale (Falke).

In the Central and Variscan province a lower part
with grey to vary-coloured sediments relatively reach
in fossils (Lower Rotliegend = Autunian) can be distinguished from an upper part with a predominance of
red, mostly barren sediments (Upper Rotliegend). This
can be stated for the occurrences in Czechoslovakia
(Holub), Central Germany and South Germany (Falke),
and France (Feys). Both stages are separated mainly
with the aid of the Saalian phase. Its stratigraphic
position is still controversial even at the locus
typicus - the Halle porphyry complex in Central Germany.
Apart from tetrapod tracks there exist no further
palaeontological proofs for the division into these
two stages. According to investigations in France and
Central Germany a subdivision for the Lower and Upper
Autunian seems to be possible only with the help of the
macro- as well as microflora. Up to now a correlation
of the individual sequences between the basins based on

arguments is not possible. There also exist only few
data on the genesis of the sediments.

In this connection the first review given on the
occurrences in Spain (Virgili) is of interest. Their
general composition and also particular data as the
accumulation of carbonate rocks in the Lower Autunian
can be compared with other occurrences in the extensive
Variscan province. Nevertheless, in Spain a palaeontologically well-founded subdivision of the whole sequence
is also very difficult as well as the question which
section of the uppermost Permian profile can be regarded as a terrestrial facies of Zechstein age since
these occurrences are mostly unconformable overlain
by the Bunter. The appearance of volcanic rocks in the
younger sediments of the whole sequence is also of
interest. In the Esterel Massif, in southern France,
they exist in a stratigraphic level which according to
micro-palaeontologic investigations belong to the Zechstein. Further research on the occurrences in Spain
should also give suggestions for investigations in
other basins of the Variscan province.

AUSTROALPINE VERRUCANO OF SWITZERLAND

R. Dössegger

Swiss Federal Institute of Technology,
Geology Department, Sonneggstrasse 5, Zürich *

ABSTRACT. The Permian rocks of four Austroalpine regions of
Switzerland (Engadine Dolomites, Silvretta nappe, Bernina nappe,
Err-Julier-zone) are described and compared with the Permian
rocks of the South-Alpine realm.

1. INTRODUCTION

The Permian rocks of Switzerland have been described earlier in
detail (Trümpy 1966, Trümpy and Dössegger 1972). Since that time,
new results have been only published on Permian rocks from the
Austroalpine nappes of Switzerland, so that we may renounce to
recapitulate facts on the Permian of the whole of Switzerland. In
this paper new results from the Val Müstair Verrucano will be discussed and compared with results of investigations (partly still
going on) in the other regions of the Austroalpine realm of
Switzerland. The Austroalpine nappes are tectonically the highest
(and consequently they have been the southernmost resp. south-
easternmost before the Alpine folding) of the Alps, they border
paleogeographically to the south on the South-Alpine realm (leaving out possible late Alpine displacements along the Insubric
line). In Switzerland, these Austroalpine nappes are only present
in its southeastern corner, in Graubünden. They are subdivided
into a Lower Austroalpine (originally lying north or north-west)
and into an Upper (1) Austroalpine. The Upper Austroalpine comprise

* Present address
(1) see Trümpy and Haccard (1969)

H. Falke (ed.), The Continental Permian in Central, West, and South Europe, 123-136. All Rights Reserved.
Copyright © 1976 by D. Reidel Publishing Company, Dordrecht-Holland.

the décollement thrust-sheets of the Northern Calcareous Alps
(now to the N) and the "Central Austroalpine "units, which are in
relation with large basement thrust masses. The original relation
of the Northern Calcareous Alps and the Central Austroalpine are
in discussion.
In the Austroalpine of Switzerland we can separate a series of
detrital and volcanic rocks between metamorphosed and folded basement rocks (incl. isolated outcrops of supposed Carboniferous)
and the carbonatic Triassic. Here we use for this series the term
"Verrucano" - in the sense of a lithostratigraphic group -.
Against this use of the term "Verrucano" many arguments can be
advanced - expecially the fact that there is little age and facies
correspondence with the type locality near Pisa -, but until now
there does not exist a correct lithostratigraphic group name,
which could replace this term which the Swiss geologists have used
since Studer (1872). For that reason we can still assent to the
opinion of Zoeppritz (1906, p. 19):
"Unter dieser Bezeichnung werden hier eine Folge von roten und
grünen Konglomeraten, Sandsteinen und Schiefern zusammengefasst,
die überall wo sie vorhanden sind, durch ihre lebhaften Farben
ein leicht kenntliches Grenzglied zwischen den älteren, kristal-
linen Schiefern und der Trias darstellen. Die Bezeichnung Verrucano ist, wie Milch (1892) in ausführlicher Weise dargelegt hat,
von den verschiedenen Autoren in der verschiedensten Weise ange-
wendet worden. Was mich bestimmt, trotzdem diesen Ausdruck beizu-
behalten, das möchte ich fast mit einem Scherz ausdrücken, dass
es auf eine Definition mehr auch nicht mehr ankommt. Denn ander-
erseits bietet der Name "Verrucano" doch den Vorzug, dass jeder
Geologe, der ihn hört, sich sofort eine Bildung wesentlich kla-
stischer Natur darunter vorstellt, die er wenigstens ungefähr an
die Schwelle von Paläozoikum und Mesozoikum zu stellen gewohnt ist".

2. DESCRIPTION OF FOUR AUSTROALPINE REGIONS

In the following the four regions, marked on figure 1, will be
discussed. These regions have been situated approximately from
the south to the north before the Alpine folding.

2.1 Verrucano of Val Müstair (Engadine Dolomites, Central Austro-
       alpine to the south of the Engadine fault)

Several tectonic elements, which have moved differentially in
partly unknown directions and lengths during the Alpine folding,
participate in the framework of the Engadine Dolomites: Crystallines and/or sediments of the Campo-, Ortler-, S-charl-Sesvenna-

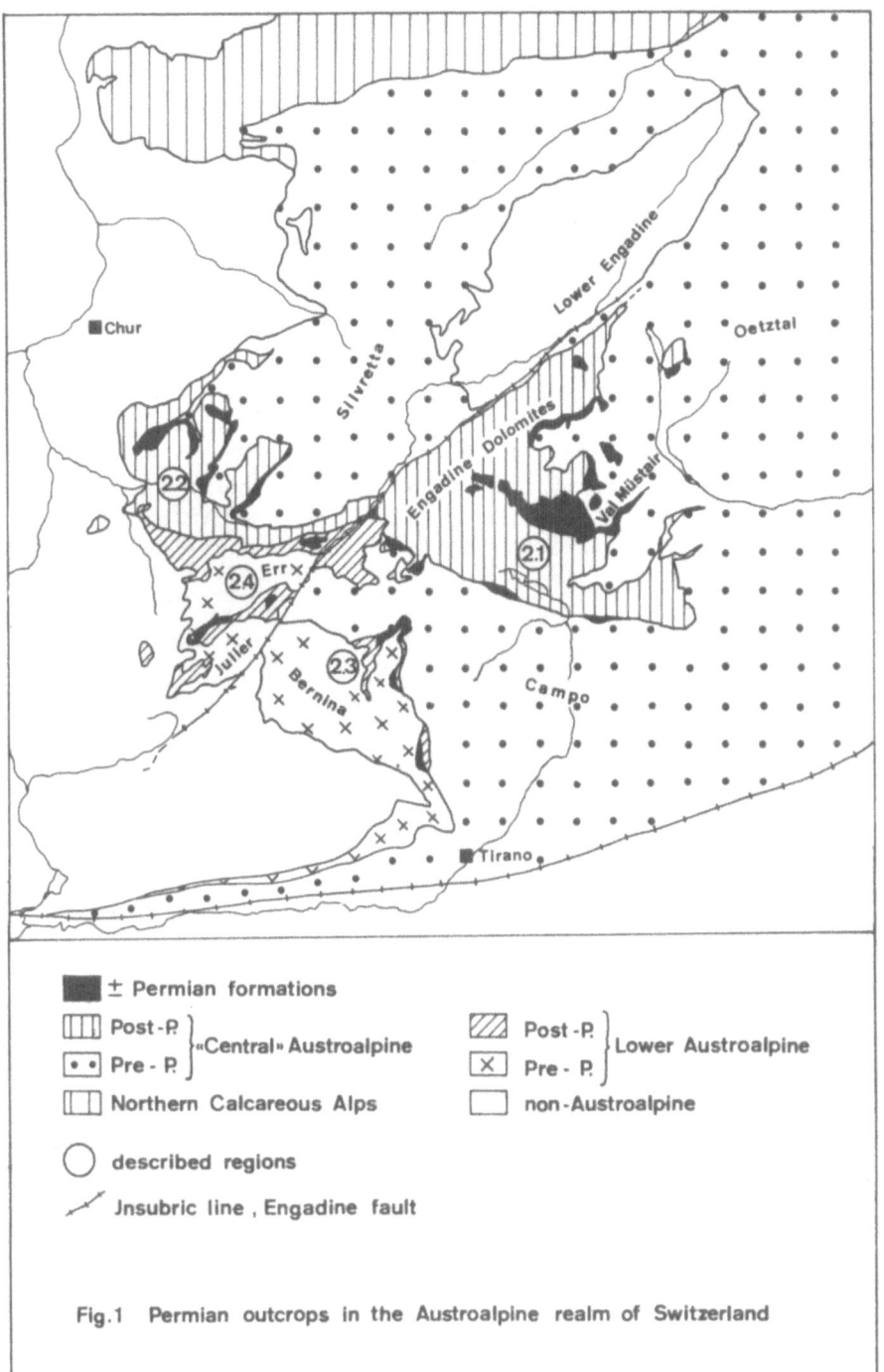

Fig.1 Permian outcrops in the Austroalpine realm of Switzerland

and Oetztalelements. For this reason the paleogeographic arrangement of these elements cannot yet be reconstructed in a completely
satisfying way. All the same, we can draw a fitting picture of
the environment at Verrucano times:

The detrital and volcanic sequences between metamorphosed
Pre-Permian basement (including scarce and uncertain Carboniferous)
and the Middle Triassic carbonate rocks are subdivided into the
lithostratigraphic group of the Val Müstair Verrucano (Ruina
Formation in the lower part and Chazforà Formation in the higher
part) and into the Fuorn Formation (in place of "Tiroler Buntsandstein") (see figure 2).

The Ruina Formation (measuring 400 to 500
metres) has been sedimentated in a trench (Val Müstair trench)
above a crystalline basement, which does not show any indications
of previous weathering. This trench runs (today) from WNW to ESE
and has a probable width of 15 to 20 km. The sediments of the
Ruina Formation consist of alternations of dark violet and green
conglomerates and green-grey to purple sandstones. The sandstones
have components of mostly volcanic (rhyolitic) material, the
conglomerates are composed of a matrix, which consists of the
same material as the sandstones, and of frequent volcanic pebbles
(stretched, with proportions of up to 10:1 between long and short
axis) as well as rounded pebbles of crystalline basement rocks.
Mostly there are no observable sedimentary structures in the
conglomerates and the sandstones, but rarely we can see a grading
or a crossbedding of conglomerate layers.

The above-lying Chazforà Formation is much
more widespread than the Ruina Formation and is absent only in
one region of the Engadine Dolomites. Its basal contact to the
Ruina Formation is sharp and often tectonically overprinted.
Outside the Val Müstair trench it lies directly on fresh crystallines, on a lateritic (2) weathering zone of the crystallines or
on rhyolitic volcanics. The Chazforà Formation consists of irregular alternations of conglomerates, sand- and siltstones with sedimentary structures of undoubtedly fluvial origin. Its colour is
mostly green-grey and rarely red. In the region of the Val Müstair
trench, where the Chazforà Formation reaches its greatest thickness
(about 500 metres), we can distinguish between two zones:

_____

(2) "lateritic" is used here for an intensively red coloured
soil-development, in which feldspar and micas are completely
disintegrated, but where the quartz is not yet dissolved.

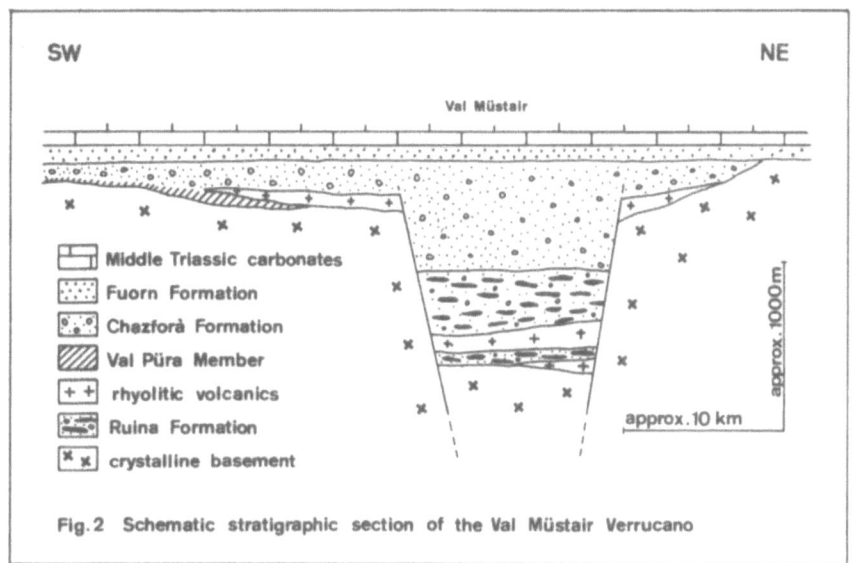

Fig. 2 Schematic stratigraphic section of the Val Müstair Verrucano

_ a central zone with sediments of river channels with few
  sediments of floodplains or oxbow lakes.

- a marginal zone, which consists in its lower part of the same
  sediments as in the central zone and in its higher part of
  sediments of floodplains or alluvial flats.

An exeption in the sedimentary character of the Chazforà
Formation makes an intensively red coloured breccia with extremely
angular quartz pebbles (Val Püra Member) which could be the result
of a sedimentation by sheet flooding in local depressions, the
source area having been affected by lateritic weathering.

Both formations comprise some layers of v o l c a n i c s
of rhyolitic character. Two layers of rhyolites of different age
have been found in the lower part of the Ruina Formation, they
reach their greatest thickness of about 150 metres in the vicinity
of the northeastern margin of the Val Müstair trench. The vents
may be connected with faults bordering the trench. Whether these
volcanics were initially normal rhyolites, tuffs or welded tuffs
cannot be established today, because of the (early?) alpine
tectonism. The great lateral distribution and the relatively small
thickness make a true lavaflow improbable. More rhyolites (partly
certain welded tuffs) and tuffites have been discovered outside of
the Val Müstair trench. They lie between the crystalline basement

and the Chazforà Formation or are associated with the Val Püra
Member within the Chazforà Formation (in this case the Chazforà
sediments s.str. set in with a marked disconformity over the
volcanics). The correlation of these volcanics is difficult;
two possibilities can be considered:

- the volcanism of the regions outside the Val Müstair trench
  (at the base of the Chazforà Formation) has more or less the
  same age as that one within the Ruina Formation. In this case
  the lower boundary of the Chazforà Formation would be heterochroneous, and the Val Püra Member would be a time equivalent
  of the Ruina Formation (marginal facies?). A regional change
  of sedimentary environment could have taken place above the
  volcanics with the fluvial Chazforà sedimentation.

- The volcanism of the regions outside the Val Müstair trench is
  younger as that one within the Ruina Formation. In that case
  the lower boundary of the Chazforà Formation would be homo-
  chroneous and would correspond perhaps to a regional change
  of the sedimentary environments.

A decision between these two possibilities cannot be taken at
this moment, even though the absence of traces of volcanic activity within the Chazforà Formation in the region of the Val Müstair
trench makes the first possibility more probable. Rare pebbles
of intermediate magma type, found in conglomerates of the Chazforà Formation, could be indicative of an initially present, but
today no more outcropping, volcanism of intermediate type.

    To the formations of the Val Müstair Verrucano and local
directly to the crystalline basement succeed brown and yellow
sandstones with increasing carbonate contents, cellular dolomites
and dolomites with decreasing detritus content (F u o r n
F o r m a t i o n). The Fuorn Formation contains, even in its
lower part, a fauna of poorly preserved lamellibranchs of marine
character. The distribution of the thickness of this formation
shows no more a maximum in the region of the Val Müstair trench,
in contrast to the formations of the Val Müstair Verrucano. On
the other hand the carbonate containing sedimentation of the
Fuorn Formation of this region is interrupted by a recurrence
of conglomerates and sandstones in a facies similar to that of
the Chazforà Formation. Whether the reason was a short regression
during the transgression of the Tethys, increasing supply of
detritus out of the hinterland or a fossil barrier has not been
yet decided. The base of the carbonate rocks of the Middle
Triassic is more or less concordant to the Fuorn Formation, but
partly heterochroneous. The age of the lowermost carbonates would
be Middle Anisian (Brachiopods, Crinoids), an other fossil find

(Conodonts) proves an Anisian age for at least the higher part of
the Fuorn Formation.

The genesis of the described formations can be characterized
as follows:

Ruina Formation: On the edges of a slowly sinking trench, rhyolitic
volcanoes were active locally. Their partly pyroclastic products
were sedimented in the Val Müstair trench as well as in its
environs. The trench was filled by volcanic detritus, which was
probably transported mainly by mud-flows (ev. by true lahars?),
partly directly from the volcanoes. The pebbles of crystalline
basement rocks (rounded, relatively large) could come from the
region of the trench-edge, and the quartz pebbles (well rounded,
relatively small) could have been brought from more distant regions
by fluvial transport. The volcanism would have ended towards the
upper boundary of the Ruina Formation or within the Chazforà
Formation (see discussion above).

Chazforà Formation: Some crystalline high zones, which show no
signs of a pre-Chazforà weathering, must have been eroded before
being covered by sediments. In other regions (perhaps high zones
too, but less exposed to the erosion) lateritic weathering zones
developed; their debris has been partly accumulated in local
depressions by sheet-floods (Val Püra Member). Over a part of
these regions have settled the predominantly pyroclastic products
of the rhyolitic volcanism (see discussion above). Then the fluvial
sediments of the Chazforà Formation s.str. follow; their vertical
development points at a gradual flattening of the relief. The
well established maximum of thickness of the Chazforà Formation
in the region of the Val Müstair trench must be explained with
a persisting activity of the trench.

Fuorn Formation: The transition from the Chazforà Formation to
the Fuorn Formation (transgression of the Tethys) seems to be
without a greater hiatus. The fact that the Fuorn Formation
spreads out over the last crystalline areas can hardly be explained as result of a longer interruption of the sedimentation
or even of a phase of folding; on the contrary it should be a
"concordant" sedimentation over a more or less flat topography
with filled depressions and eroded hills. The clear heterochrony
of the upper boundary of the detritic sedimentation and its
partly warranted Anisian age can be used as arguments for this
theory. This facts should indicate that at least parts of the
Chazforà Formation are also of Lower Triassic age.

2.2 Permian of the Silvretta nappe (Central Austroalpine in the north of the Engadine fault)

Permian (and Mesozoic) sediments have been preserved from erosion only in the northwestern corner of the Swiss part of the Silvretta nappe, in synclines formed during the (early?) Alpine folding (Ducan- and Landwasser-synclines). This review is based mainly on diplomapapers of students of the Geological Department of the ETH in Zürich, but also on publications by Brauchli (1921), Eugster (1923) and Grauert (1969).

Over a crystalline basement (metamorphosed and folded in the Variscan phase; cooling under 300° 300 my ago) are found some isolated black graphitic conglomerates and schists, which could be Carboniferous. But normally a volcanic series (acidic lava type; welded tuffs are partly proved) follows over the crystalline basement. These volcanics have a greatly variing thickness with an average of about 500 metres; but locally we can observe as much as 1600 metres (Bellaluna rhyolite) and on the other hand the volcanics can be lacking. An intensively red tuffite, which measures 80 to 100 metres and contains much components of rhyolitic volcanics and less fragments of the crystalline basement, is attached to this volcanics. Isolated pebbles (diameter up to 20 cm) consist of rhyolitic material.

The following detrital series of considerable thickness (200 to 500 metres) begins above the tuffite with a transition zone of some metres and above the crystalline (with lateritic weathering) and the volcanics with a disconformity and a basal conglomerate. It is composed of conglomerates, sand- and silt-stones with sedimentary structures, which point at a fluvial origin of these sediments. The pebbles consist of vein quartz, volcanic and crystalline material, their frequency decreases upwards. Parallel to this, the rock colour changes from the predominantly red in the lower part to green in the higher part. Brown carbonatic sandstones (measuring about 100 metres) follow with apparent conformity. They contain in their lowermost part single layers of conglomerates without carbonate, which look like conglomerates of the lower formation. On the other side carbonate rocks (cellular dolomites and sandy dolomites) increase in their upper part and lead over to the Anisian carbonates.

This sequence of the Silvretta nappe can be compared very well with that of the Engadine Dolomites, and we can use the same names of lithostratigraphic formations:

- the basal series with rhyolitic volcanism and conglomeratic
  tuffites corresponds to the Ruina Formation.

- the middle detrital series of predominantly fluviatile rocks
  is identical to the Chazforà Formation.

- the upper brown carbonate-containing series at the passage to
  the Middle Triassic carbonates correspond inits turn to the
  Fuorn Formation.

It must be noted, however, that the "Buntsandstein" of the older
authors dealing with the Silvretta nappes comprises both the
middle and upper formation, whereas in the Engadine Dolomites
the term was restricted to the Fuorn Formation. Thus, the Chazforà
Formation was regarded as " upper Verrucano" in southern Graubünden, as "lower Buntsandstein" in central Graubünden.

2.3 The southern Bernina nappe (Lower Austroalpine to the
    south of the Engadine fault)

The Lower Austroalpine in the south of the Engadine fault is
split up in three nappes; the sediments belonging to them are
found in the intervening synclines. The region of the highest
sedimentary zone (Region of Piz Alv) should be discussed here
briefly (Schüpbach 1969).

     With a disconformity over metamorphosed crystalline, which
contains apart older also granites of Upper Carboniferous and
Lower Permian age (Age of Zircons:295 - 260 my) follow rhyolites
and andesites with a thickness of some tens of metres. Brown
carbonatecontaining conglomerates and sandstones and green
siltstones without carbonate lie above, they measure 10 metres
at most. With a small zone of cornieules begin the Middle
Triassic carbonates.
In this region of the Lower Austroalpine appear thus (with
exeption of the volcanics) no equivalents of the Val Müstair
Verrucano. The detrital series, only few metres thick and already
containing carbonate, can be compared perhaps with the Fuorn
Formation of the Engadine Dolomites.

2.4 Err-Julier-zone (Lower Austroalpine in the north of the
    Engadine fault)

The Lower Austroalpine crystalline in the north of the Engadine
fault (frontal parts of Lower Austroalpine) has been split up

into different parts, which are separated by partly synclinal,
but strongly sliced sedimentary zones (Err-, Julier-nappe).
The strong Alpine deformation makes it difficult to reconstruct
the paleogeographical relationships. This description of the
(eventually) Permian rocks is based on the studies of
Cornelius (1935), Heierli (1955) and on recent diplomatheses of
students of the Geological Department of the ETH in Zürich
(Finger a.o.).

Above the gneisses and crystalline schists of the basement
we find, unfortunately without identifiable age relationships:

a) black conglomerates, sand- and siltstones and rare limestones,
   which could represent the Carboniferous (thickness about
   40 metres).

b) two layers of so-called "Nair-porphyroid"(rhyolitic volcanics,
   locally evidenced structures of welded tuffs), each measuring
   up to 250 metres. The lower one has been stressed intensively
   and is recrystallized. The two layers of rhyolitic volcanics
   are separated by andesitic volcanics (about 50 metres) which
   show a strong foliation (Vairana schists) and by purple
   tuffites (about 150 to 200 metres thick) with a gradual development from the lower Vairana schists (Sprenkelschiefer).
   The tuffites consist of a quartz-rich matrix and have light,
   mostly greenish oval spots, which have a composition different
   from that of the tuffites, (pebbles of tuff or volcanic glass?).
   In some places similar schists are found above the upper layer
   of "Nair-porphyroid" as well. But normally, 10 metres of
   sericite schists, which contain pebbles of "Nair-porphyroid",
   follow above the upper "Nair-porphyroid". Their lower boundary
   is gradual and they represent the products of working up of
   the "Nair-porphyroid". With a sharp boundary follow the
   cornieules and carbonates of the Middle Triassic.

c) a breccia, lacking in matrix; the pebbles of which are composed
   predominantly of crystalline basement material (Lavatèra breccia).
   Its relation to the series a,b or d are still not clear, this
   breccia seems to lie partly below or within or above the
   volcanics.

d) in the Albula-zone (probably belonging to the Err-nappe):
   above the crystalline basement follow green-grey and red
   quartz-conglomerates, sand- and siltstones (about 70 to 150
   metres) and above that green, red or brown, partly carbonatic
   sand- and siltstones (30 to 50 metres)without a sharp boundary
   to the Middle Triassic carbonates.

Cornelius (1935) supposes an Early Paleozoic age for the volcanics of the series b and for the series c, chiefly because of differences in the deformation and folding style between these rocks and the overlying Triassic carbonates. Finger (1974) on the other hand supposes apparently an exclusively Alpine deformation and so a Postvariscan age of these rocks. Because the crystalline basement of the Austroalpine nappes should have been subjected to the same tectonical history, and as Grauert (1969) has proved a high-grade Caledonian thermometamorphism, contemporaneous and later intrusions of intermediate to acidic lavas as well as a Variscan dynamothermometamorphism for the Silvretta crystalline, the hypothesis of Cornelius about the Early Paleozoic age seems to be improbable. Indeed, if we compare the rocks of the Err-Julier-zone with the described Central Austroalpine regions, we see a great analogy to the strong volcanism of the Ruina Formation, so that we can accept a Postvariscan (Early Permian?) age of these series in all probability. Equivalents to the Chazforà and Fuorn Formation seem to be practically absent, with the exception of the detrital series of the Albula-zone, which show great analogy to the Chazforà and Fuorn Formation of the Central Austroalpine regions.

3. SUMMARY AND COMPARISONS

The Post-Variscan sedimentary sequence begins in three of the described regions with volcano-detrital sediments, which have been deposited at least partly in intramontane trenches, and with volcanics of mainly acidic and more rarely intermediate lava type (Ruina Formation and relatives). We can scarcely tie together these three trench-areas paleogeographically, so that we must estimate at least three independent sedimentary trenches in the small Swiss part of the Austroalpine realm. For the trench of the Val Müstair (Engadine Dolomites) we could show an extend of WNW to ESE. For paleogeographic comparisons we have to remember the possible Post-Permian rotation (counter-clockwise, about 30 to 50 degrees) of the whole Austroalpine and South-alpine plate, so that the Val Müstair trench could have run approximately E-W (or NE-SW). In both Central Austroalpine regions the fluviatile sediments of the Chazforà Formation spread out over these basal formations and cover vast crystalline areas, which partly show lateritic weathering. Local high zones in the Central Austroalpine and commonly the Lower Austroalpine realm do not become covered by sediments and represent apparently denudation areas (exception: Albula-zone). The transition to the carbonatic Triassic sedimentation takes place in the Central Austroalpine realm with about 100 metres carbonate-containing

detrital rocks (Fuorn Formation), which spread out over the last
crystalline areas. The supply of detritus continues here into
the Lower Anisian. In the Lower Austroalpine, the carbonatic
facies sets in above a transition zone measuring only a few
metres, which resembles to the Fuorn Formation and which consists of the products of local desintegration of the basement
rocks.

The strong volcanic influence in the basal formations
(Ruina Formation and relatives) could be compared with the
volcanics of the region of Bolzano or with the volcano-detrital
series of the region of Passo di Maniva. In analogy to these
regions of the Southern Alps we can accept for our basal formations an age of Upper Early Permian. The striking spreading
out of the Chazforà Formation, and the regional change in the
sedimentary character at the base of this formation could so be
interpreted as result of the Saalic phase. The fluviatile-
detrital Chazforà Formation corresponds in its lithology perfectly to the Verrucano Lombardo of Central Lombardy, and the
lack of corresponding rocks in the Lower Austroalpine region
possesses its analogy in the Southern Alps in the region of
Lugano. In the Austroalpine realm of Switzerland we find so
(today) from south to north a similar development as in the
Southern Alps from east to west. Unlike the Southalpine realm
(Assereto et al, 1974) there does not seem to be a great hiatus
between the Chazforà Formation and the Fuorn Formation in the
Central Austroalpine realm. The detrital supply should have been
going on locally some times in spite of the transgression of the
Tethys. Similar developments are known in the southwest of the
Northern Calcareous Alps (Krabachjoch, Hirsch 1966) and in the
Southern Alps (Val di Non, Kreis 1970). For the Central Austro-
alpine region of Switzerland would be valid rather the model of
Riehl-Herwisch (1972) with a slowly northward wandering facies
of the Arenarie di Val Gardena (and so of course of the Verrucano
Lombardo too).

Upon the age of the Chazforà and Fuorn Formation in the
Swiss part of the Austroalpine realm we can only speculate.
Only the upper boundary of the Fuorn Formation is more or less
dated as lying within the Anisian. The boundary between Chazforà
and Fuorn Formation could lie chronostratigraphically somewhere
between the Permian/Triassic boundary and the Anisian. The
Chazforà Formation could have an age of Upper Permian and/or
Skythian.

Going out from our regions we cannot say much about the
climate of the Permian. Before the beginning of the Chazforà

sedimentation the climate must have allowed a lateritic
weathering and during the Chazforà sedimentation rainfall must
have been sufficient to account for fluviatile transport.

BIBLIOGRAPHY

ASSERETO,R.,BOSELLINI,A.,FANTINI SESTINI,N.&SWEET,W.C.(1974):
  The Permian-Triassic boundary in the Southern Alps (Italy).
  In:The Permian and Triassic systems and their mutual
  boundary (p. 176-199).- Canad.Soc.Petroleum Geologists.

BRAUCHLI,R.(1921): Geologie der Lenzerhorngruppe. - Beitr. geol.
  Karte Schweiz (N.F.) 49/2.

CASSINIS,G.(1966): Rassegna delle Formazioni Permiane dell'alta
  Val Trompia (Brescia). - Atti Ist. geol. Univ. Pavia 17,50-66.

CORNELIUS,H.P.(1935): Geologie der Err-Julier-Gruppe, 1.Teil:
  Das Baumaterial (Stratigraphie und Petrographie, excl. Quartär).
  -Beitr. geol. Karte Schweiz (N.F.) 70/1.

DOESSEGGER,R.(1974): Verrucano und "Buntsandstein" in den
  Unterengadiner Dolomiten. - Unpubl.Diss.Eidg. Tech.Hochsch.
  Zürich.

EUGSTER,H.(1923): Geologie der Ducangruppe. - Beitr. geol.
  Karte Schweiz (N.F.) 49/3.

FINGER,W.(1972): Geologie der Val d'Agnelli und des Vairana-
  Kessels (Oberhalbstein, Graubünden). - Diplomarb.Eidg.Tech.
  Hochsch. Zürich.

GRAUERT,B.(1969): Die Entwicklungsgeschichte des Silvretta-
  kristallins auf Grund radiometrischer Altersbestimmungen.
  - Diss.Univ.Bern.

HEIERLI,H.(1955): Geologische Untersuchungen in der Albulazone
  zwischen Crap Alv und Cinuos-chel.- Beitr.geol.Karte Schweiz
  (N.F.)101.

HIRSCH,F.(1966): Etude stratigraphique du Trias Moyen de la
  région de l'Arlberg. - Diss.Eidg.Tech.Hochsch.Zürich.

KREIS,H.H.(1970): Sedimentologische Untersuchungen des
    "Unteren Muschelkalks" (Anis) im Bereiche des Gampenpasses
    (Südtirol). - Festbd.geol.Inst. zur 300-Jahr-Feier Univ.
    Innsbruck,139-164.

RIEHL-HERWISCH ,G.(1972): Vorstellungen zur Paläogeographie
    des Verrucano. - Verh. geol. Bundesanst.(Wien)1972,97-102.

SCHUEPBACH,M.(1969): Der Sedimentzug Piz Alv - Val da Fain.
    - Unpubl.Diplomarb.Eidg.Tech.Hochsch.Zürich.

TRUEMPY,R.(1966): Considérations générales sur le "Verrucano"
    des Alpes Suisses. - Atti Symp.Verrucano Pisa,Soc.Toscana
    Sci.Nat.,212-232.

TRUEMPY,R.&DOESSEGGER,R.(1972):Permian of Switzerland. In:
    Falke,H.(Ed.): Rotliegend,Essays on European Lower Permian
    (p.189-215). - Int.sediment.petrogr.Ser.15.

TRUEMPY,R.&HACCARD,D.(1969): Réunion extraordinaire de la
    Société Géologique de France:Les Grisons,14.-21.9.1969.
    - Bull.Soc.géol.France 9,329-396.

ZOEPPRITZ,K.(1906): Geol. Untersuchungen im Oberengadin
    zwischen Albulapass und Livigno. - Ber.natf.Ges.Freiburg i.Br.16,
    164-231.

FACIES ZONES OF THE PERMIAN IN THE EASTERN ALPS*

E. Erkan

Montan-Universität Leoben, Austria

ABSTRACT. Detailed studies of the Permian beds in the
eastern parts of the Eastern Alps show that marine
ingression upon the continental area took place in
two steps after the Variscian orogenesis. Therefore
the paleogeographical situation of the East Alpine
Permian is as well described for a lower as for an
upper part of the Permian. The slight thickness of
several of the Eastern Alpine Permoskythian profiles
is due to the lacking of the lower part of the
section. The formation of uranium and gypsum deposits
as well as their occurence are described.

The Permoskythian sediments of the Eastern Alps
might be considered as being of great interest in
several respects. They are of economic importance due
to the occurences of uranium, gypsum, copper and iron
and of great tectonic interest as the study of the
Permoskythian is practically a study of the Eastern
Alps' tectonics. Many tectonical problems of the
Eastern Alps, however, can be solved only by means of
a classification of the problems of Permoskythian
stratigraphy.

In the present paper the author tries to contribute to the solution of the Permoskythian problems.
It is tried to give a reconstruction of the original

---

*This work has been supported by Fonds zur Förderung
der wissenschaftlichen Forschung, Austria.

H. Falke (ed.), The Continental Permian in Central, West, and South Europe, 137-147. All Rights Reserved.
Copyright © 1976 by D. Reidel Publishing Company, Dordrecht-Holland.

area of sedimentation as well as a synopsis of the
paleogeographical results. A through discussion of
all the details is reserved to a separate paper.

The tectonical scheme of the present paper is
based upon a publication of Tollmann (1965).

In the present paper I am distinguishing between
two main groups of the Permoskythian, i.e. (I) a
first one of great thickness and (II.) a second one
with slight thickness (see Fig. 1).

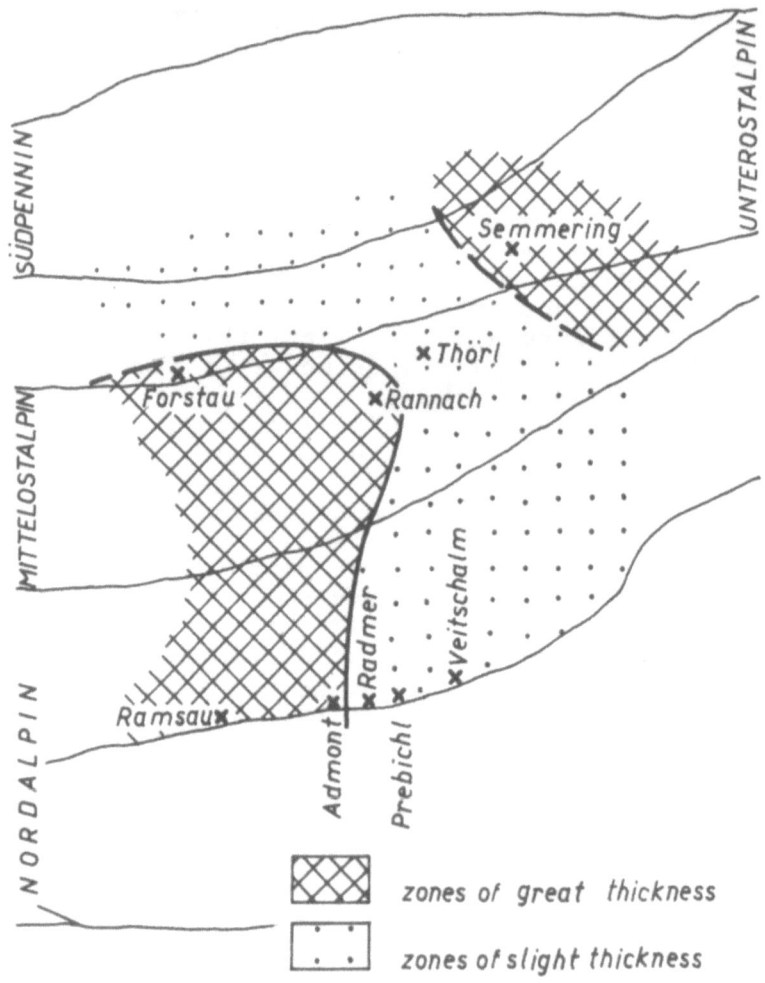

FIG.1. Permoskythian facies zones in the eastern part
of the Eastern Alps (Erkan, 1975)

I.) The first group reaches at least a thickness of 500 m, sometimes of 1000 m (for instance the Rannach-series of the Liesing-valley, Styria) at the southern border of the Calcareous Alps. The eastern-most occurence is to be found in the surroundings of Admont, and reaches to the Dachstein-area and still further westwards. Studies of literature revealed the fact that it is ending W of Fieberbrunn where the development of the second groups is beginning. Very probably this scheme, originally introduced for the eastern section of the Eastern Alps, is applicable for the whole Eastern Alps. The Semmering area in the East is belonging again to the first group. The area in between is considered to belong to the second group.

In the surroundings of Admont an area of interference of both groups is developed. NW of Admont, N of the Enns valley, the Permoskythian reaches a thickness of 600 m. SE of Admont, however, the same series reaches only about 250 m.

The series with great thickness comprises at the base either conglomerates and breccias of slight thickness (Filzmoos) or these are lacking (Rannach series, Middle Austro-Alpine and Semmering area Lower Austro-Alpine). In the second case the separation of the Permoskythian and the Older Paleo-zoic sometimes is very difficult. In the area W of the Dachstein for instance, according Bittner (1884) and Trauth (1925), a separation of Young Paleozoic series and Graywacke-zone is not possible.

II. The Series with slight thickness mostly reaches a thickness between 100 and 150 m, but - according to the relief of the Variscian orogene - sometimes beds with a thickness of about 300 m are developed.

In contrary to the first group (I) the second one (II) generally comprises a considerable series of conglomerates at its base ("Prebichl-conglomerates"). Hitherto, no Prebichl-conglomerates were found in the first group, neither in the Upper-, nor in the Middle- or the Lower Austro-Alpine. On the other hand, the sediments of the series with slight thickness always are lying discordant upon the basement, while the series of great thickness normally are concordant on the basement. The Permian of the first group domprises mainly grayish sediments while the Permian of the second group consists mainly of reddish beds.

Hitherto it was not possible to distinguish between the Permian and the Triassic section of the Permoskythian beds of the Eastern Alps Due to this fact a lot of local names for these beds were introduced.

The Seisian (=Lower Skythian) is characterized by a fauna comprising Claraia clarai (Emmr.), while the Campilian (=Upper Skythian) contains the guide fossil Naticella costata MSTR..

The stratigraphic position of all occurences of these guide fossils hitherto known first was registered in the Eastern Alpine profiles. Then these horizons have been compared and/or parallelized with the nonfossiliferous beds according to their lithology. One of the results gathered was that the Skythian parts contain marly layers of different thickness, in contrary to the lying Upper Permian beds, which consist in their upper parts of clays and gypsum but are lacking carbonates. The green clays of such gypsum bearing beds contain at Ramsau, (Styria) spores of an Upper Permian age (letter of W. Klaus, unpublished data). According to E. Pak (1974) these beds are, according to the determination of their sulphur isotopes, of Permian age at Wörschach further east.

All these hints allow, according to our present knowledge, a separation of Permian and Skythian beds of each other. A more detailed paper dealing with this problem is in print.

Gypsum bearing beds of the Upper Austro-Alpine mostly are responsible for the development of slices and faults (southern Dachstein area, northern surroundings of Gollrad, (Styria)). Permian beds without gypsum are present too, e.g. in the Prebichl area and the Veitschalm (Styria). The latter one is interesting, because of a complete profile, beginning with the transgression contact and ending with lime-stones and dolomites of middle Triassic age. Neither faults nor thrust-levels are recognizable. The logic conclusion therefore is that in the Veitschalm area Permoskythian and Middle Triassic are belonging to the same unit and are not divided by overthrust-plane as it was indicated by Tollmann (1967).

The parallelization of different Permoskythian
profiles shows the fact that in the group with great
thickness this great thickness of sediments is due
to the fact that sedimentation started earlier. At
the base of this (I.) group normally graphit-bearing
phyllites are developed. These beds are considered
to be of Upper carboniferous to Lower Permian age.
Furthermore, the lying parts of the Permoskythian are
epizonal metamorphosed in the middle parts of the
Northern Calcareous Alps. In the southern part of the
Dachstein area chloritoid-bearing horizons have been
discovered for the first time. Of course this metamorphose is of Alpidic age. In the same area, two
tuffitic horizons, only a few (5-7) centimetres
thick, are present. Up to now the acidic volcanics
and tuffites in the Eastern Alps have been considered
as having originated by the Saalian phase, a
hypothesis without any paleontological proofs.
According this problem I just want to point out the
fact that there is another tuffitic horizon present
in the Eastern Alps which is of Upper Permian age
according to palynological results (W. Klaus,
unpublished data).

Tectonically very interesting is the Rannach-
series of the area of the Liesing-valley, Styria.
There the contact of Rannach-series and Seckau
gneiss* is clearly a tectonic one (S "Rossschwanz",
SW "Ripplmauer"). This observation conflicts with
the opinion adopted up to now that the Rannach-
series has been deposited primarily upon the Seckau
gneiss. The zone of (tectonic) contact consists of
a quartzband with a thickness of 2-6 m, containing
several lenses of Seckau gneiss. The Rannach series
doesn't contain any components comparable with the
Seckau gneiss. At the zone of contact the Rannach
series is cut obliquely by the Seckau gneiss.

The Rannach series reaches up to the Middle
Triassic. The Sulzbach window shows the Middle Austro-
Alpine Rannach series which containing Upper
Skythian to Anisian marbles. These marbles contain
basic volcanics which have been metamorphosed to
greenschists. However, different Upper Permian basic
volcanics and tuffites are known. The volcanics of
the Middle Austro-Alpine are younger than Middle
Triassic. Another occurence of basic volcanic rocks
cutting obliquely across (probably Lower) Permian

---

* as several new outcrops have shown

quartzites I found in the Dachstein area.

It is wellknown that the Skythian beds of the
Eastern Alps are containing marine faunas. According
to my observation, the Permian sandstones and
conglomerates often are showing graded bedding. This
is due to an aquatic sedimentation. Aeolian sediments
may show a graded bedding too, but ruly in a much
smaller range of grain size. Flute casts are known
too, e.g. from Altenmarkt/Rax, Lower Austria.

F. Kümel (1954) described first remains of
"Pecten" of Permian sandstones in the Dachstein area.
The presence of these shells indicates a marine
environment. In the Permian of the Northern
Calcareous Alps, below the gypsum, limestones and
dolomites of considerable thickness are known
(S Admont, Schüssergraben/Radmer, Oberort S Hochschwab, according W. E. Petrascheck (unpublished)
from Faschinggraben/Radmer, too.

All these limestones and dolomites linked by
transitional beds, are joined with the hanging
gypsum bearing beds, are considered to be of marine
origin. The dolomites of Radmer contain sedimentary
pyrite with a considerable amount of organic matter,
i. e. they have been deposited in a calm, reducing
environment. Permian Uranium-ores have been sedimented in a similar reducing environment. They
contain up to 1,1 % Cu. The content of Uranium
reaches a maximum concentration of 1-2 %, while the
content of Thorium is smaller than 0,004 %. The
relation of U/Th points to a marine sedimentation too.

The paleogeographical situation principally was
a different one in the Lower and the Upper Permian.
In the Lower Permian the Variscan Lineaments and
relief were determining the distribution of facies.
In the Upper Permian, however, the Alpine E-W
direction dominates the paleogeographical situation,
but often with remains due to the Variscan direction
too.

After the Variscan orogeny and continental period
the ingression of a sea situated SE of the East
Alpine area towards the N, respectively NW in the
Semmering area. Therefore we have to distinguish two
big areas in the Lower Permian (see Fig. 2).

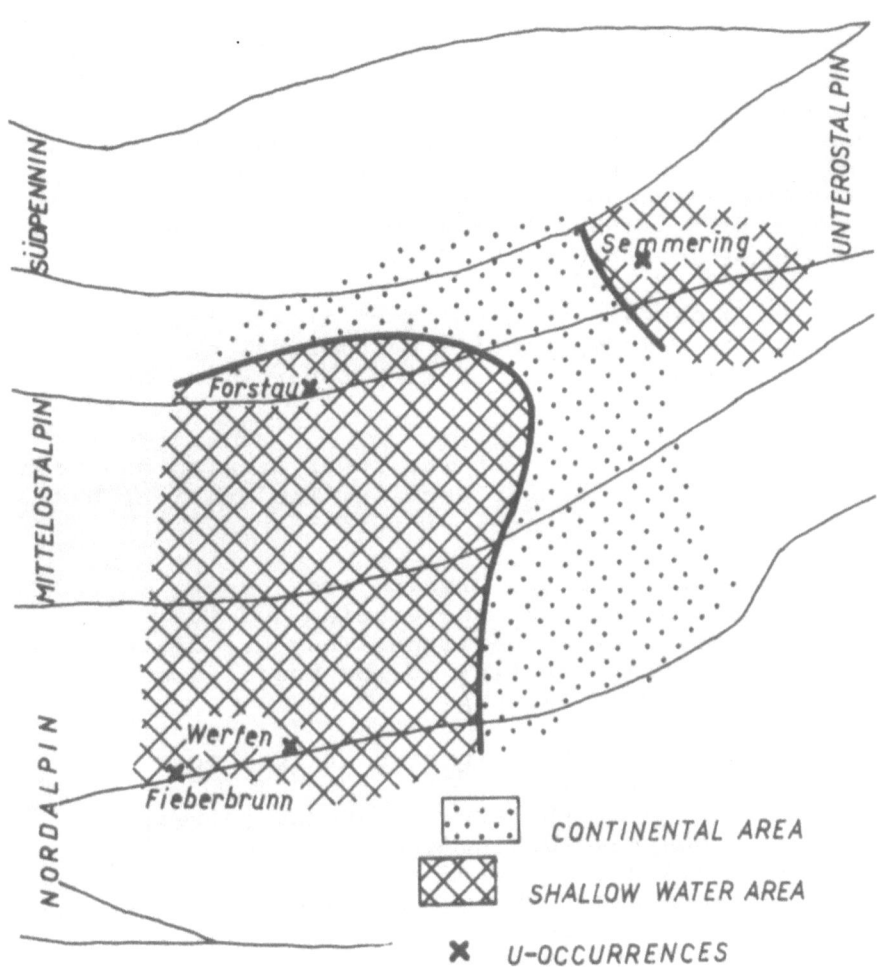

FIG.2. Lower Permian facies zones in the eastern
       part of the Eastern Alps (Erkan, 1975)

    The first region, the Continental area, consisted
of Limestones, porphyroids and granits, rather
resistable against erosion (Silurian/Devonian limestones of the Reichenstein group S of Prebichl;
porphyroids of Radmer).

    The second region is a shallow water area, whose
bottom consisted of phyllitic rocks, sericite-

quartzites a.o., relatively easily erosible
(Phyllites and sericite-quartzites near Ramsau and
Filzmoos).

The younger Paleozoic beds of the shallow water
area begins with Upper Carboniferous or Lower Permian.
They contain at their base conglomerates and lie
transgressively upon quartzites, sericite schists and
sericite quartzites which have been variscic
metamorphosed. The basal conglomerates of the Younger
Paleozoic series consist of rocks of these Older
Paleozoic metamorphic rocks.

In the Dachstein area and W of it the Younger
Paleozoic has been epizonally metamorphosed. This
Alpine metamorphose is decreasing towards the hanging
Werfen beds.

The continental area of the Lower Permian has been
transgraded in the Younger Permian. The Prebichl beds
are the basal conglomerates of this ingression.

In the Lower Permian, Uranium ores occur, according
to their high mobility of the Uranium and its
compounds, in different horizons. They are distributed
in N - S direction, parallel to the Lower Permian
shore-line (see Fig. 2).

The paleogeographic situation of the Upper
Permian is shown in fig. 3.
The N-S direction of the gypsum-lagoons (directed
northwards, open towards the south) is due to the
Variscan Lineation (Oberort S Hochschwab, Trübenbach,
Erlafboden, Annaberg). The E-W arrangement of all the
lagoons, however, is caused by the Alpine direction.
The conclusion that the Juvavic salt deposits origi-
nated south of that lagoons therefore seems to be
obviously.

Uranium ores of the Middle and Upper Permian
(Tweng, Thörl and Rannach) are restricted to a facies
of intercalating clays, clayish sandstones and
sandstones. Between this facies and the gypsum
bearing facies a flat ridge is postulated, consisting
of carbonate bearing sandstones (Middle Austro-
Alpine at and SW of Murau). In this region no
Uranium deposits are to be expected.

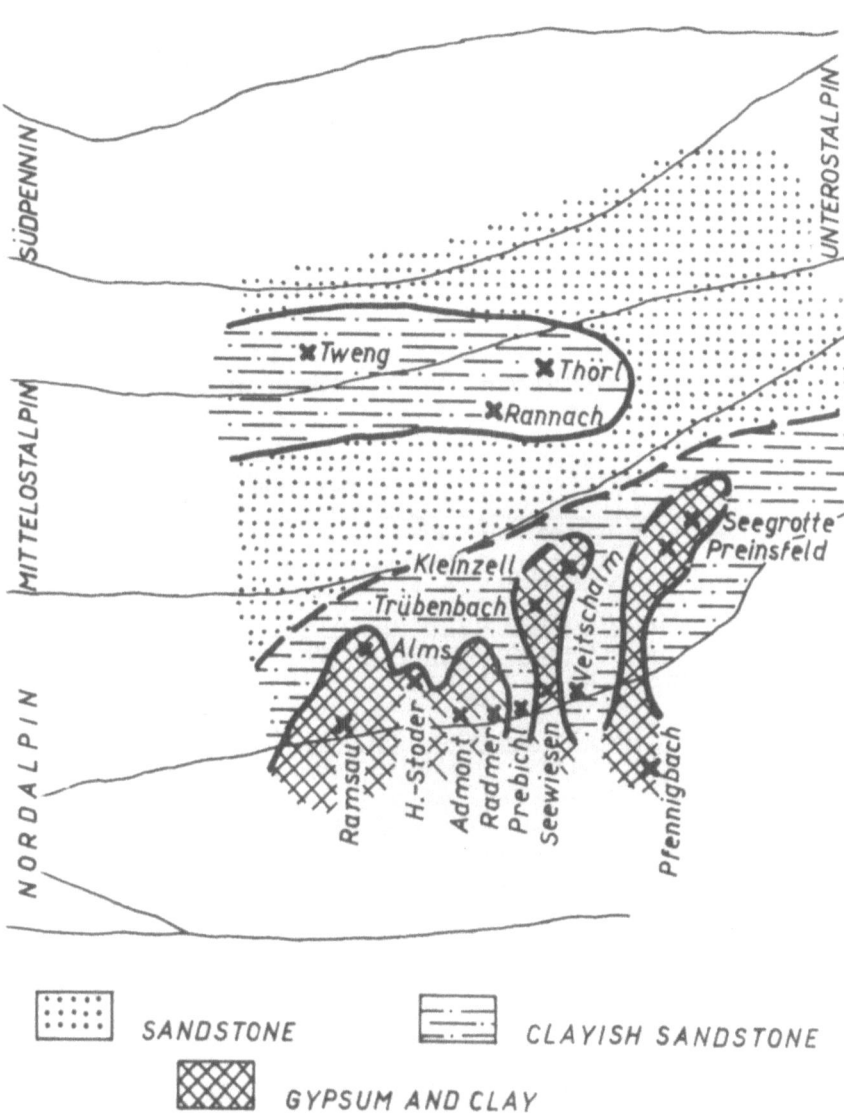

FIG. 3. Upper Permian facies zones in the eastern
part of the Eastern Alps (Erkan, 1975)

REFERENCES

1.   W.E. Petrascheck, E. Erkan and K. Neuwirth,
      Permo-Triassic Uranium ore in the Austrian
      Alps - paleogeographic control as a guide for
      prospecting, Proc. Symposium of IAEA in Athens,
      Vienna 1974.

2.   A. Tollmann, Das Permoskyth in den Ostalpen
      sowie Alter und Stellung des "Haselgebirges",
      N.Jb.Geol.Paläont.Mh., 1964, 270-299, 1 Abb.,
      Stuttgart 1964.

3.   A. Tollmann, Faziesanalyse der alpidischen
      Serien der Ostalpen, Verh. Geol. B.-A., Sonder-
      heft G, S. 103 - 133, Wien,1965.

4.   O. Schulz and W. Lukas, Eine Uranlagerstätte in
      permo-triadischen Sedimenten Tirols,
      Tschermaks Miner. u. Petrogr. Mitt. 14,
      213 - 231, Springer Verlag, 1970.

5.   F. Trauth, Geologie der nördlichen Radstädter
      Tauern und ihres Vorlandes, Dkschr. Ak. Wiss.
      Wien., m.-n.Kl., 100, 101 - 212, Wien,1926.

6.   W. Klaus, Möglichkeiten der Stratigraphie im
      Permoskyth, Mitt. Ges.Geol. Bergbaustud., 20,
      33 - 34, Wien, 1972.

7.   E. Pak, Schwefelisotopenuntersuchungen am
      Institut für Radiumforschung und Kernphysik I,
      Anzeiger Österr. Akad. Wiss., 1974.

8.   A. Tollmann, Tektonische Karte der Nördlichen
      Kalkalpen, Mitt. Ges. Geol. Bergbaustud. Wien,
      18 (1967), 207 - 248, Taf. 8, Wien, 1968.

9.   K. Metz, Die Geologie der Grauwackenzone von
      Mautern bis Trieben, Mitt. Reichsst. Boden-
      forsch. Wien, N.F. 1 , 161 - 220, Wien 1940.

10.   O. Ganss, F. Kümel and E. Spengler,
       Erläuterungen zur geologischen Karte der
       Dachsteingruppe, Wiss. Alpenvereinsh., 15,
       Geol. Karte 1 : 25 000, Innsbruck, 1954.

11.   G. Haditsch and H. Mostler, Mineralisationen
       im Perm der Ostalpen, Carinthia II,
       164./84. Jahrgang, 63-71, Klagenfurt, 1974.

12.   H.W. Flügel, Einige Probleme des Variszikums
       von Neo-Europa, Geol. Rundschau, 64 , H. 1,
       Ferdinand Enke Verlag, Stuttgart, 1975.

LE PERMIEN DES PREALPES LOMBARDES ORIENTALES[(+)]

G. CASSINIS et G. PEYRONEL-PAGLIANI

Institut de Géologie de l'Université de Pavie
et Institut de Minéralogie, Pétrographie et
Géochimie de l'Université de Milan

ABSTRACT. On the basis of new geological and petrographical researches throughout the Palaeozoic cover between the Trompia Valley and Croce Domini Pass, the authors summarize the characteristics of the lithostratigraphical units. These are from bottom to
top: "Basal Conglomerate", "lower quartz-porphyries", Collio Formation with some rhyolitic lavas interstratified, Dosso dei Galli
Conglomerate, Auccia Volcanites and M. Mignòlo Sandstone (Lombardy Verrucano).
   The succession above described is typical of eastern Lombardy, where it is chiefly noted for the fossils found in the Collio,
including vegetables and vertebrate footprints from the Rotliegende.
   This is followed by a synthesis concerning the principal stra
tigraphical characteristics westwards, in the Camonica Valley, and
eastwards, in the Caffaro Valley.
   The investigations on the geographical limits and the frequent stratigraphical variations of the Collio basin has made it
necessary to provide a general description of the Permian in the
ncighbouring localities. These are present in the western region
of the Chiese and between this river and the Rendena Valley.
   A chronostratigraphical classification of the Lombardy formations has been attempted on the basis of the recent palaeonto-

---

(+) L'étude sur le terrain a été effectuée par G.Cassinis grâce
à l'aide financière du Conseil National des Recherches. En ce qui
concerne la rédaction de la présente étude, G.Cassinis est respon-
sable des observations de caractère litho- et chronostratigraphique et paléogéographique; quant aux données pétrographiques sur
les roches volcaniques, nous les devons à Madame G.Peyronel-Pagliani

H. Falke (ed.), The Continental Permian in Central, West, and South Europe, 148-168. All Rights Reserved.
Copyright © 1976 by D. Reidel Publishing Company, Dordrecht-Holland.

logical examinations of some German specialists (H.Haubold, W.Remy). We have also described briefly the principal events in the whole Collio basin.

Comme on peut le déduire en jetant un coup d'oeil sur la bibliographie de cette étude et de nos précédentes, le Permien des Préalpes lombardes, situées à l'est du Val Camonica, a attiré, dès le milieu du siècle passé, l'attention d'une multitude de chercheurs. Surtout depuis que Suess y observa, en 1869, la présence du Rotliegende, grâce à l'étude, accomplie par Geinitz, de quelques restes végétaux découverts dans le Collio du haut Val Trompia. Ce signalement permit, en effet, de clarifier, non seulement dans les Préalpes brescianes, mais aussi dans les régions limitrophes, le tableau chronostratigraphique encore obscur des unités comprises entre le socle cristallin et le Servino du Trias inférieur. Toutefois, comme nous aurons l'occasion de le spécifier plus loin, il y eut, plus tard, de nombreuses incertitudes quant à l'acceptation inconditionnelle de cette corrélation.

Par suite, tant pour les raisons indiquées que pour la puissance atteinte, le Permien du Val Trompia est de très loin le plus classique du secteur lombard étudié. C'est pourquoi il nous semble opportun de commencer par lui cette revue à caractère régional.

I. Val Trompia

Il faut préciser, d'abord, que la succession stratigraphique qui est illustrée ci-après inclut non seulement cette vallée, mais aussi la zone de séparation des eaux entre le Val Camonica et le Val Caffaro, qui coïncide, de fait, avec le parcours de la route directe vers le Passo di Croce Domini.

A partir des terrains plus anciens, on note donc ce qui suit:

a) "Conglomérat Basal" (Auct.) - Cette formation, dont la dénomination est désormais commune chez les géologues lombards, apparaît constituée par des conglomérats grossiers rouge-brun, avec des éléments de roches métamorphiques et de quartz ainsi que, secondairement, de grès. Elle affleure, de façon marquée, uniquement à l'ouest de la vallée du Stabil Solato, où elle est délimitée audessous par le socle cristallin, à travers une nette discordance angulaire, et au-dessus par des vulcanites. Sa puissance est aux alentours de 25 m.

b) "Porphyres quartzifères inférieurs" (Auct.) - Fait suite régulièrement, au-dessus, une succession de roches ignées à qui les auteurs, sur la base de la position stratigraphique, ont depuis longtemps réservé la dénomination de "Porphyres quartzifères inférieurs". Au début, elle est pour la plupart du temps caractérisée par des tufs multicolores, semi-cohérents, de puissance irrégulière (5 à 20 m) et souvent insignifiante.

Les lithotypes successifs correspondent à des roches volcani-

ques compactes, morphologiquement très prononcées, à structure
porphyrique, de couleur surtout rougeâtre, violacée, grise. La
stratification est imprécise ou presque; toutefois, localement,
comme aux Corni del Diavolo, on peut observer une disposition en
triple paroi des roches en question.

La masse de fond, où ressortent les structures pseudo-fluida-
les particulières des ignimbrites, prévaut toujours sur la fraction cristalline; celle-ci est représentée, dans l'ordre, par des
phénocristaux (ou phénoclastes) de quartz, du K-feldspath (généralement sanidine, parfois amplement albitisée), de la biotite. Les
xénolites sont fournies par des fragments de métamorphites et de
vulcanites, ponces ou "porphyrites", ces dernières des roches dont
la structure est microlitique, avec de petits cristaux rectangu-
laires de plagioclase.

Le chimisme peut se rapporter à celui d'un magma aplit-grani-
tique selon Niggli; d'après la classification de Streckeisen basée
sur la norme AMS de Rittmann, on peut les définir comme des ignimbrites rhyolitiques à tendance alcaline (ou alcali-rhyolites). La
teneur extrêmement élevée en K_2O dans ces lithotypes est caractéristique; cette teneur est accompagnée d'une valeur exceptionnel-
lement faible en CaO, que l'on doit attribuer à l'intense transfor-
mation deutérique hydrothermale-pneumatolytique qui, localement,
a blanchi la roche. Caractère sérial pacifique moyen.

La puissance de cette unité semble varier de 30 à plus de
100 m.

c) Formation de Collio (pro parte) - Les types lithologiques
déjà décrits sont recouverts par des tufs cristallino-vitreux passant parfois à des cinérites multicolores, stratifiés, avec des
intercalations de conglomérats et de grès surtout rougeâtres, dont
l'origine est en partie alluviale.

Les tufs se distinguent des ignimbrites sous-jacentes,en particulier par l'apparition d'un plagioclase répandu chez de grands
individus,très frais et toujours très acide (An 8 à 10%). Les éléments lithiques sont surtout donnés par des vulcanites, avec des
ponces, des ignimbrites et des roches d'aspect "porphyritique"
dont nous avons déjà parlé et dont la structure est pseudo-trachytique.

Quant au chimisme, il est proche du granit-engadinitique.
Par rapport aux manifestations précédentes, les tufs examinés jusqu'à présent offrent une acidité plus faible et une plus grande
richesse en sodium; le caractère hyperaluminifère, déjà observé
dans les ignimbrites, persiste. En suivant Streckeisen, ils font
partie, eux aussi, du domaine des rhyolites, avec un caractère sérial pacifique moyen.

Les intercalations terrigènes signalées dans ces tufs de transition au Collio typiquement sédimentaire se composent de grès et/
ou de rudites lithico-feldspathiques. Sur la base de la systémati-
que de Folk, elles peuvent être définies comme étant des rhyolites-
rudites (ou arénites) feldspathiques passantes, à hauteur des ni-

veaux moins grossiers où le quartz et le feldspath dominent sur
les fragments de roche, à arkoses arénites lithiques. Le plus caractéristique de ces bancs détritiques est celui qui est situé immédiatement au-dessous des sédiments de la Formation de Collio.

La puissance de l'unité tufacéo-épiclastique considérée est,
en moyenne, de 80 m environ.

La partie inférieure du Collio alluvio-lacustre est représentée, du bas vers le haut, par des grès et des siltites gris-vert,
des argilites-grès fins rouge-brun, des argilites et des siltites
gris-noir. Elle constitue l'horizon caractéristique de la formation tant par sa continuité latérale, que, surtout, pour le fait
de trouver à hauteur de ses assises supérieures la flore permienne
déjà citée.

Les types lithologiques en question sont généralement marqués
par des feuillets et par une nette stratification.

Les distinctions chromatiques observées à l'échelle macrosco-
pique ne sont toutefois pas confirmées par les résultats des analyses effectuées sur un certain nombre d'échantillons (I). Les
grès et les siltites sont représentés, de fait, surtout par des
arkoses. La fraction détritique se compose de feldspaths représentés par des plagioclases (albite-oligoclase) et des orthoses, de
quartz anguleux d'origine tant volcanique que métamorphique, et
de fragments de roches généralement volcaniques, de structure différente, parmi lesquels on note la structure pseudo-trachytique ou
microlithique déjà signalée. La matrice, plutôt abondante, est donnée surtout par des minéraux argileux de type illitique, recristal-
lisés partiellement en de grosses feuilles micacées claires parfois réunies en bandes ou festons ondulés.

Les argilites sont, elles aussi, caractérisées par l'abondance d'illite, à laquelle s'associe généralement, en sous-ordre, de
la chlorite et, parfois, de la kaolinite en pourcentage très faible ou en traces, avec une composition minéralogique pratiquement
constante, et intéressées par des phénomènes de recristallisation
identiques à ceux des lithotypes décrits ci-dessus.

La présence, dans les niveaux supérieurs, d'éléments lithiques blancs irréguliers, apparaît avoir une relation avec un début
de manifestation d'une seconde activité volcanique. Selon toute
probabilité, ces éléments particuliers, constitués de feldspaths
(albite-oligoclase) associés à du quartz et, parfois, à de la calcite, sont, en effet, attribuables à des cendres et à des lapilli
qui ont été, eux aussi, soumis à des processus de recristallisation
avec une formation d'albite néogène.

La portion du Collio sédimentaire examinée atteint, le long

(I) G.Cassinis remercie vivement son ami, le Professeur L.Matta-
velli, de l'Agip Mineraria, de lui avoir permis de profiter, dans
la forme jugée la plus adaptée aux caractéristiques de cette étude,
des résultats de quelques-unes de ses recherches pétrographiques
et diffractométriques entreprises sur le Permien brescian.

de la route Maniva-Croce Domini, la valeur de 200 m. A l'occident,
par contre, elle s'amenuise jusqu'à disparaître soudain à hauteur
de la Vallée haute de Cigoleto, au contact d'une faille.

d) Laves rhyolitiques intercalées dans le Collio ("Porphyres
quartzifères supérieurs" Auct.) - Vers le haut, les susdites argilites et siltites gris foncé incluent un ou plusieurs épanchements
volcaniques. Celui qui a le plus de continuité est le supérieur,
dont la puissance est de 15 m environ.

Les roches en question, qui peuvent être comparées à des laves qui se sont déposées dans un milieu aquatique, se présentent
à structure porphyrique, de couleur grisâtre et privées de stratification. Le fond a une structure qui va de la micro- à la cryptocristalline; il est constitué, en grande partie, par de l'albite
et du quartz. La fraction cristalline est fournie par le quartz,
le plagioclase albitique souvent intensément séricitisé, parfois
remplacé par de l'albite néogène en damier ou entouré d'une auréo-
le plutôt étendue de cette dernière. L'orthose est très rare et
parfois profondément altérée. La biotite est souvent en feuilles
de dimensions importantes, complètement décolorées ou chloritisées.
Les xénolites représentées par des fragments de roches du Collio
sédimentaire sous-jacent, des métamorphites et des vulcanites
sont fréquentes.

Le chimisme peut se rapporter à l'engadinitique sodique selon
Niggli. La classification exacte pour des roches transformées aussi intensément n'est pas très facile. D'après Streckeisen, sur la
base de l'AMS de Rittmann, ces lithotypes devraient être interpré-
tés comme des alcalis-trachytes avec quartz; selon la classification de O'Connor, il s'agirait plutôt de lithotypes dont les
points représentatifs tombent à la limite entre le domaine des
rhyolites et celui des kératophyres quartzifères (ces derniers pouvant être définis comme des rhyolites fortement auto-métamorpho-
sées). Les données pétrochimiques typiques pour ces laves sont la
teneur élevée en Na_2O par rapport à K_2O (k = 0,22 à 0,27) et la
valeur médiocrement élevée en SiO_2 et en CaO. Le caractère sérial
est pacifique moyen.

e) Formation de Collio (pro parte) - Au-dessus de ce dernier
épanchement volcanique, font suite surtout des grès et des siltites verts alternant avec des très minces mais fréquentes manifestations pyroclastiques que l'on peut reconnaître grâce à la présence des éléments blancs déjà cités.

La fraction détritique des roches verdâtres en question n'a
pas, par ailleurs, subi de différences substantielles par rapport
à celle que l'on a observée dans les roches terrigènes fines pré-
existantes; c'est pourquoi on peut, dans l'ensemble, les définir
comme des arkoses lithiques. Il existe, naturellement, toute une
gamme de variations graduelles entre des dépôts de cette nature et
ceux qui sont typiquement volcano-clastiques.

L'épaisseur totale de ces lithotypes est de 20 m environ.

L'apparition successive, dans le Collio, de grès dont le grain est généralement gros, et celle encore plus haute de conglomérats, pousse à reconnaître dans la formation une portion supérieure aux caractères nettement plus grossiers que ceux relatifs à sa portion inférieure.

Les grès, passant à des siltites, sont surtout gris et verdâtres, en strates massives. Sur le vu de quelques échantillons examinés au microscope, on ne relève pas de variations pétrographiques importantes, comparativement à ceux qui sont directement sous-jacents. Il s'agit, en effet, d'arkoses et d'arkoses lithiques.

Les siltites et les argilites superposées, gris foncé, feuilletées, avec des débris végétaux, reproduisent elles aussi, nécessairement avec une plus faible composant détritique, la même composition minéralogique.

L'épaisseur globale de ces deux lithozones peut être estimée à 50 m.

La Formation de Collio se termine, à hauteur du susdit tracé où se trouve sa coupe-type, par une puissante succession composée essentiellement de grès et de conglomérats. Ces lithotypes, qui constituent, de fait, une unité de transition à la formation sus-jacente du Conglomérat du Dosso dei Galli, se présentent, en général, avec des couleurs mal définies et avec des éléments de roches métamorphiques, volcaniques, quartzeuses et sédimentaires (celles-ci appartenant au Collio), avec un faible degré d'arrondissement et de sphéricité, et des dimensions pouvant être considérables. La stratification est, en majeure partie, massive, avec des bancs fréquemment en lentille.

L'épaisseur de cette dernière unité est sujette à des variations. Le long de la coupe-type du Collio, on peut l'évaluer à I50 m environ.

f) <u>Conglomérat du Dosso dei Galli</u> - Cette formation est constituée sur le flanc oriental de ce Dosso, où se trouve la coupe-type, de conglomérats puis, en second lieu, de grès et de siltites, et est divisible en deux parties.

La partie inférieure, dénommée Membre de la "Pietra Simona", est composée principalement de grès et de siltites rougeâtres, plus ou moins bien stratifiés. La caractéristique la plus marquante est donnée par la présence de structures vermiculaires ("röhrigen Wülsten" ou "budellature" <u>Auct.</u>), dues probablement à l'activité d'organismes limivores.

La partie supérieure est constituée par des conglomérats grossiers, avec des éléments de vulcanites, de schistes cristallins et de quartz, immergés dans une matrice gréseuse rouge sombre. Les fragments de métamorphites, surtout, représentent, en raison de leur abondance et de leur évidence, une note très particulière de cet horizon supérieur. La stratification est imprécise ou formée de gros bancs.

Le "Conglomérat" en question, pratiquement réduit à la seule portion supérieure, affleure, en Val Trompia, uniquement dans le

groupe des Colombine et dans la zone comprise entre le Passo delle
Sette Crocette et la Foppa del Mercato, se surposant, dans la première de ces deux localités, au Collio inférieur, tandis que dans
la seconde aux tufs qui lui sont sous-jacents.

La puissance de la formation varie, donc, de 0 à environ 300
m.

g) Vulcanites d'Auccia - Toujours à hauteur de la route Maniva-Croce Domini, où se trouve la coupe-type, on peut distinguer
dans cette unité volcanique trois parties: celle du milieu, contient des bancs d'ignimbrites rhyolitiques violacées, à structure
porphyrique très évidente, qui représentent le faciès le plus caractéristique et le plus constant de la formation.

Dans ces lithotypes, la fraction cristalline est essentiellement formée de quartz, de plagioclase de type oligoclase-andésine
ou, également, plus acide (albite = I0% An), de K-feldspath (sanidine), qui peut toutefois manquer dans quelques niveaux, de biotite. Les xénolites sont assez fréquentes et sont représentées par
des vulcanites et, plus rarement, par des métamorphites. Le fond
présente une importante variété de structures, dont les plus fréquentes sont celles microfelsitique et sphérolitique; les typiques
structures ignimbritiques sont évidentes.

Le chimisme de ces roches est très proche de celui de type
yosemit-granitique; il en est de même pour leur nature rhyolitico-
alcaline avec des caractéristiques petrochimiques semblables à
celles des ignimbrites sous-jacentes au Collio qui, cependant, mon-
trent une teneur plus élevée en SiO_2 et un caractère potassique
beaucoup plus accentué. Le caractère sérial est faiblement pacifique.

L'épaisseur des Vulcanites d'Auccia oscille entre 0 (Val
Trompia) et I30 m environ (dans la coupe-type).

h) Grès du M. Mignòlo (Verrucano Lombard) - Il fait suite, à
l'unité précitée -qui représente le dernier acte du volcanisme
permien acide dans la région- une formation constituée de grès et,
dans une plus faible mesure, de siltites et de conglomérats, de
couleur rouge, qui a pris, dans la littérature, des noms variés
comme ceux de "Verrucano", de "Arenarie di Val Gardena", etc. La
stratification est généralement à bancs ou imprécise.

A côté d'éléments de quartz blanc ou rose, on remarque parfois dans la formation, des éléments représentés, surtout, par des
roches volcaniques. Il ne manque d'ailleurs pas l'occasion de relever, à certains niveaux, la présence, due à des processus d'érosion, de nombreux et gros fragments lithiques des siltites rouge
sombre qui se trouvent intercalées ou alternées avec les grès. Les
micas sont assez diffus.

La matrice, composée de matériau pélitique associé à des oxydes de fer, constitue en moyenne un faible pourcentage du volume
de la roche. Celle-ci apparaît le plus souvent recristallisée en
de minuscules produits micacés.

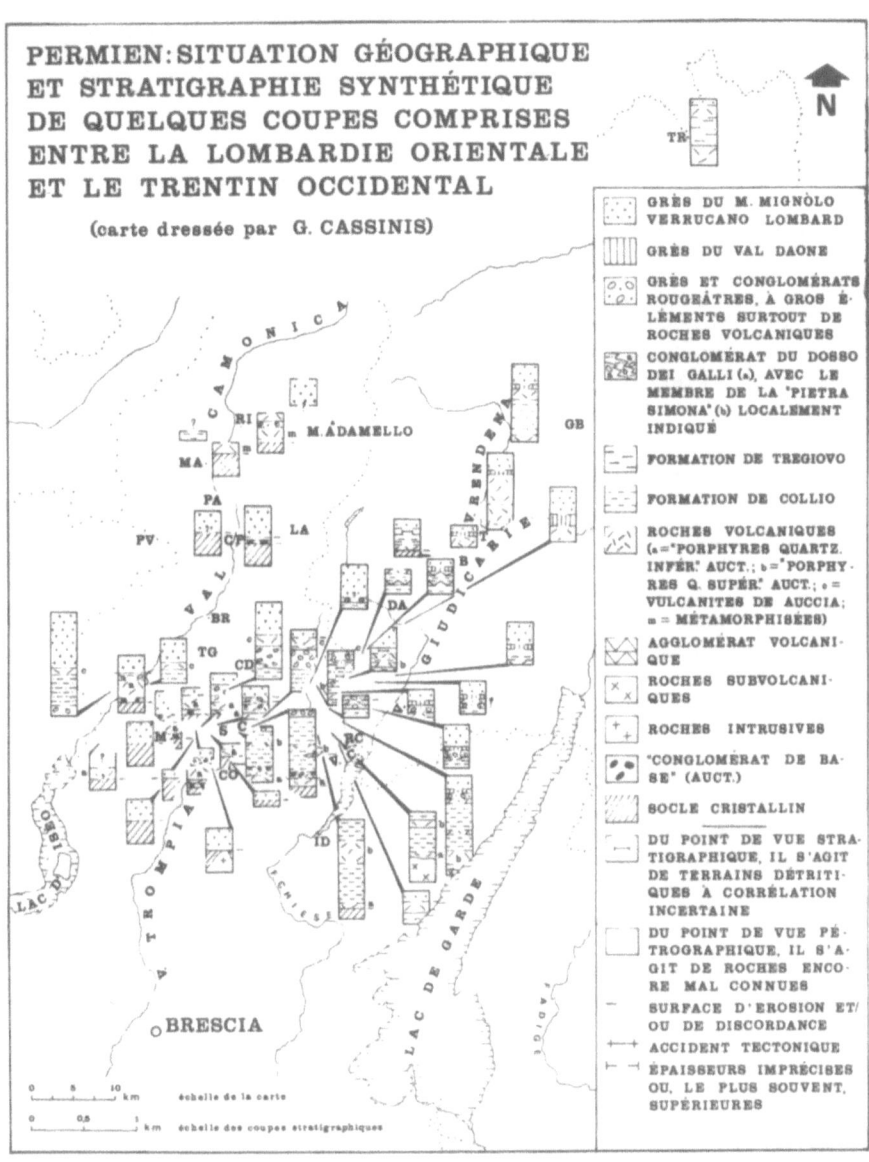

Fig.I – Cette carte est simplement destinée à placer les ensembles
lithologiques les plus évidents, d'âge permien, établis dans la ré-
gion étudiée, afin de faciliter la lecture du texte. Situation de
quelques lieux indiqués: B=Breguzzo, BR=Breno, C=M.Colombine, CD=
P.C.Domini, CO=Collio, CP=C.di Ponte, DA=V.Daone, GB=G.Brenta, ID=
Idro, LA=L.Arno, M=M.Muffetto, MA=Malonno, PA=Paisco, PV=P.Vivione,
RC=Riccomassimo, RI=Rino, S=P.Sette Crocette, T=Tione, TG=T.Grigna,
TR=Tregiovo, V.C.=V.Caffaro.

Sur la base de la classification de Folk, l'étude locale d'u-
ne coupe représentative de ces roches détritiques rouges, qui fer-
ment la succession stratigraphique permienne, a permis d'y reconnaître la présence de lithoarénites-rudites feldspathiques et d'ar-
koses-arénites lithiques.

L'unité a une puissance d'environ 300 à 400 m. Elle s'étend
vers le sud, en Val Trompia, sur des terrains toujours plus anciens, et les érode progressivement et entièrement, jusqu'à repo-
ser sur le socle cristallin.

Mélangées à ce socle, on trouve également de petites masses
épiplutoniques granitiques qui passent à des types adamellitiques
et granodioritiques, ou de diorites quartzifères (Giuseppetti,
I959). Elles ont métamorphosé, de façon limitée, les roches traver-
sées, tandis qu'elles ne semblent pas avoir donné lieu à des phénomènes considérables ou évidents de contact sur le Grès de M. Mignòlo, qui les recouvre. Les roches intrusives en question doivent
donc être considérées comme étant plus anciennes que ce grès. Elles font vraisemblablement partie d'un cycle magmatique hercynien.

2. Val Camonica

En direction du Val Camonica, la succession stratigraphique
permienne, ci-dessus illustrée, subit, par rapport au groupe des
unités sous-jacentes au Grès du M. Mignòlo (Verrucano Lombard),
des modifications.

Nous pouvons distinguer trois zones entre le Passo di Croce
Domini et le Lac d'Iseo. Une zone septentrionale, caractérisée par
une gamme de terrains similaires à ceux du haut Val Trompia. Une
zone centrale, constituée, en partant du bas, par le "Conglomérat
Basal", fréquemment intéressé par des structures vermiculaires
semblables à celles de la "Pietra Simona"; puis par les "Porphyres
quartzifères inférieurs" suivis, mais pas toujours, des tufs qui,
normalement, les recouvrent dans le Val Trompia; et, enfin, sauf
à proximité du Val del Re, par le Conglomérat du Dosso dei Galli.
Une zone méridionale, mise en évidence par le contact immédiat entre le Verrucano et le cristallin ancien; ici, toutefois, entre
Gratacasolo et Fraine, réapparaît, délimité généralement par des
failles, un affleurement représenté par les "Porphyres inférieurs"
qui se trouvent, probablement, interposés entre les deux dernières
unités précitées.

Par contre, sur le versant opposé du bas Val Camonica, la succession permienne est plus régulière, car de Boario Terme à Rondi-
nera, elle est représentée par le Conglomérat du Dosso dei Galli,
par les Vulcanites d'Auccia et par les conglomérats et les grès
rouges sus-jacents (Verrucano Lombard). En particulier, en ce qui
concerne la première formation, elle apparaît généralement constituée par le Membre de la "Pietra Simona" qui, contrairement à la
région située plus à l'est, occupe une position supérieure au sein
de cette formation et atteint la puissance remarquable de 300 m.

Passant au moyen Val Camonica, autour des contreforts occiden-

taux du massif de l'Adamello, le Permien est représenté presque u-
niquement dans les cartes officielles de ce que l'on appelle le
Verrucano Lombard. Les dépôts détritiques de cette unité sont toujours au contact direct du socle cristallin.

Au-dessus de Paspardo, il a été observé entre autres, toujours
dans le Verrucano, quelques fragments foncés qu'Accordi (1953) in-
terpréta comme des schistes argileux que l'on peut rapporter au
Collio; d'après le même auteur, ils proviendraient d'orient, étant
donné que l'épaisseur de la formation augmente graduellement d'est
en ouest (300 à 400 m au nord du M. Colombè, I000 m environ vers
Schilpario).

Le long de la rive septentrionale du Lac d'Arno, on attribue,
en outre, au Permien des cornubianites bigarrées brun rougeâtre
et vertes, avec des niveaux de dolomies cariées friables qui, si
elles devaient effectivement faire partie -comme il est indiqué
dans la feuille I9 Tirano de la carte géologique italienne au
I00 000e- de la Formation à Bellerophon, donneraient lieu à l'affleurement plus occidental de cette unité commune dans la région
située surtout à l'est de l'Adige (2).

Sur le versant opposé du moyen Val Camonica, le Verrucano Lombard se prolonge de Capo di Ponte vers le Mont Elto, et se dispose
en bande, encore au contact du cristallin, le long du flanc droit
du Val Paisco et atteint le Passo del Vivione.

On est donc contraint d'affirmer, sur la base des renseignements rapportés, qu'il n'existe pas, à l'heure actuelle, d'éléments
permettant d'être amenés à soutenir une continuation du bassin intérieur du Val Trompia dans ce secteur intérmediaire de la vallée
considérée.

Plus au nord, la succession permienne vient résumer un plus
grand développement. Mais celle-ci, généralement marquée par des
roches volcaniques métamorphosées ("porphyroïdes" ou "schistes por-
phyriques" Auct.), localement superposées par des couches ou des
bancs pouvant se rapporter à la Formation de Collio et par le Verrucano Lombard, appartient déjà au secteur oriental du bassin tar-
do-hercynien des Alpes bergamasques.

3. Val Caffaro

Retournant au bassin brescian du Val Trompia, celui-ci est caractérisé, vers est et en direction du moyen et du haut Val Caffaro, par une succession stratigraphique qui ne diffère pas beaucoup

---

(2) Sur la base des récentes connaissances acquises sur la stratigraphie à l'est du massif de l'Adamello, par exemple dans le groupe
de Brenta et dans les Giudicarie, il semblerait plus facile d'accep-
ter l'hypothèse que les niveaux en question correspondraient à une
unité située au début, ou presque, de la Formation de Werfen ou de
Servino (équivalant, peut-être en partie, à l'horizon de Tesero) et
estimée par certains auteurs d'âge triasique inférieur.

de celle qui a été examinée à hauteur de la route Maniva-Croce Domini. Les faits les plus intéressants sont fournis par le plus
grand nombre de coulées rhyolitiques intercalées dans le Collio
et par des variations de faciès et d'épaisseur au sommet de la
formation du même nom.

Dans le bas Val Caffaro, une étude, publiée depuis peu par
les auteurs de cette relation, a dénoncé la continuation latérale
des "Porphyres inférieurs" déjà décrits, contenant localement en
plus, parmi les phénocristaux, un plagioclase albitique très frais,
qui ne rentre pas, au contraire, dans les lithotypes examinés à
l'ouest. Dans la même zone, entre autres, les témoignages volcaniques d'âge moins ancien se font nettement plus imposants à la suite de l'existence, supposée, d'une zone d'adduction. Il s'agit de
masses subvolcaniques qui apparaissent pénétrées dans le cristallin et dans la succession permienne de base du Val Trompia, et de
épanchements volcaniques bien plus puissants que ceux précités.

Il n'existe, dans les laves de deux groupes, aucune diversité
tant dans la structure -toujours porphyrique-, que dans la composition minéralogique. La masse de fond, toujours subordonnée à la
fraction cristalline, a une structure allant de la micro- à la
cryptocristalline, parfois fluidale pseudo-trachytique dans les
laves des masses subvolcaniques. La fraction cristalline est toujours constituée par du plagioclase oligoclasique ou albitique,
de couleur parfois rosée macroscopiquement, du quartz et par de
la biotite. Il manque le K-feldspath en phénocristaux, alors qu'il
est constamment présent sous forme de microcline très repandu dans
la masse de fond, avec la chlorite, le quartz et l'albite. Les xénolites ne sont pas très fréquentes; elles sont représentées par
des métamorphites et des roches "porphyritiques".

En ce qui concerne le chimisme, toutes les laves qui s'intercalent dans le Collio peuvent être classifiées comme des rhyolites,
à tendance surtout alcaline et dont la composition est très proche
de celle des types magmatiques, selon Niggli, alcalo-granitiques
et engadinitico-granitiques. Les laves hypogées sont, elles aussi,
des rhyolites, mais avec une tendance quartzo-trachytique alcaline
et avec un chimisme proche de celui des magmas nordmarkitiques.

L'absence locale du Conglomérat du Dosso dei Galli et des Vulcanites d'Auccia semble cependant se préciser entre la ligne faî-
tière du moyen Val Caffaro et le Rio Riccomassimo. En général, ici,
le Collio supérieur, c'est-à-dire celui suivant la coulée la plus
importante de laves rhyolitiques interstratifiées, est caractérisé par des lithotypes plus ou moins semblables à ceux des monts
du haut Val Trompia; mais ils ne sont pratiquement pas cartographa-
bles, surtout parce que les épisodes terrigènes à grain fin, en
devenant assez nombreux, compliquent cette tâche. Normalement, la
succession vers le haut est la suivante: un banc rhyolitique (I3 m);
des grès et des siltites rouge-violacé, parfois tufacés, qui sont
accompagnés de conglomérats de la même couleur, constitués presque
uniquement d'éléments volcaniques plutôt arrondis (60 m); des conglomérats d'un grain plus fin que les précédents et des grès de

couleur claire avec des éléments lithiques de diverse nature (I5 m); enfin, les grès rouges de M. Mignòlo (Verrucano Lombard).

4. Val Giudicarie inférieur

La succession, que nous venons de donner, du Val Riccomassimo se poursuit jusqu'au T.Sorino ou peu au-delà. Sur les monts, à droite du Chiese, toutes les manifestations laviques et pyroclastiques permiennes prennent, en outre, un développement plus accentué qu'à ouest, simultanément avec l'apparition de dépôts sédimentaires plus puissants ou souvent grossiers au sein du Collio.

Toutefois, en de nombreuses zones, le Collio dénote, au contact de la succession stratigraphique (plus ou moins complète) qui le délimite supérieurement, la présence d'un horizon caractéristique à grain fin, riche en végétaux et localement semblable à celui, bien plus connu sous l'aspect paléo-botanique, qui distingue, à occident, la portion inférieure de la formation.

En outre, comme il ressort du tableau stratigraphique de la Fig.I, dans les coupes examinées au bord septentrional des terrains paléozoïques reliés directement au bassin du Val Trompia, le passage aux grès rouges du Verrucano subit d'ultérieures modifications. Les causes de ces variations ne nous sont pas encore très claires; c'est pourquoi nous renvoyons leur approfondissement à un prochain travail.

Par ailleurs, l'étude générale du bassin du Val Trompia nous pousse à effectuer quelques observations même en dehors des limites géographiques imposées par le titre de cette étude, précisément vers nord-est, dans le but de connaître les caractéristiques les plus évidentes relatives aux successions stratigraphiques permiennes dans les différentes localités, et de risquer quelques tentatives de corrélation entre elles pour en examiner les éventuelles affinités.

Ainsi, en Val Daone, le Permien est représenté, du bas vers le haut, par des vulcanites verdâtres ou rouge-violacé, par des roches détritiques claires semblables à celles observées plus au sud jusqu'au Val Riccomassimo, et par des grès rouges du Verrucano.

Le Permien réaffleure plus au nord, entre le Val Bondone et le Val Breguzzo. Localement, au-dessus du socle cristallin, il a été découvert des roches volcaniques (45 m) suivies de lithotypes fins (I5 m env.) rappelant les caractères du Collio, et de lithotypes plus grossiers (50 m env.), qui passent au Verrucano Lombard.

5. Val Rendena

A hauteur du grand ensemble éruptif permien qui court des environs de Tione jusqu'au-delà de Massimeno, les recherches effectuées ont permis de confirmer la présence non seulement d'affleurements de Collio argileux ou gréseux, plus ou moins vastes, émergeant à différents niveaux, mais également d'un horizon détritique lithologiquement semblable à celui du Val Daone, situé, lui aussi,

sur des roches volcaniques et au-dessous des grès rouges du Verrucano ou du Val Gardena.

D'après Trevisan (1939), cet horizon apical, appelé sous le
nom de "Anagénites", serait assez continu. Par conséquent, compte
tenu de son importante distribution sur l'arête, que l'on peut é-
valuer, à vol d'oiseau, à 40 Km environ, et des aaractères lithologiques particuliers qui le distinguent, il est jugé opportun de
l'indiquer stratigraphiquement sous le nom de "Arenaria della Val
Daone" (Grès du Val Daone), car il est plus puissant dans cette
zone (100 m env.) et qu'il est signalé depuis longtemps dans la
littérature (p.ex. Dozy, 1935).

Toutefois, malgré les affinités signalées, les caractéristiques d'ensemble du Permien du Val Rendena s'écartent beaucoup de
celles qui sont connues dans les terrains contemporains, situés
plus au sud - sud-ouest, surtout à partir du Val Daone.

TENTATIVE DE CLASSIFICATION CHRONOSTRATIGRAPHIQUE

Le seuls fossiles du Permien de la Lombardie orientale proviennent surtout de la Formation de Collio du haut Val Trompia et
des monts situés immédiatement au nord de ce dernier. Il s'agit
de restes végétaux et, secondairement, de traces de vertébrés et
de valves de lamellibranches lacustres.

En 1869, quelques pièces furent rassemblées par Suess et classifiées par Geinitz: Walchia piniformis, Walchia filiciformis,
Noeggerathia expansa ?, Schizopteris fasciculata, Sphenopteris
tridactylites, Sphenopteris Suessi, Sphenopteris oxydata ? (écrits
dans leur diction originale); elles portèrent à attribuer la formation en question au Rotliegende.

Par la suite, ayant eu l'opportunité d'examiner quelques plantes recueillies par Dozy (1935), Jongmans se prononça en faveur
d'un âge permien inférieur à cette dèrnière unité.

Plus spécifiquement, en 1939, Heritsch, dans un mémoire concernant le Carbonifère et le Permien de l'Europe sud-occidentale,
jugea devoir la rapporter au Rotliegende moyen.

Après la dernière guerre en Europe, surgirent, cependant, des
doutes sur la validité stratigraphique précise de la flore sus-ci-
tée, étant donné que quelques-unes de ses formes commencent pendant le Carbonifère supérieur (De Sitter et De Sitter-Koomans,
1949). Malgré cela, en raison de la simultanéité du Collio à une
activité volcanique du type surtout acide, considérée dans de nombreuses régions européennes d'âge permien, on continua, par conven-
tion, à attribuer la formation à cette période.

Les incertitudes sur le dilemme chronologique -Carbonifère
supérieur et/ou Permien inférieur- ne disparurent cependant pas
complètement; pas même après une seconde étude paléo-botanique
conduite par Tyroff, du Musée d'Histoire Naturelle de Francfort-
sur-le-Main, sur quelques échantillons déjà déterminés par Geinitz
et sur divers autres que j'ai personnellement recueillis.

Il indiqua la présence de: Lebachia laxifolia, Walchia geinitzii, Schizopteris fasciculata, Sphenopteris suessi, Strobilifer sp., Walchiostrobus ? (Cassinis, 1966 b).

L'examen du Docteur Mademoiselle Paproth, de Krefeld en Westphalie, accompli, lui aussi, sur une faune à lamellibranches d'eau douce, trouvée par Berruti à hauteur de la partie supérieure de la formation, n'apporta aucune contribution au problème de l'âge, sinon celle de reconnaître leur appartenance à la famille des Anthracosiidae (Cassinis, 1969).

A la même époque, une note de Berruti, qui fréquentait le Musée d'Histoire Naturelle de Brescia, mit en lumière divers autres matériaux fossilifères, dérivant surtout des niveaux supérieurs du Collio et non seulement -comme on pouvait le lire dans la littérature- des niveaux inférieurs, sous-jacents aux coulées volcaniques intercalées dans la formation.

Publiée en 1969, cette note de Berruti offrit à Haubold de Halle l'occasion de reconnaître, dans les photographies et les dessins relatifs aux traces de vertébrés, la présence de Dromopus lacertoides, cf. Gilmoreichnus brachydactylus, Anthichnium salamandroides, Amphisauropus imminutus, Amphisauropus latus, qui devraient permettre, d'après l'auteur, de rapporter la Formation de Collio à l'Autunien supérieur de l'Europe centrale et de l'estimer, par exemple, plus ou moins contemporaine des "Oberhöfer Schichten".

L'attribution au Permien inférieur, suggérée dans une lettre envoyée au Dr Berruti en personne, le 10 octobre 1973, fut confirmée peu après par une ultérieure révision de toutes les formes fossiles de végétaux (dont j'étais au courant) déjà recueillies au sein de cette unité, ainsi que par un examen de quelques exemplaires appartenant aux "Strates de Tregiovo" du haut Val di Non et de quelques localités peu distantes, car ces dernières strates constituent une formation permienne assez analogue à celle du Collio. D'après Remy, qui s'est occupé de cette étude, et sur la base des connaissances actuelles et des formes reconnues, il n'y aurait aucune différence chronologique sensible entre les flores des Formations de Collio et de Tregiovo, et elles appartiendraient au Saxonien.

La liste des déterminations qui m'est parvenue (6 mai 1974) comprend, sans qu'il subsiste une référence quelconque à l'une ou à l'autre formation, les fossiles suivants: Callipteris cf. nicklesi, Sphenopteris bipinnata, Sphenopteris kukukiana, Sphenopteris suessi, Lesleya sp., Lebachia hypnoides, Lebachia laxifolia, Walchia arnhardti, Walchia gallica, Walchia geinitzii, Walchia stricta, Ullmannia frumentaria.

Cette datation du Collio contredit toute affirmation précédente, car elle situe la formation à des niveaux permiens nettement plus récents que ceux qui ont été considérés jusqu'à présent. Etant donné l'énorme intérêt de cet élément, la flore examinée sera donc l'objet d'une description dans un prochain numéro de la revue "Argumenta Palaeobotanica" qui comportera, comme on peut certainement le supposer, une discussion concernant l'âge.

Donc, dans les Préalpes bresciannes, sur la foi des renseigne-

ments rapportés et des résultats obtenus jusqu'à ce jour dans les
régions environnantes, on serait porté à attribuer le groupe de
formations sous-jacentes au Grès du M. Mignòlo (Verrucano Lombard)
au Permien inférieur, exactement à un intervalle de temps qui s'in-
sère dans l'Autunien supérieur et/ou dans le Saxonien. Si ce dernier étage (= Rotliegende supérieur) devait exister, il ne serait
cependant limité qu'à une de ses parties -vraisemblablement la
partie inférieure-, compte tenu du fait que les "grès rouges", du
type Val Gardena, ont mis en évidence, bien qu'à l'est de la Lombardie, de nombreux témoignages paléontologiques considérés d'un
âge qui est encore en partie saxonien et en partie plus récent. Le
Grès du M. Mignòlo se serait, par conséquent, constitué entre la
partie la plus élevée du Permien inférieur et le Permien supérieur.

Naturellement, ce tableau à peine tracé, représentant une éta-
pe des recherches en cours, est susceptible d'ultérieures modifications. Ainsi, bien qu'il n'ait pas encore été recueilli une preuve
quelconque dans la région examinée, contrairement à d'autres zones
des Alpes méridionales, il a déjà été signalé la présence d'une
lacune entre le Permien et le Trias inférieur (Assereto et coll.,
I972).

BREVE HISTOIRE GEOLOGIQUE ET CONSIDERATIONS CONCLUSIVES

La Lombardie orientale fut caractérisée, surtout dans le Permien, par l'existence d'un bassin intérieur qui partait du haut
Val Trompia vers le Val del Chiese. L'histoire de ce bassin du Val
Trompia, ou de Collio, a été amplement mise en évidence (Cassinis,
I966 b; Cassinis et coll., I975), en particulier pour ce qui concerne ses aspects plus méridionaux. C'est pourquoi nous renvoyons
aux chapitres correspondants des travaux déjà cités, ceux qui pourraient être intéressés à connaître les détails. Cette étude trace-
ra donc les faits géologiques les plus marquants et les problèmes
qui restent à résoudre.

Sur la base des observations effectuées, après l'événement métamorphique hercynien, il apparut, surtout à proximité du Val Camonica, quelques irrégularités dans le paysage qui donnèrent lieu à
des conglomérats alluviaux (Conglomérat Basal), peut-être liées à
la manifestation prochaine d'une activité volcanique. Celle-ci fut
tout d'abord caractérisée par des dépôts tufacés, qui ne se trouvent cependant que localement. Mais avec l'arrivée d'une ou de plusieurs étendues ignimbritiques -dont on ignore, jusqu'à ce jour,
la provenance- cette activité prit un développement nettement plus
important, que l'on peut attribuer à toute la région comprise entre les vallées de l'Oglio et du Chiese. Cette première séquence
volcanique se conclut partout par des manifestations pyroclastiques qui semblent s'être générées par des ouvertures très proches.

Selon toute probabilité, c'est au cours de cette période que
se forma le Bassin de Collio. Celui-ci fut marqué, dans sa phase
initiale, par des dépôts fluvio-lacustres à grain fin, qui évoluè-

rent également vers des conditions de type marécageux.

Vers la fin de ces dépôts, survint une seconde activité volcanique, représentée par des laves et des fragments pyroclastiques. Elle s'exerça surtout vers l'orient, car les manifestations relatives sont plus nombreuses et puissantes dans cette direction et qu'il en a été découvert quelques conduits d'alimentations tant dans l'environs de Bagolino qu'en Val Giulis. Cette activité fut accompagnée, généralement, de processus de sédimentation plus énergiques que ceux qui se sont produits précédemment. Il semble donc évident de supposer qu'il y ait eu un lien entre ces deux phénomènes et que la vitesse de la subsidence, due à l'évacuation des laves et des éjecta, puisse avoir subi un sollecitation plus forte.

Le fait d'avoir remarqué, par la suite, que les laves se sont extrudées dans le Collio dans des aires non proprement périphériques au bassin, peut en outre suggérer l'hypothèse qu'il existait une masse magmatique active sous une bonne partie des affleurements paléozoïques actuels. Certainement, comme le long du Rio Secco, au nord de Bagolino, la présence de ces corps volcaniques hypogés donnait lieu à des irrégularités dans le fond du bassin, et provoquait, par suite, des déséquilibres dans les quantités d'accumulation des sédiments. Malgré cela, le tableau général qui ressort de l'ensemble des observations effectuées au sein du bassin, amène à reproposer l'idée déjà formulée par De Sitter et d'autres auteurs, c'est-à-dire qu'il était caractérisé par des flancs raides, souvent délimités par une ou plusieurs failles et par des aires intérieures plates ou légèrement déclives.

A ce propos, il vaut la peine de rappeler que l'importante ligne tectonique du Val Trompia, d'âge alpin, s'est justement délimitée le long du saut structural existant entre le bassin du Collio et la dorsale montagneuse qui le délimitait au sud, c'est-à-dire, probablement, le long d'un système de failles contemporaines de l'activité volcanique permienne. A l'est de Bagolino, une preuve ultérieure naît du fait que cette ligne montre clairement s'être établie en coïncidence d'une bande linéaire verticale de laves, vraisemblablement débordantes en surface, par qui auraient été alimentés les coulées intercalées dans le Collio.

Etant donné les nombreuses structures que l'on découvre avec une certaine continuité dans les sédiments du Collio (rides, fissures par dessèchement, empreintes de vertébrés et de gouttes de pluie, etc.), cette fosse fut occupée par des eaux plutôt basses et souvent consacrée à l'émersion. Signe que la subsidence était contrebalancée, avec une certaine régularité, par la sédimentation.

Il semble, en outre, intéressant de souligner que l'immaturité minéralogique et texturale des grès et des conglomérats du Collio -que nous avons déjà mise en évidence- dénoncerait l'existence d'un transport rapide et d'un climat modérément semi-aride, qui ne auraient généré que de faibles changements, sur les particules les plus fines également.

Selon les recherches déjà nommées de Mattavelli, qui sont encore inédites, les modifications post-sédimentaires subies par la susdite formation, dans son aire-type, pourraient, dans son ensemble, être rapportées à une diagénèse avancée, mais ne rentreraient pas dans le domaine des processus métamorphiques. Cette constatation est surtout prouvée par le fait que la fraction détritique, associée à celle argileuse, conserve sa structure originaire et ne démontre pas avoir subi de phénomènes significatifs de recristallisation.

Avec la survenue du Conglomérat du Dosso dei Galli, le paysage subit probablement d'ultérieures transformations. D'après la distribution des affleurements relatifs à cette formation, on pourrait supposer, par exemple, que la structure positive -ce que l'on appelle "Val Camonica Ridge" des De Sitter- séparant le bassin du Val Trompia de celui des Alpes bergamasques, passait à l'ouest du cours inférieur de l'Oglio, plutôt qu'à l'est comme il ressort de la littérature.

Ainsi que nous l'avons dit, le passage aux formations détritiques sus-jacentes se présente avec des modalités différentes. Il suffit, pour cela, d'observer les coupes columnaires de la Fig. I. Seule la région à l'ouest du Caffaro, jusqu'au Val Camonica, est marquée par le dépôt d'une unité volcanique surtout ignimbritique, dont on ne connaît cependant pas, comme pour celle de base de la succession permienne, l'aire de provenance. Les zones restantes, jusqu'à Tione, sont caractérisées, par contre, par une multiciplité de faciès terrigènes, tant grossiers que fins. Par ailleurs, des manifestations de type volcanique semblent s'y substituer ou s'y ajouter localement.

La présence presque constante du Grès du Val Daone marque pratiquement dans toute cette aire orientale la fin de tout dépôt sédimentaire déjà cité ou la fin de l'activité volcanique permienne. Il s'agit, dans tous les cas, dans les zones où ce "Grès" a été observé jusqu'à présent, d'une masse de détritus, toujours immédiatement antérieure aux grès rouges du Permien plus jeune.

Par conséquent, à la suite des renseignements ci-dessus reportés et de la coexistence de tectoniques tardo-hercynienne et alpine plutôt compliquées et souvent superposées entre elles, on hésite encore à exprimer une opinion précise sur la direction structurale du bassin étudié et sur les limites géographiques qu'il avait atteint. La seule affirmation que nous nous sentons en mesure de soutenir est que ce bassin a pris place en marge ou à proximité du vaste plateau porphyrique du Trentin-Haut Adige.

Sur la base des récentes recherches sur le complexe volcanique permien de la Lombardie orientale, il nous semblerait en outre possible de déclarer qu'il y ait une corrélation entre les vulcanites lombardes et la partie la plus élevée du susdit plateau. Par contre, il serait encore prématuré, à l'état actuel des connaissances, d'établir une comparaison entre les caractères de composition des vulcanites se rapportant aux deux régions.

L'apparition des "grès rouges" supérieurs représente un événement commun à toutes les successions permiennes de la Lombardie

orientale. Elles se sont formées tant à hauteur des dépressions
qu'à des reliefs préexistants. C'est pourquoi, en comblant les
premières, ils progressèrent sur les seconds en les démantelant
ou en les "fossilisant". Il s'ensuit que ces grès possèdent ini-
tialement des âges différents selon la zone et reposent, le plus
souvent, en discordance sur les terrains sous-jacents.

En ce qui concerne les monts au nord du haut Val Trompia,
l'étude de la coupe-type du Grès du M. Mignòlo -qui serait, d'a-
près le schéma de la Fig.2 donné récemment par Assereto et coll.
(1972) un faciès particulier du "lithosome" baptisé sous le nom
de Verrucano Lombard- a conduit à souligner leur état d'immaturité du point de vue tant texturale que minéralogique.

Une confrontation préliminaire entre les détritus qui constituent les grès rouges en question et ceux de quelques types représentatifs de la Formation de Collio, permet toutefois de relever
qu'elles sont moins immatures, car le quartz est moyennement plus
élevé dans les premières, tandis que les pourcentages de feldspaths
et de fragments de roches sont plus bas. Du point de vue textural,
également, les grès du M. Mignòlo présentent un classement plus
grand, un arrondissement meilleur et une sphéricité plus parfaite
par rapport aux constituants du Collio.

Sous l'aspect paléogéographique, la présence de feldspaths,
même s'ils sont assez altérés, et de fragments de roches volcaniques dénoterait, en outre, un transport relativement court et un
climat de type semi-aride. Ce fait serait d'ailleurs confirmé par
les oxydes de fer diffus dans la matrice, qui indiqueraient la
prédominance de conditions oxydantes dans l'aire des dépôts ori-
ginaires.

Pour conclure, les caractéristiques pétrographiques et sédimentologiques du Grès du M. Mignòlo (Verrucano Lombard) semble-
raient indiquer un milieu de sédimentation continentale du type
fluvial ou de plaine alluviale. A ce milieu fit suite, dans le
Trias inférieur, l'arrivée de la mer du Servino.

BIBLIOGRAPHIE

Accordi B. (1953) : Geologia del Gruppo del Pizzo Badile (Adamel-
        lo sud-occidentale). - Mem. Ist. Geol. Univ. Padova, vol.18,
        58 p., Padova.
Ardigò G. (1955) : Geologia della regione fra il Sebino e l'Eri-
        dio. IV. La porzione nord-occidentale. Introduzione. Strati-
        grafia. - Atti Ist. Geol. Univ. Pavia, vol. 5 (1951), p. 65-
        82, Pavia.
Ardigò G. et Boni A. (1952) : Sulla stratigrafia del massiccio
        delle Tre Valli Bresciane.-Boll. Serv. Geol. Italia, vol.
        74 (1953), n° 2, p. 323-333, Roma.
Assereto R. et Casati P. (1965) : Revisione della stratigrafia
        permo-triassica della Val Camonica meridionale (Lombardia) -
        Riv. Ital. Paleont. Strat., vol. 71, n° 4, p. 999-1097, Mila-

no.

Assereto R. et Casati P. (1966) : Il "Verrucano" nelle Prealpi lombarde. - Atti Symp. Verrucano (Pisa-Settembre 1965), p. 247-265, Pisa.

Assereto R., Bosellini A., Fantini-Sestini N. et Sweet W.C. (1972) : The Permian-Triassic Boundary in the Southern Alps (Italy). - Bull. Can. Petrol. Geol., vol. 20, p. 176-199.

Berruti G. (1969) : Osservazioni biostratigrafiche sulle formazioni continentali pre-quaternarie delle Valli Trompia e Sabbia. II. Sulla fauna fossile della Formazione di Collio (alta Val Trompia). - Natura Bresciana (Ann. Mus. Civ. St. Nat.), an. 5, n° 6, p. 3-32, Brescia.

Boni A. (1943) : Geologia della regione fra il Sebino e l'Eridio. Introduzione. Bibliografia tettonica. Parte prima. La porzione centrale. - Atti Ist. Geol. Univ. Pavia, vol. I, 143 p., Pavia.

Boni A. (1952) : Geologia della regione fra il Sebino e l'Eridio. III. Il margine orientale. a) Stratigrafia. - Atti Ist. Geol. Univ. Pavia, vol. 5 (1951), p. 13-64, Pavia 1965.

Boni A. et Cassinis G. (1973) : Carta geologica delle Prealpi Bresciane a Sud dell'Adamello. Note illustrative della legenda stratigrafica. - Atti Ist. Geol. Univ. Pavia, vol. 23, p. 119-159, Pavia.

Cassinis G. (1966 a) : Rassegna delle formazioni permiane dell'alta Val Trompia (Brescia). - Atti Ist. Geol. Univ. Pavia, vol. 17 (1965-66), p. 51-66, Pavia.

Cassinis G. (1966 b) : La Formazione di Collio nell'area-tipo dell'alta Val Trompia (Permiano inferiore bresciano). - Riv. Ital. Paleont. Strat., vol. 72, n° 3, p. 507-588, Milano.

Cassinis G. (1968) : Sezione stratigrafica delle "arenarie rosse" permiane presso il Passo di Croce Domini (Brescia). - Atti Ist. Geol. Univ. Pavia, vol. 19, p. 3-14, Pavia.

Cassinis G. (1969) : Appunti su una fauna a lamellibranchi non marini rinvenuta nel "Collio" trumplino (Paleozoico sup. bresciano). - Atti Ist. Geol. Univ. Pavia, vol. 20, p. 82-86, Pavia.

Cassinis G., Origoni-Giobbi E. et Peyronel-Pagliani G. (1975) : Osservazioni geologiche e petrografiche sul Permiano della bassa Val Caffaro (Lombardia orientale). - Atti Ist. Geol. Univ. Pavia, vol. 25, p. 17-71, Pavia.

Dal Cin R. (1972) : I conglomerati tardo-paleozoici post-ercinici delle Dolomiti. - Mitt. Ges. Geol. Bergbaustud. (Verrucano-Symp., Wien 1969), vol. 20, p. 47-74, Wien.

De Sitter L.U. et De Sitter-Koomans C.N. (1949) : The Geology of the Bergamasc Alps. Lombardia. Italy. - Leid. Geol. Meded., vol. 14 B, p. 1-257, Leiden.

Dozy J.J. (von) (1935) : Ueber das Perm der Südalpen. - Leid. Geol. Meded., vol. 7, n° 1, p. 42-61, Leiden.

Folk R.L. (1968) : Petrography of sedimentary rocks. - Vol. de 170 p., Hemphill's (Drower M. University Station), Austin (Texas).

Geinitz H.B. (1869) : Ueber fossile Pflanzenreste aus der Dyas
    von Val Trompia. - N. Jb. Min. Geol. Paleont., p. 456-461,
    Stuttgart.
Giuseppetti G. (1959) : Ricerche petrografiche sull'alta Val Trom-
    pia. Le formazioni eruttive di Val Navazze e di Val Torgola.
    - Atti Ist. Geol. Univ. Pavia, vol. 9-10, p. 3-222, Pavia.
Heritsch F. (1939) : Karbon und Perm in den Südalpen und in Süd-
    osteuropa. - Geol. Rund., vol. 30, n° 5, p. 529- 588, Stutt-
    gart.
Klaus W. (1963) : Sporen aus dem südalpinen Perm (Vergleichsstudie
    für die Gliederung nordalpiner Salzserien). - Jb. Geol.Bundes.,
    vol. 106, n° 1, p. 229-361, Wien.
Leonardi P. (1951) : Orme di tetrapodi nelle Arenarie di Val Gar-
    dena (Permiano medio-superiore) dell'alto Adige sud-orienta-
    le. - Mem. Ist. Geol. Min. Univ. Padova, vol. 17 (1951-52),
    23 p., Padova 1953.
O' Connor J.T. (1965) : A classification for quartz-rich igneous
    rocks based on feldspar ratios. - U.S. Geol. Survey Prof.
    Paper, 525-B, p. 79-84, Washington.
Peyronel-Pagliani G. (1965) : Studio petrografico delle vulcaniti
    della "Formazione di Collio" in alta V.Trompia (Brescia). -
    Rend. Ist. Lomb. Sc. Lett., Cl. Sc. (A), vol. 99, n° 1, p.
    148-174, Milano.
Peyronel-Pagliani G. et Clerici-Risari E. (1973) : Le ignimbriti
    paleozoiche costituenti la formazione "Vulcaniti di Auccia"
    (Permico bresciano). - Atti Ist. Geol. Univ. Pavia, vol. 23,
    p. 160-169, Pavia.
Rau A. et Tongiorgi M. (1972) : The Permian of Middle and Northern
    Italy. - En : Rotliegend. Essays on European Lower Permian
    (éd. par H. Falke), Int. Sed. Petr. Ser., vol. 15, p. 216-
    280, E. J. Brill, Leiden.
Rittmann A. (1973) : Stable Mineral Assemblage of Igneous Rocks.
    A method of calculation. - Vol. de 262 p., Springer Verlag,
    New York.
Streckeisen A.L. (1967) : Classification and Nomenclature of I-
    gneous Rocks. N. Jb. Min. Abh., vol. 107 (2 et 3), p. 144-
    240, Stuttgart.
Suess E. (1869) : Ueber das Rothliegende im Val Trompia. - Sitz.
    K. Ak. Wiss., Math.-Nat. Kl., vol. 59, p. 107-119, Wien.
Trevisan L. (1939) : Il Gruppo di Brenta (Trentino occidentale). -
    Mem. Ist. Geol. R. Univ. Padova, vol. 13 (1938-39), 128 p.,
    Padova.
Ulcigrai F. (1969) : Geologia dei dintorni di Tregiovo (Trentino-
    Alto Adige). - St. Trent. Sc. Nat., s.A, vol. 46, n° 2, p.
    243-300, Trento.

Cartes géologiques
(outre celles comprises dans les travaux précités)

- Carta geologica delle Tre Venezie, F° 35 Riva, au 100 000ᵉ,

I948.

- Carta geologica delle Tre Venezie, F° 20 M. Adamello, au
  I00 000e, I953.
- Carta geologica d'Italia, F° I9 Tirano, au I00 000e, I969.
- Carta geologica d'Italia, F° 34 Breno, au I00 000e, I970.
- Carta geologica delle Prealpi Bresciane a Sud dell'Adamello,
  au 50 000^e. - Atti Ist. Geol. Univ. Pavia, vol. 22, pl. 4
  et 5, Pavia I972.

SEDIMENTATION, CLIMATE AND DEVELOPMENT OF LANDFORMS IN THE POST-
HERCYNIAN NORTH TUSCANY: A CONTRIBUTION

A. Rau[o] and M. Tongiorgi[oo]

[o] Centro di Minerogenesi,petrogenesi e tettogenesi dello
Appennino settentrionale del C.N.R.,Pisa (Italy)
[oo] Istituto di Geologia dell'Università di Pisa,Via Santa
Maria 53, Pisa (Italy)

ABSTRACT - In North Tuscany the remains of the Hercynian mountain
chain are covered in places (Massa metamorphic Unit) by fluvio-lacustrine deposits with layers of coal of the Stephanian and Autunian Ages, as determined from the rich fossil flora. These flora
suggest a hot-humid intertropical-type climate. The sedimentological features furthermore point to a remarkable instability of the
Hercynian relief in wich were found the Permo-Carboniferous sinking
basins. During the Lower Permian this instability increased further,
thus causing a corresponding increase in the erosion on the reliefs
and in the amount of clastic material deposited in the sinking basins between the mountains. In the upper part of the Autunian the
lacustrine basin of San Lorenzo in the Pisan Mountains is completely filled in. At this point (Saalian movements) the relief under-
goes a sharp rejuvenation: the deposit of coarser detritic material
(Asciano Formation) is a result of the morphological variations
following on the increase in the vertical movements. However the
climate by now has changed to a drier sub-arid type. On the Pisan
Mountains this variation can be seen in the sedimentological features
of the red fanglomeratic deposits of Asciano Formation. The Post-
Hercynian sedimentation on the Pisan Mountains ends with the Asciano Formation: in this area there are none of the fine-grained deposits which, in other areas, cover the arid lowlands at the end of
the Permian.

H. Falke (ed.), The Continental Permian in Central, West, and South Europe, 169-180. All Rights Reserved.
Copyright © 1976 by D. Reidel Publishing Company, Dordrecht-Holland.

1. Introduction

In North Tuscany the formations belonging to the Paleozoic "Tuscan"
basement and their stratigraphic cover of clastic Permo-Carboniferous and Triassic sediments outcrop over a wide area in the Apuan
Alps. Moving up one of the valleys which from the Tyrrhenian coast
lead right into the core of the Apuan Alps as, for example, the Fri-
gido Valley, we can see some very beautiful natural sections which
prove that these formations form the basement for two different tectonic units, both of which are slightly metamorphic: that of Massa
and the underlying Autochthon.
In the autochthonous core of the Apuan Alps the carbonatic formations of the Upper Triassic lie transgressively over the folded
Hercynian basement. Only in places, at the bottom of the Noric dolomites, some Triassic quartzite and conglomerate lenses can be seen.
In the Massa Unit, however, the most complete stratigraphical sequence allows a quite detailed reconstruction to be made of the period between the last Hercynian foldings and the beginning of the
Alpine sedimentation cycle.
The Massa Unit is a long continuous belt stretching down the western
edge of the Apuans from the Paleozoic and Triassic outcrops in the
North, belonging to the Eastern promontory of the Gulf of La Spezia,
to the Pisan Mountains in the South [15].
From among the outcrops of the Massa Unit we shall take as an example those of the Pisan Mountains which, studied recently by the same authors, are certainly the best-known.

2. Buti Banded Phyllite and Quartzite Formation

The oldest formation outcropping in the Pisan Mountains is that of
the Banded Phyllite and Quartzite of Buti; this is a complex of sericitic-chloritic schists with quartz and albite of varying grain
size, which show traces of a metamorphic pre-Alpine event dated at
275±12 MY [2] . This event probably correspond to the last folding
phase of the Hercynian orogenesis, i.e., the Asturian. In fact,the
strong angular unconformity between this formation and the overlying
Permo-Carboniferous and Triassic sediments leads to the identification within the basement of the Tuscan units, of the remains of an
eroded Hercynian chain. Similar formations can also be seen in the
Apuan region and in several areas south of the Arno river.

3. San Lorenzo Schists Formation

The first post-Hercynian sediments in the Pisan Mountains are the
San Lorenzo Schists which have undergone a very slight Alpine metamorphism; these detritic deposits of a continental environment
vary considerably in grain size and have a high content of material
of organic origin. In what is probably the oldest part of this formation (Villa Massagli) the massive, dark grey to black silty shales predominate. In what is held to be the intermediate part of the
formation (Villa La Valentona, Traina,Monte Togi) the main deposits
are similar to those of Villa Massagli but more argillaceous and
fissile; dark,lustrous sericitic shales, frequently associated with
thin graphitic layers are also present. In the upper part of the
formation (Monte Vignale, Sasso Campanaro, high part of Monte Togi),
there are interbeddings of quartzose-micaceous sandstones and also
layers of conglomerates. One of the few good exposure points of this
part of the formation is along the Via Pari, near Monte Vignale.
Here it is possible to see rhythmic sequences each of a few metres
thickness and comprising (from bottom to top): a thin graphitic
layer, lustrous finely-foliated graphitic shales, blackish phyllites
with vegetal traces and finally a thick layer of coarse-grained micaceous sandstone or of quartzose conglomerate. The coarser layers
have poorly definable cross-bedding.
Purple colourings can be seen, in some places, in the uppermost
part of the San Lorenzo Schists: originally the reddish tones which
are considered all over as being characteristic of the Permian deposits, must have made their first appearance in these layers.
The age of San Lorenzo Schists has been determined from the well-
known fossil flora [7] even though there is still some uncertainty
as to the age of the oldest outcropping layers (Westphalian D: Trevisan [17] ; more probably Stephanian A: Remy, oral communication).
However,the presence of the Autunian in the layers on the Via Pari
and at Sasso Campanaro has been proved without any doubt by the floristic association with Lebachia piniformis and Callipteris conferta. In the older layers near Villa Massagli Canavari [3] found,along
with the vegetal fossils, traces of limnic bivalves (Anthracosia).
In 1892 the same Canavari [4] at Traina found two insect remains which
he generically assigned to the sub-family of the Blattinariae;then,
during an excursion with W.Remy, the authors of this paper discovered, near Sasso Campanaro and therefore in the highest part of the
San Lorenzo Formation, a small fauna with Pelecypods and Estheriae
which was unfortunately difficult to recognize.

The sedimentation environment of the San Lorenzo Schists is certainly continental. It could be imagined as a subsident basin towards
which was carried the detritic material resulting from erosion of
the nearby Hercynian reliefs. It still has to be decided whether it
was a paralic basin or a marshy depression on the inside of the reliefs.

Although, as the marine fossils accompanying the carboniferous phyllites in Tuscany south of the Arno lead us to suppose, the sea-coast
was not very far off, it must also be said that no proof at all exists in the Pisan Mountains that there ever was a marine transgression, even temporary, in this area. The Pelecypods found so far,and
also the Phillopods, certainly lead to deduce a limnic environment.
The one conclusion does not automatically exclude the other, but
the nearness of the reliefs (there is no Permo-Carboniferous in the
southern part of the Pisan Mountains nor in the Apuan massif even
where there is still evidence of the Permian), the reduced thickness
of the San Lorenzo Schists, the absence of marine fossils and carbonate interbeddings, all lend support to the limnic basin hypothesis.
In this sunken area there was a delicate balance, periodically broken and re-formed, between the vegetational growth, the detritic
material brought down from the steep slopes of the reliefs and the
basin's subsidence. This balance can be seen in the rhythms typical
of deposits filling subsiding basins in very unstable areas. The coal
layers are evidence of the vegetational growth on marshy ground;
the coarse material fails to reach the flooded zone through the fil-
ter of vegetation. The later sinking of the basin, unsuccessfully
filled in by detritic material, led to complete lacustrine conditions. The clay sediments, with a few vegetal remains and bivalve
shells, are deposited beneath a greater thickness of water. When
the vegetation barrier between the limnic basin and the surrounding
steep slopes is also broken on the basin edges, new coarse material
begins to fill the basin. Thus the vegetation is re-planted and the
cycle begins again. In this type of environment the fossil flora
composition only partly reflects the original local floristic association. The present fossil deposits are in fact the result of a
mixing of vegetal remains which were for the most part brought along
with the water. Therefore, along with the plants from the flooded
lowlands there are species from a drier environment which originally grew on the dry slopes of the surrounding mountains.

Therefore we believe that the difference in the floristic composition in the carboniferous deposits on the Pisan Mountains cannot be
considered as representative of the climatic variations during the
Stephanian, especially as the aquatic flora deposits are found in-

terbedded with layers where the predominant species is that typical
of a much drier environment. On the whole the carboniferous flora
in the Pisan Mountains in many ways suggests a humid intertropical
climate, similar to that in which forests of the watery environment
of coastal marshes are now developing. That is the same climate as
seems to be suggested by the flora from most of the European deposits of the same period [13].

The paleomagnetic measurements carried out in various regions of
southern Europe seem to confirm this interpretation. In fact a pa-
leoparallel map could be drawn showing that during the Carboniferous
the studied region lay between the paleoequator and 10°S [10].

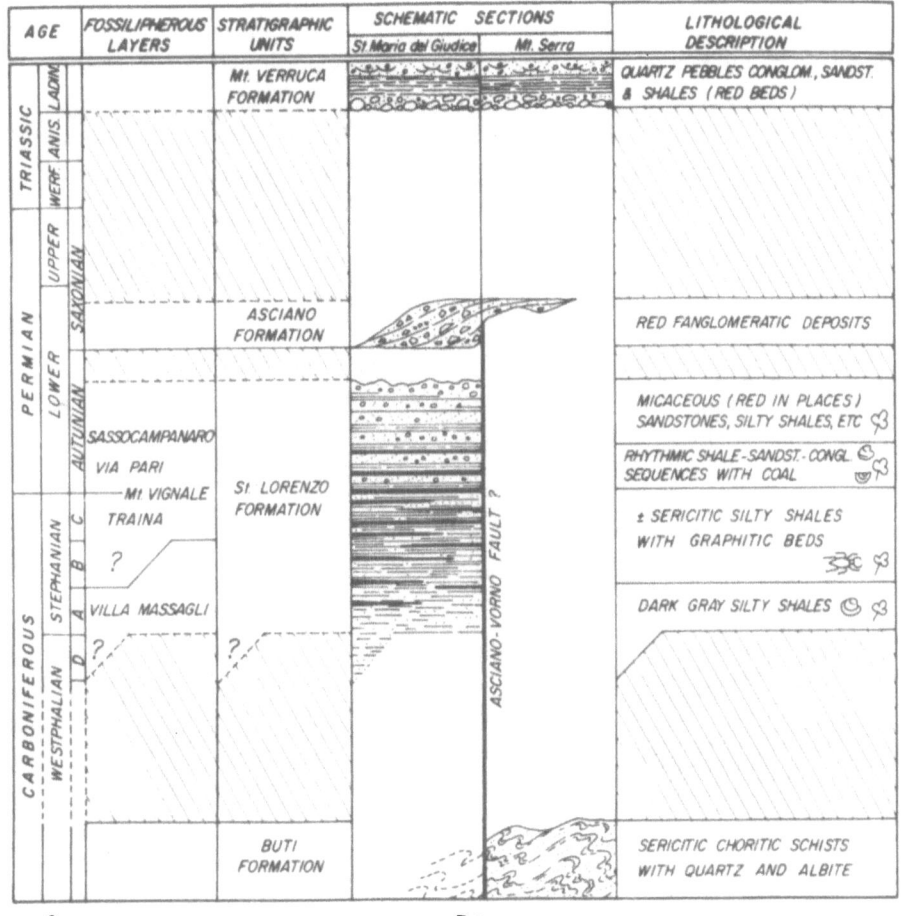

Fig. 1

Nor does appearance of dry environment flora with <u>Lebachia pinifor-</u>
<u>mis</u> at the base of the Permian, a bit higher, the appearance of the
first red interbeddings, seem to us definite proof of a variation
in climate. The coarser grain size of the Autunian sediments would
suggest an increase in the vertical movements which were to reach
their peak towards the end of the Lower Permian in the so-called
Saalian phase. The rejuvenation of the relief led on the one hand
to a stronger erosion and, consequently, a more rapid filling in of
the basins; on the other hand there was also a general uplift of
the Hercynian reliefs. Therefore the drier type of flora predominated in an area where the flooded sunken areas were gradually disap-
pearing. Finally, we believe that the prevalent climate in our region from the end of the Westphalian to the Autunian was usually
a hot-humid one.

These events were not necessarily a result of regional climatic variations during the early Permian. In fact, the local environmental
variations associated with vertical movements of the earth's crust
could account for the water level changes during the Carboniferous,
the increase in erosion, the filling up of the intramontane basins
and the disappearance of the humid forest during the Autunian [8,13] .

4. Asciano Breccia and Conglomerate Formation

Asciano Formation is made up of detritic deposits of mainly purple-
coloured phyllites including many angular, or sub-rounded, fragments
of different rocks. As the argillaceous matrix decreases and the
grain size increases we come to the breccia proper or, where the
rounded elements are in the majority, massive layers of conglomerates. However, even in beds where the coarse clasts are predominant,
there is still plenty of the matrix. In some places between the
clasts there are angular blocks of up to about a metre in diameter,
enclosed by a finer grained sediment. Sediment sorting of Asciano
Formation is very limited. Therefore where texture is concerned
the facies lies within the sphere of fanglomeratic deposits [11] or
"paraconglomerates" in Pettijohn's classification.

As regards petrography the matrix is a sericitic-chloritic phyllite
with ematite; the clastic elements seem to be made up exclusively
of silicic rocks with, in particular, numerous sericitic quartzites
with albite along with the quartz and,often, phyllite pebbles.Many
of these clasts show the characteristic banding of the Buti rocks
from which they seem to have originated. Taken as a whole this is a

very immature sediment. Asciano Formation varies greatly in thickness; in places it is entirely missing, in other places we might
estimate an original thickness of between 50 to 80 m.
Although no fossil have ever been found in Asciano Formation its
stratigraphical position would suggest its possible deposition period as lying between the end of the Autunian and beginning of the
Mid-Triassic. We shall later show how this period can be defined
more precisely.
The immaturity of the sediment, its limited or non-existent sorting,
its purple-colouring caused by the large amounts of iron oxides,
its varying thickness and lack of a well-defined stratification
all suggest a continental deposition in a relatively warm and sub-
arid climate with occasional humid periods characterized by short
violent rainstorms. As Asciano deposit was probably preceded by
tectonic movements (stratigraphical break and disconformity at the
base), which rejuvenated the Hercynian reliefs, these deposits probably correspond to the local deposition en masse of detritic material on the borders of the uplifting areas. The latter had no
vegetation and underwent intense erosion with only slight weathering;
during the rainy periods the detritic covering was dislodged and,
after a short journey and almost no sorting at all, was re-deposited
on the edges of the old Permo-Carboniferous basin of San Lorenzo
Schists which had by then been filled in.

5. Conclusions

a) The Hercynian diastrophism - We have already shown how the "Hercynian unconformity" has left evidence in our zone, thus proving
that the latter was affected by diastrophism before the end of the
Westphalian. In the Upper Carboniferous, after the Asturian phase,
the Pisan Mountains area formed part of an emerged continent, probably connected with the Hercynian continent which stretched northwards beyond the Alps to include the mid-European block [14] .

b) The Post-Asturian sedimentary cycle - The San Lorenzo Schists
outcropping in the northern part of the Pisan Mountains (Santa Maria del Giudice Unit: see in Rau and Tongiorgi [15]) are proof that
at the end of the Carboniferous there was an intramontane subsiding
basin in which fluvio-lacustrine sediments were deposited. In the
southern part of the Pisan Mountains, on the other hand,(Monte Ser-

ra Unit), there was probably a relief subjected to erosion [15],
and other reliefs further North in what is now the metamorphic
core of the Apuan Alps. In Central-South Tuscany the Permo-Carboniferous is characterized by paralic facies comprising layers with
marine fossils and layers with terrestrial plants (Iano [12,20];
Island of Elba [16,18]). Further South in the Monticiano-Roccastrada mountain chain in the same period the open sea facies prevailed
[5,6,9].
Therefore a coast-line can be recognized in Tuscany south of the
Arno. This coast-line can be joined, along an approximative NNE-
SSW direction, to that lying between the Dolomites and the Carnia.
During the sedimentation of the San Lorenzo Schists the climate
was probably a humid tropical one but, as we have seen, the vegetational growth was mostly tied to the variations in the amount of
sedimentation which, in its turn, was connected with the vertical
movements which repeatedly modified the morphological equilibrium
of the still unstable Hercynian chain.

c) The Saalian movements - As the instability of the Hercynian continent worsened during the Autunian, so the grain size increased
in the deposits on the upper part of the San Lorenzo Schists in the
Pisan Mountains (sandstone with Lebachia piniformis).This increase
had in fact led us to suppose a certain rejuvenation of the relief
as a result of the uplifting then in course.
Whereas in other regions the eopermian movements led to a deepening
of some of the carboniferous basins or the formation of new subsidence basin, the Pisan Mountains area in the Upper Autunian was an
uplifted zone; even in San Lorenzo basin, filled in for the most
part, the sedimentation diminished and disappeared entirely.
The sudden appearance on the top of San Lorenzo Schists of reddish
fanglomeratic deposits of Asciano Formation, that strongly contrast
with the underlying dark schists and well-stratified sandstones,
testifies to the great changes that took place in the sedimentation
after the Upper Autunian. In fact, the sedimentation was affected
by the variation of two decisive factors:the relief and the climate.
The rejuvenation of the relief was accompanied by a much drier sub-
arid climate. In these conditions the debris covering which had
formed on the Hercynian reliefs, below what was the tropical forest
before its disappearance, underwent an erosion which also reached
the old basement rocks.
The stratigraphical break separating the Asciano fanglomerates from
the San Lorenzo Schists certainly correspond to a period of increa-

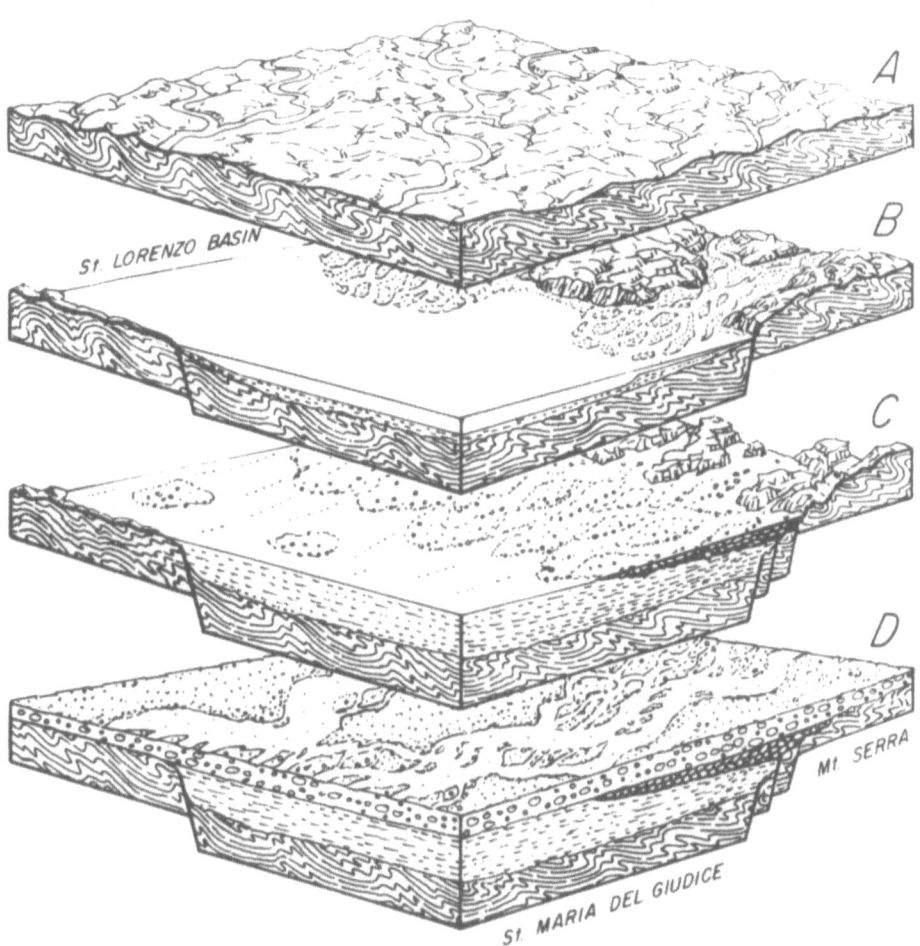

Fig. 2. Scheme of the paleoenvironmental evolution of the Massa
metamorphic Units in the Pisan Mountains area during the Upper
Carboniferous, Permian and Lower-Middle Triassic Ages. A-Upper
Westphalian; B-Stephanian, Autunian: fluvio-lacustrine deposits
(San Lorenzo Schists Formation); C-Permian: post-saalian fanglo-
meratic deposits (Asciano Breccia and Conglomerate Formation);
D-Middle Triassic: beginning of a new Alpine, sedimentary cycle,
fluvial (braided streams)deposits of Mt.Verruca Formation.

sed tectonic movements. Such a period of intense tectonic activity
is generally called the "Saalian phase" and is often associated
with an acid-type volcanism.

The particular distribution of Asciano Formation, which is ammassed on the edge of San Lorenzo basin and then thins down rapidly until it eventually disappears above the old rocks of the substratum (Monte Serra Unit), but also towards the centre of the basin, leads to the hypothesis that it was deposited at the foot of fault escarpments. In fact, the presence of large blocks of old rocks enclosed within the Asciano fanglomerates would suggest an intense erosion undergone by the steep slopes. Finally it seems logical to conclude that at the end of the Autunian there was a new sharp movement in the fractures which probably formed the boundary to San Lorenzo subsidence basin.

On the Pisan Mountains there are no traces of the magmatic phenomenon which often accompanies the movements of the "Saalian phase". However, in other parts of Tuscany porphyries and the arkosic products of their weathering have been found beneath Triassic deposits. These magmatic-originating products were commonly considered to be of Permian [1,14]. However Vai [19] recently contested this age as regards the outcrops in the Apuan Alps. Nevertheless the Permian age as referred to the porphyries and arkoses in the stratigraphical sequences of Iano and the Island of Elba seems still to hold valid [1].

The frequent porphiric pebbles found in the basal conglomerates of the Triassic transgression in the Pisan Mountains (Verrucano s.s.) could therefore be proof of a magmatic activity in neighbouring areas, perhaps connected with the "Saalian phase" tectonic activity.

The fact that no volcanic products, or volcanic rock clastics, have been found in Asciano Formation could be explained by a local supply of these deposits from an area with no volcanic activity.

On the other hand all evidence points to their being syntectonic deposits; in other words Asciano fanglomerates, in the light of present knowledge, could be considered of the same period as at least a part of the Saalian movements and therefore no later than the last Permian acid magmatic activity. They could therefore be of the same period as, for example, the coarser clastic sediments (Dosso dei Galli Conglomerate) of the Collio Permian basin whose facies are also very similar.

d) The Post-Saalian landscape - After the Saalian movements the complete peneplanation of the Hercynian relief began. The Pisan Mountains area throughout the Upper Permian will remain a relief in the late-Hercynian landscape, predominated by erosion. In fact the deposits of Asciano Formation brought to an end the Post-Hercynian sedimentation. There are none of the finer deposits usually found covering the arid lowlands at the end of the Permian. However, in the Pisan Mountains on the denuded surface of the old eroded and by non stable relief, we can see the beginning of the long sub-aerial reworking of the detritic material later to be added to the basal conglomerate deposits of the Alpine sedimentary cycle (Mt.Verruca Formation: Verrucano s.s.).

REFERENCES

1. Barberi F.,1966,I porfiroidi della Toscana e la loro posizione stratigrafica,in:Atti Symposium sul Verrucano,Pisa-settembre 1965,ed.by Soc.Tosc.Sci.Nat.,34-54,Pisa.

2. Borsi S.,Ferrara G.,Rau A.and Tongiorgi M.,1967,Determinazione con il metodo Rb/Sr dell'età delle Filladi e Quarziti listate di Buti (Monti Pisani),Atti Soc.Tosc.Sci.Nat.,Mem.,A,73,632-646, Pisa.

3. Canavari M.,1891,Due nuove località nel Monte Pisano con resti di piante carbonifere,Atti Soc.Tosc.Sci.Nat.,Proc.Verb.,7,217-2i8,Pisa.

4. Canavari M.,1892,Insetti del Carbonifero di San Lorenzo nel Monte Pisano,Atti Soc.Tosc.Sci.Nat.,Proc.Verb.,8,33-34,Pisa.

5. Cocozza T.,1965,Il Carbonifero del Gruppo Monticiano-Roccastrada (Toscana),La Ricerca Scientifica,35,(II°A),38pp,Roma.

6. Cocozza T.,Lazzarotto A.and Vai G.B.,1974,Flysch e molassa ercinici del torrente Farma (Toscana),Boll.Soc.Geol.It.,93,115-128, Roma.

7. De Stefani C.,1901,Flore carbonifere e permiane della Toscana, Pubbl.R.Ist.Studi sup.pratici e di perfez.,Sez.Sci.Fis.e Nat., 212pp,Firenze.

8. Fabre J.and Feys R.,1962,Reflexions sur la genèse des bassins houillers et la théorie bio-rhexistasique,in:Biogéographie du Permo-Carbonifère et genèse des charbons,C.R.Somm.des Séances de la Societé de Biogéographie,335,336,337,4-13,Paris.

9. Giannini E.,Lazzarotto A.and Signorini R.,1972,Lineamenti di Geologia della Toscana meridionale,Rend.Soc.It.Min.e Petr.,special issue,33-168.

10.Irving E.,1964,Paleomagnetism and its application to geological and geophysical problems, Wiley and Sons, New York.

11.Krumbein W.C.and Sloss L.L.,1951,Stratigraphy and Sedimentation, Freeman,San Francisco.

12.Mazzanti R.,1961,Geologia della zona di Montaione tra le valli dell'Era e dell'Elsa (Toscana),Boll.Soc.Geol.It.,80(2),37-126, Pisa.

13.Pruvost P.,1962,La genèse des gisements houillers envisagée à la lumière de la Bio-rhexistasie,in:Biogéographie du Permo-Carbonifère et genèse des charbons,C.R.Somm.des Séances de la Soc. de Biogéographie,335,336,337,81-88,Paris.

14.Rau A. and Tongiorgi M.,1972,The Permian of Middle and Northern Italy,in:Rotliegend.Essay on European Lower Permian,ed.by H.Falke,Intern.Petrographical Series,15,216-280,Leiden.

15.Rau A. and Tongiorgi M.,1974,Geologia dei Monti Pisani a SE della Valle del Guappero,Mem.Soc.Geol.It.,13,227-408,Pisa.

16.Trevisan L.,1950,L'Elba orientale e la sua tettonica di scivolamento per gravità (con bibliografia precedente),Mem.Ist.Geol. Univ.Padova,16,30pp,Padova.

17.Trevisan L.,1955,Il Trias della Toscana ed il problema del Verrucano triassico,Atti Soc.Tosc.Sci.Nat.,Mem.,A,62,1-30,Pisa.

18.Trevisan L.,1955,La 55ª Riunione estiva della Società Geologica Italiana. Isola d'Elba,18-23 settembre 1951,Boll.Soc.Geol.It., 70(2),435-472,Roma.

19.Vai G.B.,1972,Evidence of Silurian in the Apuane Alps (Tuscany, Italy),Giornale di Geologia,38(1),349-372,Bologna.

20.Vai G.B. and Francavilla F.,1974,Nuovo ritrovamento di piante dello Stefaniano a Iano in Toscana,Boll.Soc.Geol.It.,93,73-80, Roma.

GEOLOGY AND PETROGRAPHY OF THE VERRUCANO AND PALEOZOIC FORMATIONS OF SOUTHERN TUSCANY AND NORTHERN LATIUM (ITALY)

E. Azzaro [o], T. Cocozza [oo], B. Di Sabatino [o], G. Gasperi [ooo],
R. Gelmini [ooo], and A. Lazzarotto [oo]
  [o]   Istituto di Petrografia Università di Roma
  [oo] Istituto di Geologia Università di Siena
  [ooo] Istituto di Geologia Università di Modena

ABSTRACT. This paper present: a stratigraphic outline of the
Paleozoic-Triassic Formations that crop out south of the Arno
River; a hypothesis regarding their correlation with the M.Pisano
sequence north of the Arno; and conclusions about their position
in the tectonic evolution of the Hercynian and Alpine cycles.
These conclusions were also based on a study of the metamorphism
of the region, based in particular on the b_0 value of the potassic
mica.

1. INTRODUCTION

For the past few years a series of coordinated studies on the
Verrucano of Tuscany south of the Arno River has been under way.
This paper presents the results of these studies. Although they
shed considerable new light on the subject, they are still
incomplete, and hence subject to modification (1).
The paper is divided into two parts, one on the stratigraphy
and one on the petrography of the area. In the former, T. Cocozza
and A. Lazzarotto present the data on the Montagnola Senese, the
Monticiano-Roccastrada Ridge and Boccheggiano, while G. Gasperi
and R. Gelmini cover the Argentario, the Uccellina Mountains,
Mount Leoni and the Romani Mountains. The petrographical part is

---

This work has been supported by the "Consiglio Nazionale delle
  Ricerche".
 [o] Present address
(1) The used lithological terms refer to former nature of sediments
   and not to present conditions; all considered rocks are
   epimetamorphic (low metamorphic stage).

H. Falke (ed.), The Continental Permian in Central, West, and South Europe, 181-195. All Rights Reserved.

by E. Azzaro and B. Di Sabatino under the direction of Prof. G.C.
Negretti. The conclusions emerged from discussions between all of
the researchers.

In our stratigraphic outline, a tentative correlation with
the Monte Pisano succession studied by A. Rau and M. Tongiorgi is
proposed, regarding which we refer you to their recent detailed
monograph (9).

2. OUTCROPS OF MONTICIANO-ROCCASTRADA AND BOCCHEGGIANO (T.C. and A.L.)

The oldest paleontologically dated sediments of the northern
Appenines crop out in the Farma Valley, which sharply cuts across
and through the Monticiano-Roccastrada ridge. Another important
outcrop of Paleozoic rocks is in the area of Boccheggiano (Fig.1).
In these areas we have been able to reconstruct the following
sequence from top to bottom (Fig.2):
  a) Ferriera Formation
  b) Boccheggiano Shales
  c) Carpineta Formation
  d) Farma Formation
The Triassic Verrucano Group overlies these in nonconformity.

2.1. Farma Formation

The best exposures of this formation are along the bed and
banks of a short stretch of the Farma Stream south of the hamlet
of Solaia. It consist of a rhythmic sequence of graded pelitic-
arenaceous strata, lead-grey in color. At the base of the strata,
evidence of slumping and marks, especially groove-casts are found.
Prevalently calcareous olistostromes and olistoliths are included
in the formation. The former consist of mostly calcareous clasts
in a calcarenitic matrix with marine fauna.

The calcarenites are also present in lenses and flames within
the pelitic-arenaceous strata. Most of the carbonatic clasts are
pelmicritic and micritic limestone with crinoids, others are oolitic
and micritic limestone with brachiopods.

Thin sections of the clasts and calcarenites revealed the
presence of Tuberitina sp., Profusulinella sp., and Pseudostaffella
sp. as well as calcispheres, endothyrida, fragments of brachiopods
(spiriferids) etc.. This association was assigned to the lowest
Bashkirian (2, 3). Large limestone olistoliths appear along the
Farma towards Bagni di Petriolo.

It is likely that the S. Antonio Limestone near Casale di Pari
also constitutes a large olistolith within the Farma Formation. It
is a black limestone, wich has been partially dolomitized or
mineralized in the upper part and bears echinids, crinoids,
bryozoans, brachiopods, gastropods, calcareous algae, and

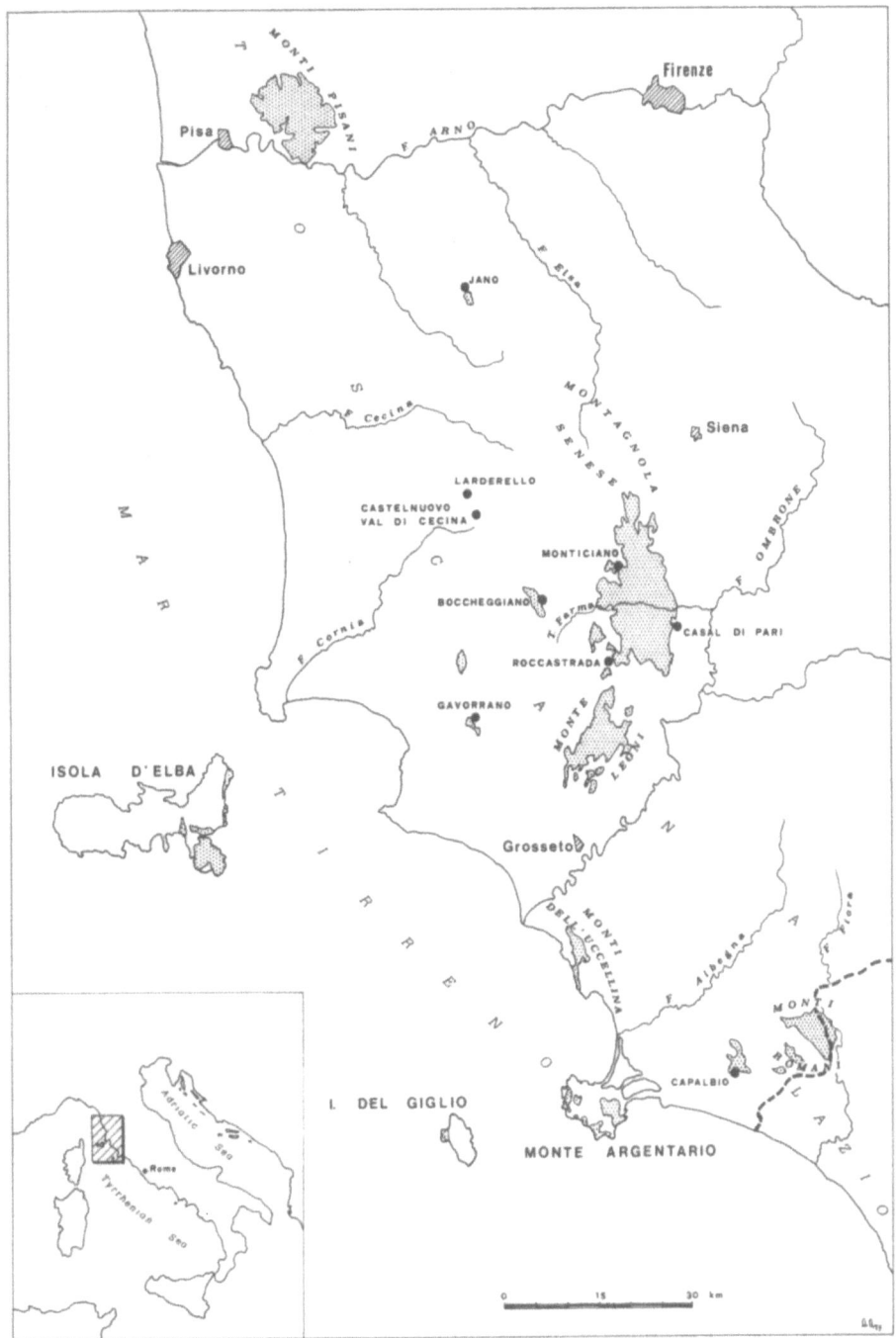

Fig. 1. Location of Verrucano and pre-Verrucano rock outcrops (dotted).

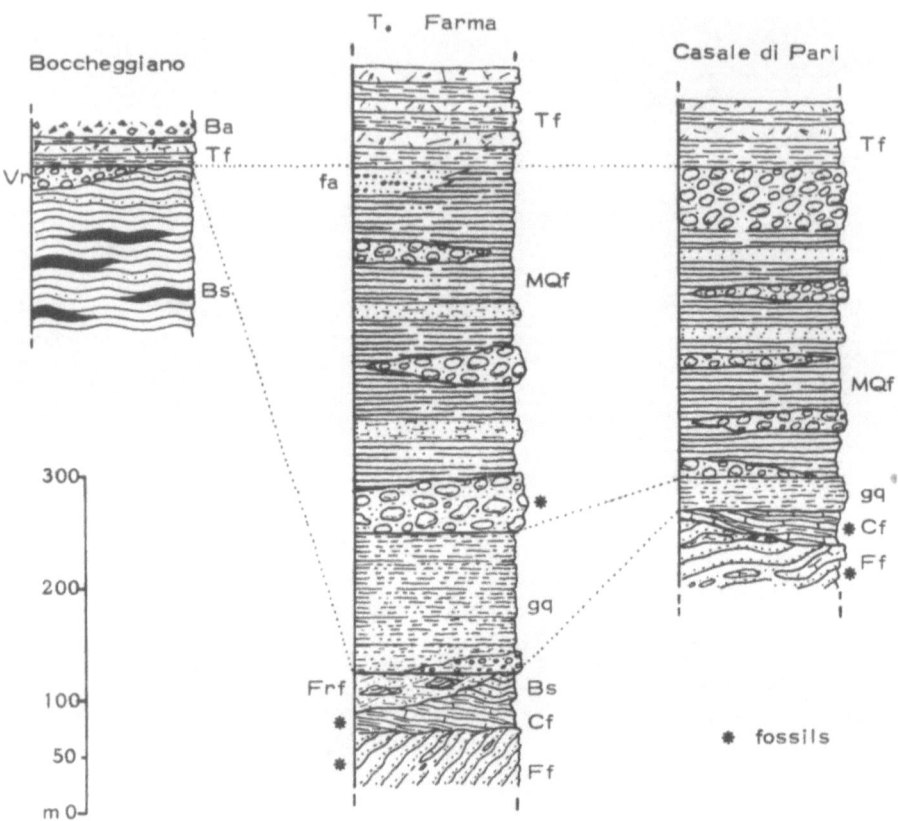

Fig. 2. Ba: Burano anhydrite — Tf: Tocchi formation —Vr: Verrucano Group.

fa: fine-grained anagenites
MQf: M. Quoio formation   } Verrucano Group of T. Farma
gq: green quartzites   and Casale di Pari

Frf: Ferriera formation — Bs: Boccheggiano shales (an: anhydrite and
dolomite lenses) — Cf: Carpineta formation — Ff: Farma formation.

foraminifera of the Moscovian [3, 4].

On the basis of these fossils, the Farma Formation has been
dates as Middle Carboniferous (Fig. 4).

2.2. Carpineta Formation

This formation consist of black argillite and siltite with
intercalations of crinoidbearing limestone, which crop out
especially along the Farma between Solaia and Bagni di Petriolo.

The "Spirifer-bearing Shales" that overlie the S. Antonio Limestone
in nonconformity (separated from it by a shaley-arenaceous crust)
have also been included in this formation (2, 3).

Redini (5) assigned the spirifer horizon to the Upper
Carboniferouus-Lower Permian, and, at least tentatively, we have
assigned all the out-crops of the Carpineta Formation to this age.

2.3. Boccheggiano Shales

The most extensive outcrop of this formation is in the
Boccheggiano area, where it is several hundred meters thich (6).
Along the Farma, there is a very small outcrop east of Ferriera.
To date it has not been possible to observe the contact of this
formation with the underlying one.

Dark grey shale with shiny surfaces and grayish green shale
and quartzite predominate in the Boccheggiano Shales. Towards the
top dark grey quartzose conglomerates with some white quatz
elements in an abundant quartzitic matrix also appear.Thin sections
show the shale to consist of a rapid alternation of quartzose and
sericitic laminae with thin lenticular levels of graphite
interposed (7).

Numerous drills for mining use in the Boccheggiano area have
revealed intercalations of anhydrite and dolomite, often in
alternating strata. These intercalations are rarely more than ten
meters thick (6).

To date the Boccheggiano Shales has yielded neither fossil
flora nor fauna. Palinological research has also been fruitless.
The only indication for dating is provided by the anhydrite, which
suggests a correlation with the evaporitic facies of the Zechstein
(Upper Permian).

2.4. Ferriera Formation (new name)

Small outcrops have been found on the Argentario and in the
Valley of the Farma near Ferriera (F° 120 of the Istituto Geografico
Militare, Table III NE-Chiusdino).

In the former area, they overlie the Monte Argentario
Sandstones (8) and form a layer several meters thick, which can be
seen from the path that leads from the panoramic road up to the
Antico Fortino Stella.

In the type area, it overlies the Boccheggiano Shales and
unconformably passes up into the Verrucano Group. Here it consists
of ocher or orange colored sandstone and microanagenite with
intercalations or lenticular inclusions of black graphitic shale.
It is further characterized by pronounced cross-stratification.

Other outcrops are found along the Botro del Confine, right
affluent of the Farma, and along the road that runs between Casale
di Pari and the farm of S. Antonio.

In all of the above-mentioned localities the formation overlies Paleozoic sediments in nonconformity and is overlain, also in nonconformity by the Triassic Verrucano. The maximum thickness found is slightly under a hundred meters.

No fossils have been discovered in this formation. It has been assigned to the Permo-Triassic solely on the basis of its stratigraphic position.

The isolated outcrops that have been observed to date may be the remains of a clastic sediment that originally was much more widespread and thicker, correlatable to the so-called "Verrucano" of the Alpine areas. The only other trace to date of such a sedimentary complex is the Asciano Formation of M. Pisano (9).

Such a sedimentary complex would fit Trümphy's definition of similar deposits in the Mediterranean mountain chains of Europe, interpreted as the "final product of the degradation of the Hercynian chain and its subsequent and postorogenic volcanoes" (Symposium on the Verrucano, Pisa, 1965).

2.5. Verrucano Group

This group includes eminently clastic formations of Middle Triassic age. It still has not been possible to correlate the lithostratigraphic units in the various outcrops with each other, mostly because of the great heterogeneousness of the sediment and its variable thickness (Fig. 2 and 3).

In the area of the Farma, where its thickness is considerable (600-700 meters), three lithofacies have been distinguished. From top to bottom they are:

    a) Fine-grained anagenite and violet quartzite
    b) M. Quoio Formation
    c) Green quartzite

The dominant lithofacies is the M. Quoio Formation, with a maximum thickness of roughly 300-400 meters. It consist mostly of siltite and purple sandstone intercalated with thick conglomerate beds.

The conglomerates have large clasts with a maximum diameter of 40 cm, consisting mosthy of green and purple quartzite and subordinately of limestone. The generally very abundant purple matrix is prevalenty micaceous and quartzitic.

In addition to fragments of echinoderms, gastropods and lamellibranchs (Eumorphotis ?), the carbonatic clasts have yulded a microfauna (Citaella (?) iulia, Ammodiscus (?) incertus, and probable fragments of Tolypammina sp.) indicative of the Werfenian-Lower Anisian (10).

These findings are particularly important because they confirm the Triassic age of the Verrucano in Tuscany south of the Arno, and permit its correlation (despite the variability of the lithofacies) with the M. Pisano Verruca Formation (9) as well as with the Verrucano of the Punta Bianca area (Bay of La Spezia) and the western slopes of the Apuan Mountains.

2.6. Tocchi Formation

This formation overlies both the Verrucano and the Paleozoic sediments (e.g. Boccheggiano Shales) and consists of sediments laid down in a marine-lagunal environment (1, 7).

Since, on the whole, the Verrucano can be-considered a continental deposit (despite brief marine episodes indicated, for example, by the thin carbonatic beds on M. Argentario), the Tocchi Formation, though characterized by a certain variability of facies, represents the first real Alpine marine transgression exposed in Tuscany south of the Arno. In this respect it can be correlated to the M. Serra Quartzites in the M. Pisano area, at the base of which the first marine fossils (Lower Carnian) of the post-Hercynian succession appear.

3. OUTCROPS OF MOUNT ARGENTARIO, THE ROMANI MOUNTAINS AND MOUNT LEONI (G.G. and R.G.)

In southern Tuscany and a small portion of northern Lazio (Fig.1), the Verrucano Group transgressively overlies the Paleozoic rocks (Monte Argentario Sandstones (7, 8)) and is followed by the Tocchi Formation and the Burano Anhydrite (Calcare Cavernoso).

3.1. Monte Argentario Sandstones

Paleozoic outcrops have been found on M. Argentario (8, 11) at Capalbio, in the Romani Mountains (7),at Gavorrano (12), and M. Leoni (13). The most extensive outcrops are those of the Romani Mountains (roughly 40 km^2).

Lithologically these outcrops are the same, and closely resemble those of the Island of Elba (14;15) , Iano (16, 17) and the Carpineta Formation of the Farma Stream (2, 3).

They consist of irregular alternations of sandstone, siltite and shale, generally grey or blackish in color, deposited in a tidal environment. They are epimetamorphic and have marked slip folds. Sometimes the sandstone predominates over the other lithological types, but usually they alternate in equal proportions.

The sandstone is medium-fine grained, quarzose and rich in muscovite and sericite with a micaceous quarzose cement. It is subdivided into lenticular, parallel laminae which are sometimes cross-bedded. The siltite, which is also laminated, alternates with the sandstone in layers that vary greatly in thickness (from one decimeter to many meters).The shale is less frequent than the other two lithotypes, between wich it is sometimes thinly intercalated.

The greatest thickness of these outcrops (Monte Argentario

Sandstones) has been found in the Romani Mountains and is roughly
400 meters. For the present, its age can only be deduced from
fossils in the lithologically and stratigraphically analogous
formations found outside the area we are dealing with namely, at
Iano, Elba and the Farma. On the basis of fossil flora and fauna,
these analogous formations have been referred to the Upper
Carboniferous and possibly the Permian (2, 3, 14, 16, 17, 18).

3.2. Ferriera Formation

As in Monticiano-Roccastrada area it seems that at M.
Argentario, Ferriera Formation overlies the Monte Argentario

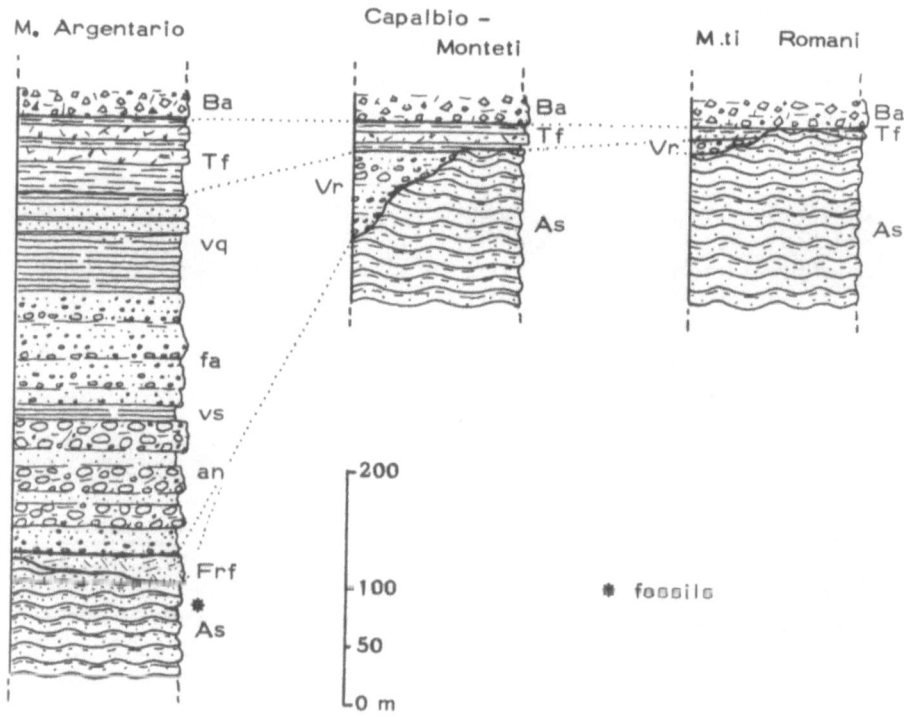

Fig. 3. Ba: Burano anhydrite – Tf: Tocchi formation – Vr: Verrucano Group.

vq: violet quartzites and shales
fa: fine-grained anagenites   Verrucano Group of
vs: violet shales   M. Argentario
an: anagenites

Frf: Ferriera formation – As: M. Argentario sandstones.

Sandstones. Ferriera Formation, over 10 meters thick, consists of yellow and purple quartzites, shales and subordinate anagenites which form lenticular, crossbedded layers.

3.3. Verrucano Group

The Verrucano Group overlies the above-described Paleozoic sediments in more or less marked angular unconformity; at M. Argentario the Verrucano sometimes overlies the Ferriera Formation.

The most extensive outcrops of the Verrucano Group are those of M. Leoni (13), the Uccellina Mountains (19, 20), and M. Argentario (8, 11), while smaller ones have been found at Capalbio and on the Romani Mountains (7) (Fig.3).

This group consists of alternations of anagenite, quartzite and shale. The anagenite is pinkish white or purple somewhat angular clasts under 15 centimeters in diameter. The clasts are of white, pink or purple quartz, quartzite and occasionally (in particular at the base of the Verrucano) tormalinolite and porphyroid. In places the anagenite forms strata over 10 meters thick; elsewhere it constitutes roughly 1-meter-thick layers and or lenses within the quartzite. The quartzite in white, greenish, pinkish-white and, more rarely, purple. Its grain is medium or large and occasionally fine grain and it is usually thinly bedded.

The shale is micaceous and varies in color from white to dark purple. It is found in strata varying in thickness from a few centimeters to many meters.

Anagenite predominates at the base of the formation, quartzite at the middle, and shale at the top.

The thickness of the Verrucano Group in the outcrops considered here varies greatly. In fact, it may decrease from almost 700 meters to just a few or taper off completely.

Going from north to south, from M. Leoni to the Uccellina Mountains and M. Argentario, the thickness diminishes from roughly 700 meters to slightly more than 300 m. From M. Argentario eastwards to the Romani Mountains the thickness of the formation decreases to zero. Similarly, to the west of M. Leoni, at Gavorrano, the Verrucano is lacking.

3.4. Tocchi Formation

The continental deposits of the Verrucano pass up into the marine sediments of the Tocchi Formation and the Burano Anhydrite. The Tocchi Formation, which like the Verrucano, is missing in certain areas, consists of fine-grained, greenish or whitish shale that is powdery to the touch, with rare intercalations of yellowish limestone. At most it is several tens of meters thick.

On the Romani Mountains as elsewhere, both the Verrucano and the Tocchi Formation are lacking, and the Paleozoic rocks are

directly succeeded by the Burano Anhydrite with only a few meters
of breccia in between. This breccia, which is almost unworked, was
derived from the degradation of the underlying rocks and is sometimes
accompanied by limonitic crusts.

4. PETROGRAPHICAL CHARACTERISTICS OF THE METAMORPHITES WITH
   PARTICULAR REFERENCE TO THE b_o VALUE OF THE POTASSIC MICA (E.A.
   and.B.D.)

Several hundred determinations of the b_o parameter of the
potassic white mica paragenetically contained in the various
metamorphic series of numerous outcrops in central-southern Tuscany
and northern Lazio were carried out.

X-ray analyses revealed a constant bimodality in both
composition and the b_o parameter of the white potassic mica present
in the metamorphic rocks as follows:

muscovites: in the oldest metamorphic levels (Permo-Carboniferous
            formation), in the anagenite, and, locally, in the
            violet shales of the Verrucano Group

phengites : overlying the above-described formations.

This subdivision is valid on a regional scale. It is also
possible to identify the geological guide levels that mark the
muscovite-phengite transition.

Sometimes it is also possible to recognize in the oldest
muscovite slates a $S_1 \cong S_0$ and a $S_2 \perp S_1$ (Carpineta Formation and
Buti phyllites and quartzites). In the former, an intense peak for
the b_o parameter of the potassic muscovite corresponds to the
$S_1 \cong S_0$, and small quantities of phengitic mica correspond to the
S_2.

The stratigraphic boundary between the two kinds of metamorphites
(as distinguished by the b_o parameter) appears to be located at the
roof of the anagenite (Verrucano), even when it is interbedded with
blastopsammitic and slaty members. The muscovite mica in the latter
suggests that the metamorphic event that affected it, is the same
as for the underlying levels, but further research is needed to
corroborate this because the coarse micaceous fraction (muscovite)
being of older origin, cannot be used for such a confirmation.
Moreover the purple shales intercalated within the anagenetic complex
sometimes contains muscovite, especially in the coarser and lower
levels, which pass up into phengite either gradually or with an
abrupt jump in the b_o value; X-rays show that the gradual transition
from muscovite to phengite corresponds to a descrease of the clastic
muscovite mica towards the roof of the anagenite.

Instead, the Tocchi Formation and the shales at the basal
part of the some were consistently shown to contain decidedly
phengitic potassic mica.

Regarding the blastopsammites and graphitic slates underlying
the anagenite, the muscovite-paragonite solvus (at 2 Kb (21))
allows us to evaluate the thermic value of the metamorphism at about

400° C. The baric value can be calculated as higher than 2 Kb and
certainly lower than 5 Kb (low metamorphic stage) (22).

The presence of phengite in the metamorphic series indicates
decidedly higher baric conditions (23, 24) as the remarkable
difference in b_0 value confirms. On the other hand, the quartz-
kaolinite association (and possibly montmorillonite) leads us to
assign a lower thermic value (very low stage) to the subsequent
metamorphic event, suc that the already existent metamorphic series
was not affected by it, chiefly because the temperature was too low
to reopen the previsions system.

5. CONCLUSIONS

On the basis of our data we have subdivided the clastic
sediments underlying the Burano Anhydrites (Upper Triassic) into
several formations and inserted them into the overall picture of
the tectonic evolution of the Hercynian and Alpine orogenic cycles
(Fig. 4). This has been done on the basis of both stratigraphic and
petrographic data. The stratigraphic sequence from top to bottom
is as follows:
Alpine cycle
        Tocchi Formation (Carnian)
        Verrucano Group (Ladinian-Anisian)
Hercynian cycle
     1. post-geosynclinal phase
        Ferriera Formation (Permo-Triassic)
     2. late geosynclinal phase (with mostly heteropic formations)
        Boccheggiano Shales
        Carpineta Formation, Monte Argentario Sandstones and Iano
        Shales (Upper Carboniferous-Lower Permian)
     3. geosynclynal stage
        Farma Formation (Moscovian-Bashkirian).
The Permo-Carboniferous sediments can be divided into two
complexes since the Carpineta Formation unconformably overlies that
of the Farma. The latter, which was involved in a tectonic phase
that is probably the Asturian, represents the Hercynian geosynclinal
complex, whereas the Carpineta Formation and the other heteropic
formations may be considered as late geosynclinal.

The Boccheggiano Shales, which overlies that of Carpineta in
the Farma Valley (though the nature of their contact is still
unclear) should also be considered as late geosynclinal.

After the sedimentation, came the general uplifting of the
region, followed by the degradation of the reliefs with the
accumulation of the post-geosynclinal sediments represented by the
Ferriera Formation. At present these only appear in small isolated
outcrops below the Verrucano Group.

In fact the Ferriera Formation, which is correlatable with the
Asciano Formation of M. Pisano, was subsequently eroded in the
Triassic to form, at least in part, the Verrucano sensu stricto, as

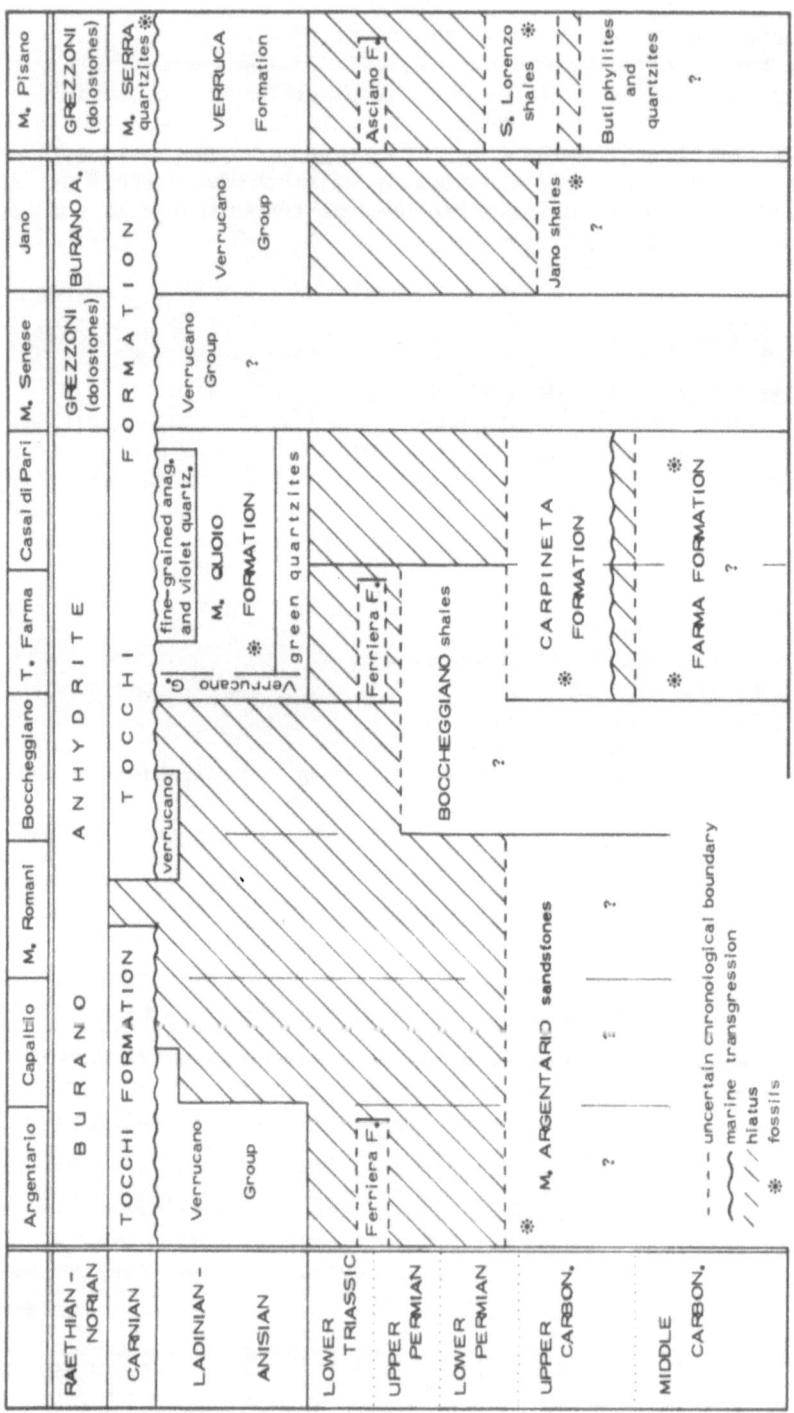

Fig. 4. Schematic diagram of stratigraphic correlations

suggested by Rau and Tongiorgi (9).

The Verrucano Group in Tuscany south of the Arno River, which has now been definitively assigned to the Ladinian-Anisian (10), represents the first sediments of the Alpine Cycle and is correlated with the Verruca Formatiìon of M. Pisano. Its greatest thickness (roughly 600-700 meters) is along the Monticiano-M. Argentario alignment, whereas to the east and west it thins out to nothing, indicating a subsident area running north-south between two high zones. Another subsident basin, also running north-south, seens to have been located to the west, from M. Pisano to the Island of Elba.

Lastly, the Tocchi Formation represents the beginning of the general marine transgression of the Alpine Cycle after the continental episode of the Verrucano. It can be correlated (because of its identical stratigraphical position and lithological analogies) with the M. Serra Quartzites, the basal part of which has yielded marine fauna of the lower Carnian.

The above-described sequence of the formations and their collocation in the Alpine and Hercynian cycles seems to be further confirmed by the study of the metamorphism of the region, based in particular on the b_0 value of the potassic mica, as will be illustrated in paragraph 4.

REFERENCES

1. E. Giannini, A. Lazzarotto and R. Signorini, Lineamenti di geologia della Toscana meridionale. Rend. Soc. Ital. Min. Petr., 27, 1972.
2. T. Cocozza, A. Lazzarotto and G.B. Vai, Flysch e molassa ercinici del Torrente Farma (Toscana). Boll. Soc. Geol. It., 93, 1974.
3. T. Cocozza, Il Carbonifero nel Gruppo Monticiano-Roccastrada (Grosseto). Ric. sci., 35, 1965.
4. M. Pasini, Archaeolythophyllum missouriensum Johnson, una Rhodophycea di interesse stratigrafico nel Carbonifero superiore della Toscana e delle Alpi orientali. Riv. Ital. Paleont. Strat., 80, 1974.
5. R. Redini, Su varie questioni geologico-paleontologiche della Catena Metallifera Toscana. Sulla età neotriassica della fauna del M. Rotondo, del M. Pisano e della fauna di Poggio Troncone, nelle Alpi Apuane. Boll. Serv. Geol. d'It.,79,1958.
6. L. Vighi, Descrizione di alcuni sondaggi che hanno attraversato lenti anidritico-dolomitiche intercalate alle fillade triassiche (Verrucano) dei dintorni di Massa Marittina (Grosseto, Toscana). Atti Symposium sul Verrucano, Pisa 1965, Soc. Tosc. Sc. Nat., 1966.
7. T. Cocozza, G. Gasperi, R. Gelmini and A. Lazzarotto, Segnalazione di nuovi affioramenti paleozoici (Permo-Carbonifero ?) a Boccheggiano e tra Capalbio e i Monti Romani (Toscana

meridionale e Lazio settentrionale). Boll. Soc. Geol. It.,93, 1974.

8. G. Gasperi and R. Gelmini, Ricerche sul Verrucano. 1- Il Verrucano del Monte Argentario e dei Monti dell'Uccellina in Toscana. Boll. Soc. Geol. It., 92, 1973.

9. A. Rau and M. Tongiorgi, Geologia dei Monti Pisani a sud-est della Valle del Guappero. Mem. Soc. Geol. It., 13, 1974.

10. T. Cocozza, A. Lazzarotto and M. Pasini, Segnalazione di una fauna triassica nel conglomerato di M. Quoio (Verrucano del Torrente Farma, Toscana Meridionale). Atti Soc. Tosc. Sc. Nat. (in the press).

11. A. Lazzarotto, R. Mazzanti and F. Mazzoncini, Geologia del Promontorio Argentario (Grosseto) e del Promontorio Franco (Isola del Giglio, Grosseto). Boll. Soc. Geol. It., 83, 1964.

12. G. Gasperi and R. Gelmini, Ricerche sul Verrucano. 3- Le successioni preverrucane e Verrucane tra Monte Bellino e Capalbio (Toscana meridionale - Lazio settentrionale). Riv. Ital. Paleont. Strat. (in the press).

13. R. Gelmini, Ricerche geologiche sul gruppo di M. Leoni (Grosseto, Toscana). 1- La geologia di M. Leoni tra Montepescali e il Fiume Ombrone. Mem. Soc. Geol. It., 8, 1969.

14. C. De Stefani, Fossili carboniferi dell'Isola d'Elba. Paleontographia Italica, 23, 1917.

15. L. Trevisan, Carta geologica d'Italia alla scala 1:100.000, F° 126 "Isola d'Elba" (II ed.). Serv. Geol. d'Ital., 1969.

16. R. Mazzanti, Geologia nella zona di Montaione, tra le valli dell'Era e dell'Elsa (Toscana). Boll. Soc. Geol. It.,80, 1961.

17. G.B. Vai and F. Francavilla, Nuovo rinvenimento di piante dello Stefaniano a Iano (Toscana). Boll. Soc. Geol. It., 93, 1974.

18. C. De Stefani, Gli schisti paleozoici dell'Isola d'Elba. Boll. Soc. Geol. It. 13, 1894.

19. R. Signorini, Descrizione geologica della parte settentrionale dei Monti dell'Uccellina presso Grosseto. Boll. Soc. Geol. It., 71, 1952.

20. R. Signorini, Descrizione geologica della parte centrale dei Monti dell'Uccellina. Boll. Soc. Geol. It., 74, 1955.

21. H.P. Eugster, A.L. Albee, A.E. Bence, J.B. Thompson and D.R. Waldbaum, The two-phase region and excess mixing properties of paragonite-muscovite crystalline solutions. J. Petrol, 13, 1972.

22. E. Azzaro and B. Di Sabatino, Prime indagini su alcuni affioramenti di serie metamorfiche dei Monti Romani (Lazio settentrionale). Period. di Mineral., 43, 1974.

23. B. Velde, Fengites micas:synthesis, stabily and natural occurrence. Am. J. Sc., 263, 1965.

24. F.P. Sassi, The petrological and geological significance of the b_o value of potassic white micas in low-grande metamorphic rocks. An application to the Eastern Alps. Tschermarks Min. Petr. Mitt., 18, 1972.

25. C. Cipriani, F.P. Sassi and C. Viterbo Bassani, *La composizione delle miche chiare in rapporto con le costanti reticolari e col grado di metamorfismo*. Rend. Soc. Ital. Min. Petr., **24**, 1968.

SUMMARY   by H. Falke

The "Verrucano" province includes the Alps as
well as the northern to central Apennines. Their Permian sediments have been discussed at the congress in
Pisa 1965 and that in Vienna 1969. The results have
been published in the "Atti del Symposium sul Verrucano", Pisa 1966, and in the "Verhandlungen der Geologischen Bundesanstalt", Vienna 1972, Vol. 1.

With regard to the western Alps from the Ligurian
Alps to the area of Brianconnais and Vanoise, the above
mentioned publications describe Permian series which
are influenced by a more or less strong metamorphism.
After deposits of an Upper Carboniferous to Permocarboniferous age with coal-seams and in part prasinites
the older Permian = the so-called Eopermian follows.
It consists of in part various coloured sequences of
quartzite, sandstones, arkoses and shales in which at
some places conglomerates as well as carbonates, but
mainly porphyroides derived from acid volcanic rocks
appear. These sediments are not developed in the
Vanoise region. Above the Eopermian the younger Permian
= the so-called Neopermian occurs above an unconformity which is supposed to reflect the Saalian phase and
to embrace a gap of unknown size. In the Brianconnais
the Neopermian is composed of a claret-coloured series
composed of conglomerates, arkoses and shales with
dacite at the base and rhyolite at the top of the
sequence. Above these deposits a conglomerate of Permo-
Triassic - respectively younger Neopermian age - occurs
which is present only locally in the area of Vanoise.
This conglomerate is called "Verrucano" especially
"Verrucano brianconnais". It results from the onlapping
onto older beds from the destruction of underlying sediments. Upwards it passes into the Triassic Werfen-
quartzite with which a new part of the geological
history of the Alps begins.

Within the Swiss Alps, especially North of the
central zone, some ENE-WSW extending occurrences of
Permian age exist. Here especially the Glarner "Verrucano" is well known. In this case the term "Verrucano",
characterizes the entire Permian sequence. The Eopermian is composed of a fanglomeratic marginal facies
(= Sernifit) and a vary-coloured as well as fine-grained
basin facies containing volcanics and carbonates. The
Neopermian lying concordantly above the deposits in the
centre of the basin consists of fanglomerates (mainly

H. Falke (ed.), The Continental Permian in Central, West, and South Europe, 196-199. All Rights Reserved.
Copyright © 1976 by D. Reidel Publishing Company, Dordrecht-Holland.

of the marginal facies), red sandstones, and slates
which are similar to the situation in the Brianconnais
unconformably overlain by a quartz conglomerate and the
Triassic sandstone. This Glarner "Verrucano" corresponds in part with Rotliegend occurrences in South
Germany. This picture of the western Alps is completed
in this book by the described occurrences of the austroalpine nappes (Dössegger) which establish a connection
with the Permian of the southern Alps.

   The last-mentioned Permian and that of the central
and eastern Alps has been dealt with in the second
above-mentioned publication. It can be divided into a
northern and southern zone. In the first-mentioned area,
in West Austria, a conglomerate-sandy to dolomitic-si-
liceous series which belongs to the Carboniferous is
followed in some places by rhyolites with tuffs and
tuffites, by conglomerates and red-coloured sandstones
of the "Grödener" type. This sequence is placed into
the Permian to Triassic. Farther to the East (in the
area Wörgl - Mitterberg) above the basement basal
breccias resp. conglomerates which passes into red
slates with tuffs and after an unconformity as Saalian
phase, conglomerates and an alternation of sandstones
and slates occur which represent the Lower and Upper
Permian. Cross bedded sandstone of permoskytic age
follow above. The last mentioned boundary is not sharp
in the area of the Central Alpine facies. Here above
the cristalline basement a sericitic-schistose to seri-
citic-quartzitic and in part coarse-grained quartzitic
series with local conglomerates is present. This has
been called by Tollmann "Alpine Verrucano". It continues
upwards into the solid, well-bedded, Skythquartzite =
Semmeringquartzite. In the further region of the Semmering in the East Alps the permoskythic sediments on
which Erkan reports in this book an older arkose-shale
- breccian-porphyroid series can be destinguished from
the overlying Semmering-quartzite the basal conglomerate
bearing section of which still belongs to the "Alpine
Verrucano". This profile establishes a contact to the
Permian in the Magdalensberg (Kärnten) highland north-
east of Klagenfurt. Here above the Carboniferous fanglomerates volcanic rocks and shales occur which are placed
into the Lower Permian. After a change in the sedimentation pattern as a concequence of the Saalian phase
a permoskytic sandstone is present which is overlain by
the Werfen beds. The Permian part of this sequence has
formerly been called "Grödener" beds.

   Herewith a term is used which as the "Grödener"

sandstone indicates a characteristic marker bed in the
Permian deposits of the South Alps. In the East, in the
Julian and Carnian Alps, this sandstone occurs after an
erosion gap with the "Tarvisio Breccie" and the below
following marine limestones with a conglomerate which
is overlain by red fine-sandy beds with argillaceous
intercalations. Due to an increase of conglomeratic
components westwards in the region of the Lombardian
Alps it passes into the "Verrucano Lombardo". The plant
fossils found in the "Dolomites" allow to determine the
age of the sandstones between Upper Rotliegend and
Zechstein and its position is obviously diachronous in
the series of strata. As a sediment of continental to
lagune-like origin it passes upwards into and inter-
fingers with the marine Bellerophon-sandstone which can
be followed to the western margin of the Dolomites.
This limestone in connection with the transgression of
the Zechstein-sea which extended to the North as well
as to the West characterizes the change in the palaeogeographic situation of the Alps leading to the events
associated into the Mesozoic history of the Alps. To
the North of the Alps in contrast to the Carnian Alps
the cristalline basement is overlain from the base towards the top by red slates, conglomerates and acid
tuffs. These sediments are believed to be of Lower Rotliegend age. They are equivalent to the deposits near
Magdalensberg mentioned above and superimposed by a
permoskytic sequence of conglomerates and intercalated
sandstones which are of the same age as the "Grödener"
sandstones but they cannot be immediately compared
with it. The Permian volcanism culminated in the Dolomites which follow in the West. Here volcanism is represented by the famous Bozen porphyry complex which
is underlain by a basal conglomerate and overlain by
the "Grödener" sandstone. At its top the "Trigovio"
shales are present which are believed to represent the
eastern continuation of the Collio-series existing in
the West and which belong to the Permian of the Lombardian Alps. The last-mentioned Permian is discussed
in great detail in the present book. Apart from the
Collio-series the younger Permotriassic "Verrucano Lombardo" is also known. It is overlain by the Servino
sandstone of Lower Triassic age. It continues to the
Lugano area in the West where besides this "Verrucano"
and a basal conglomerate a thick sequence of acid volcanic rocks with sediment intercalations occurs. Herewith the transition is reached into the already above-
mentioned Permian occurrences of the Alps in Switzerland
and the Western Alps. With the Ligurian Alps mentioned
above connection to the Apennines is given.

In the present book the Permian occurrences in Toscany of the Apennines are described north of the Arno, where they occur in continental facies and in the South where they are of marine origin. Especially for the youngest beds correlation for both facies areas has been carried out. These sediments include also the Verrucano named according to the type locality at the Monte Verruca in the Pisa mountains of the Apuan Alps. This Verrucano begins with coarse conglomerates of continental origin continuously changing into marine fossil bearing beds according to which it belongs to the Triassic.

After this summary the following unsolved problems result from the review given on the Verrucano province.

1)   In spite of repeated discussions the applicability of the term"Verrucano" has not been cleared satisfactory.
2)   In the Permoskythian the drawing of the boundary between the Triassic and Permian parts is in many places an unsolved problem.
3)   It is necessary to demonstrate by using generelly recognized arguments whether the established unconformities respectively gaps have to be co-ordinated to a Saalian phase or tectonic movements which have taken place before or afterwards.
4)   Apart from further tests to examine the possibilities of correlation of the profiles in the occurrences between each other it should be investigated whether volcanic rocks are restricted exclusively to the lower part of the Lower Permian = the Lower Rotliegend as it is the case in Central Europe or if they also occur in younger beds as it is believed for the Neopermian in the Brianconnais area of the Western Alps.
5)   Since in the Eastern Alps the transition between the continental facies in the West and the marine facies in the East takes place further tests should be made to classify the continental Permian into the marine Permian.

These raised questions give very extensive foundation for further discussions on the occasion of the Symposium in Bled (Jugoslavia) planned for 1977.

THE PERMIAN IN THE NEAR EAST

Tuvia Weissbrod

Geological Survey of Israel

GENERAL

A sedimentary sequence of Permian age is found in all the
marginal countries of the Arabo-Nubian massif. Discoveries are,
however, of comparatively recent dates and the local stratigraphy
is not always clear.

First to be known was the Permian of central and northwestern
Saudi Arabia, one of the largest exposures of marine Permian in the
world, which was described by Steineke and Bramkamp in 1952. With
the progress of geological exploration, smaller outcrops and
subsurface occurrences of Permian formations became known from
elsewhere.

Generally, The Permian sequence is 100 to 500 m thick and
consists of carbonates and shaly sediments. Nearer the massif, the
facies becomes progressively more sandy, and thicknesses diminish.
In places the Permian is missing due to stratigraphic truncation.

Permian sedimentation followed a period of intense denudation,
and Permian deposits are everywhere found overlying unconformably
older sediments, usually Lower Paleozoic or Precambrian. The
earliest Permian deposits are found in the northern part of the
region, starting with ?Lower Permian deltaic deposits. The first
deposits in the south, onlapping the massif, belong to the Upper
Permian. In spite of a certain biostratigraphic hiatus (demonstrated
only in the subsurface of southern Israel), sedimentation appears
to be continuous into the Triassic.

The present review, which has to deal with areas that are

H. Falke (ed.), The Continental Permian in Central, West, and South Europe, 200-214. All Rights Reserved.
Copyright © 1976 by D. Reidel Publishing Company, Dordrecht-Holland.

politically divided, is an attempt to form a regional concept of
the Tethyan stratigraphy of the Near East.

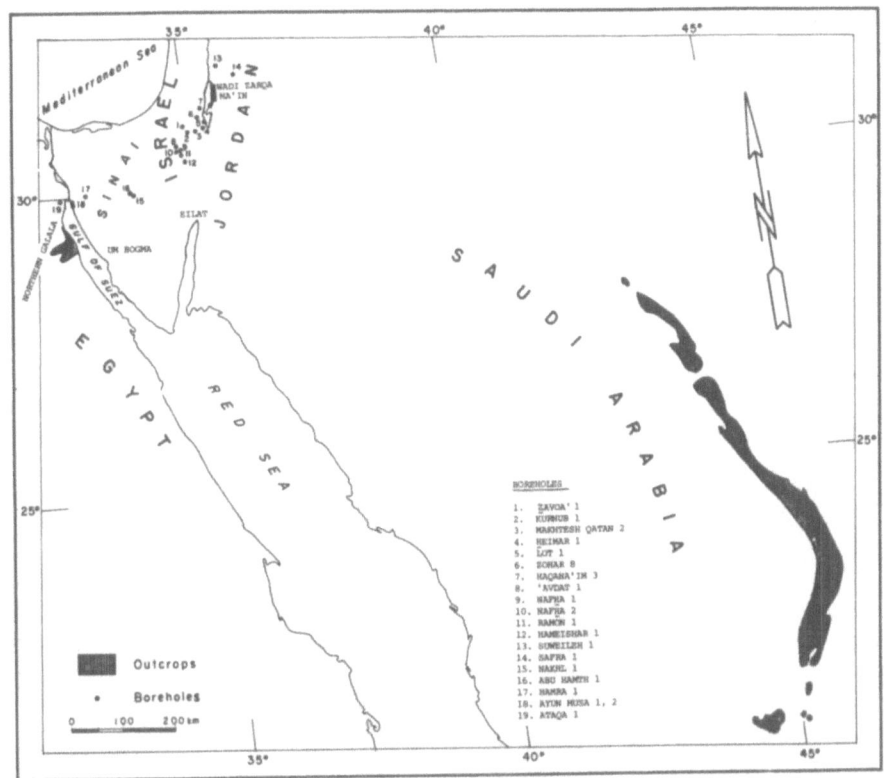

Figure 1. Occurrences of Permian strata in the Near East.

STRATIGRAPHY AND DISTRIBUTION OF PERMIAN STRATA

      1. The Permian in Israel: A Permian sequence was encountered
in twelve deep boreholes in the Negev, southern Israel (Fig. 1),
and completely penetrated in five of them. The sequence includes
carbonates, shales and sandstones ranging from 370 to 480 m in
thickness. In some places the sequence is intruded by post-Permian
magmatic sills. The Permian sediments overlie Precambrian arkoses
(Zenifim Formation) in the northern Negev, Cambrian subarkoses with
thin dolomite layers in the central Negev (Hameishar 1). In the
southern Negev (Sinaf 1 and the exposed Paleozoic of the Eilat
area) it is missing altogether due to Early Cretaceous truncation.
The Permian is everywhere overlain conformably by Lower Triassic
formations. It is nowhere exposed.

On the basis of lithology and electric markers, the Permian

is divided into three formations (Weissbrod, 1969 and in prep.),
that are recognized in all boreholes. They are, from below upward:

The Sa'ad Formation, consisting mostly of anhydrite - cemented
sandstones, with interbedded shales, partly coaly, and clays.
Thin layers of dolomite and limestone occur in some boreholes.
The formation is 100-150 m thick.

The 'Arqov Formation, consisting of alternating shale and
limestone, with a minor sandstone constituent in the northern Negev
boreholes, and sandy throughout in the southern ones. The formation
is 150-200 m thick.

The Yamin Formation, consisting mainly of limestone and
dolomite with several sandstone horizons. Shales are rare.
Thickness ranges from 120 to 150 m.

The sands of the Permian formations are mostly fine-grained
orthoquartzites; the carbonates include micrite, bio-, pelbio-
and dolomicrite, as well as sparite and biosparite. Fossils include
foraminifera and ostracods, plant remains and a fairly rich
microflora, which was studied in detail in some boreholes and which
was used to date the interval. Horowitz (1970, 1973, 1974)
examined the sequence in Zohar 8 and 'Avdat 1 boreholes and
established two palynomorph assemblage zones: a *Punctatosporites
minutus* zone that includes the Sa'ad Formation, and a
Klausipollenites schaubergeri zone that includes the 'Arqov, Yamin
and most of the overlying Zafir Formation. The first zone includes
the sporomorph types *Punctatosporites minutus, Laevigatosporites
maximus, Planisporites granifer, Triquitrites priscus* and
Reticulatosporites lorchi, which do not occur above it. On this
basis, Horowitz dated the Sa'ad Formation as Late Carboniferous
(Westphalian). However, it should be noted that Dunay (pers.
comm.) found other sporomorph types in the Sa'ad Formation of other
boreholes (Makhtesh Qatan 2, 'Avdat 1, Kurnub 1 and Ramon 1).
These include *Vittatina* cf. *costabilis, V. vittifera,
Protohaploxypinus rugalus, Guttulapollenites hannonicus,
Punctatosporites marattoides, Fimbriaesporites fimbriatus* and
others, part of which do not occur before the Permian. The Sa'ad
Formation is thus alternatively dated as Early Permian.

The *Klausipollenites schaubergeri* zone includes the sporomorphs
*Kl. schaubergeri, Aculeisporites variabilis, Falcisporites zapfei,
Sulcatisporites kraeuseli, Grandispora ullrichii, Vittatina
costabilis* and *Lueckisporites virkkiae.*

A floral break occurs in the upper part of the Yamin Formation,
where several species, including *Lueckisporites virkkiae,* disappear.
Both Horowitz and Dunay, working separately on borehole material
assign a Late Permian age to the zone up to this break

('Thuringian' acc. to Horowitz), although Dunay (pers. comm.) found Early Triassic palynomorphs in the Yamin and Zafir formations, above the floral break, and noted the disappearance of Late Permian forms.

The possibility that the upper part of the Yamin Formation and the Zafir Formation are not Permian but Lower Triassic is supported by the presence of the Lower Triassic conodonts *Pachycladina obliqua, P. symmetrica, P. tricuspidata, P. inclinata, P. longispinosa, Lonchodina nevadensis, Neohindeodella* sp. and *Hadrodontina* cf. *H. adunca* found in Makhtesh Qatan 2 borehole (cores 7 and 8, see Fig. 2) (Hirsch and Gerry, 1974).

Besides microflora, the 'Arqov and Yamin formations yielded foraminifera species (*Geinitzina, Tournayella, Globivalvulina, (?)Codonofusiella* and *Pseudovermiporella*) (Hamaoui, 1969; Derin,1972) and ostracods (*Polytilites, Sargentina, Knightina, Amptissites*) (Gerry, 1967) some of which indicate a Late Permian age.

2. The Permian in Jordan: No Permian formations have been reported to-date from Jordan, and the only information available on the Upper Paleozoic is a vague report on Lower-Middle Carboniferous microflora in the middle section of Safra 1 borehole, central Jordan (Bender, 1968).

Recent studies in Israel concerning the regional distribution of Triassic and Paleozoic deposits (Druckman *et al.*, in press; Weissbrod, in prep.) indicate that a Permian sequence is present in Jordan, both in outcrop (at Wadi Zarqa Ma'in, northeast of the Dead Sea, Fig. 1) and in the subsurface. Unfortunately, field verification is not possible at this stage. The outcrop is described by Wetzel and Morton (1959) as follows (see also Fig. 2):
>   Hathira Sandstone - Upp. Jurassic - L. Cret.
>   Humrat Ma'in Deltaic Formation - L. Triassic - 290 m.
>   Ram and Um Sahm sandstones - Paleozoic - Triassic - 100 m.
>   Qunaya Sandstone - Paleozoic - 65 m.
>   Burj Limestone - Mid. Cambrian - 40 m exposed.

Bender (1968) revised this stratigraphy, noting that the Ram and Um Sahm sandstones were defined in southern Jordan. Their recognition here is unfounded, and the sandstones thus named are more properly a part of the Qunaya Sandstone ("Massive Brownish Weathered Sandstone").

We regard the lower part of the Humrat Ma'in Deltaic Formation in this sequence as Permian. It is exposed in the Wadi Zarqa Ma'in-Wadi Mujib area and is divisible into two parts. The lower part consists of 150 m of variegated sandstones, alternating with layers of shale and micaceous sandstone. *Lingula* and other molds are found in some horizons. The upper part consists of 140 m of

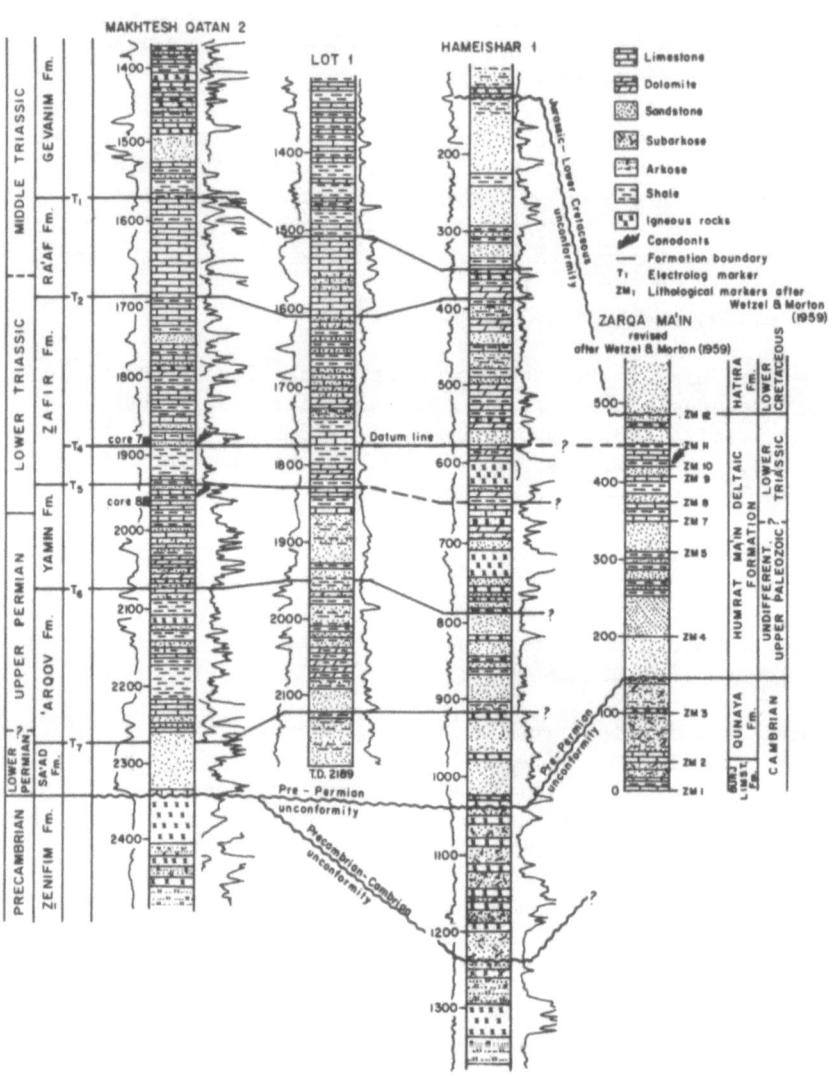

Figure 2.   Correlation of the Paleozoic and Triassic in southern
            Israel boreholes with the Wadi Zarqa Ma'in, Jordan,
            outcrop.   (from Druckman *et al.*, in press).

alternating sandy and marly limestones, with variegated, partly
clayey sandstones and shales.   Lower Triassic (Scythian) pelecypods,
brachiopods and conodonts were found here, first identified by Cox
(1932) as follows: *Anodontophora munsteri*, *Myophoria transversa*,
Pseudomonotis (Claraia) aurita, and *Lingula tenuissima*.

D. Stoppel (*in* Bender, 1968) identified the conodonts
*Pachycladina inclinata, P. obliqua, P. symmetrica, Hadrodontina
anceps* and *H. biserialis*. The conodont assemblage is similar to
the one found in the lower Zafir - upper Yamin formations in
Makhtesh Qatan 2 borehole, southern Israel (see above), suggesting
synchrony. It must be noted that correlation between the
localities is difficult due to the differences in lithology, and
the general difference in stratigraphic sequence (Fig. 2). However,
when assuming a 105 km sinistral strike slip along the Jordan -
Arava Rift Valley (Freund *et al*., 1970), the Zarqa Ma'in outcrop
should be properly placed at the approximate latitude of Hameishar 1
borehole in the Negev, and in that case the juxtaposed sequences
appear to be more similar, as are the lithologies (Fig. 2). On
the basis of such a correlation, the non-fossiliferous sandstone
that composes the lower part of the Humrat Ma'in Deltaic Formation,
between the Lower Triassic upper part and the underlying Cambrian
Qunaya Sandstone, is probably of Permian age.

Possible subsurface Permian sequences are found in Suweileh 1
borehole, some 20 km northwest of Amman, and Safra 1 borehole, some
42 km southeast of Amman (Fig. 1). The stratigraphy of the lower
section of these boreholes, based on correlation with outcrops in
southern Jordan is according to Quennell's 1951 nomenclature, and
was regarded as Cambrian both by the oil companies and Bender
(1968), although it is not supported by any paleontological evidence.
On the basis of the demonstrable similarity between the Jordanian
subsurface sequence and the subsurface Negev stratigraphy, it seems
preferable to correlate the Jordan boreholes with the latter, and
not with the south Jordanian outcrops. In doing this we have used
electrolog data rather than lithology, which is overgeneralized
in the Jordanian borehole data (Fig. 3).

Taking into account the Rift Valley strike slip displacement,
Suweileh 1 and Safra 1 were correlated with Lot 1 (west of the Dead
Sea, see Fig. 1) and Hameishar 1, respectively. The similarity
between the electric logs is striking both in pattern and in distance
between markers. The deeper section of Suweileh 1 was correlated
with sections in the deeper Negev boreholes.

Mutual biostratigraphical correlation between the Negev
boreholes points to time-equivalent sections. The correlative
sections in the central Jordan boreholes are therefore also possibly
equivalent, thus indicating that the Triassic and Permian units are
of similar lithology and thickness in south Israel and in central
Jordan.

3. The Permian in Sinai: A Permian sequence was penetrated
in five deep boreholes in central Sinai (Fig. 1), two of which
(Abu Hamth 1 and Ayun Musa 2) reached the Precambrian. The
stratigraphy of these boreholes was established by Schlatter (1951),

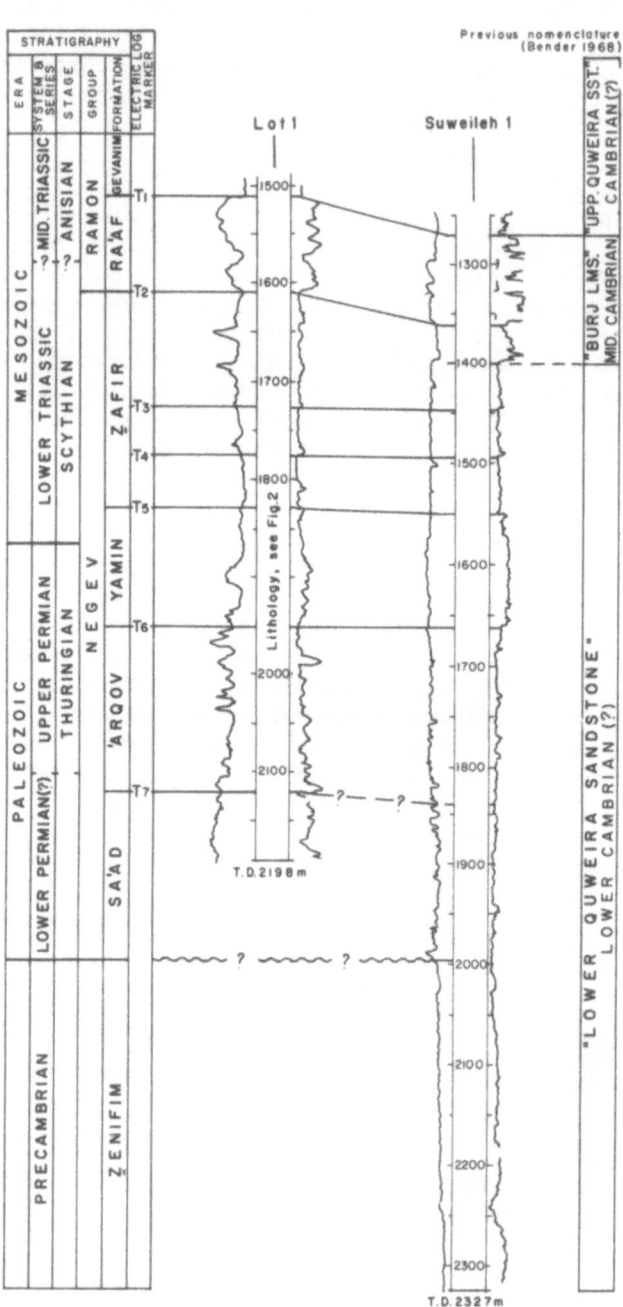

Figure 3.   Electrolog correlation between the boreholes Lot 1
            (southern Israel) and Suweileh 1 (central Jordan).

who used the nomenclature proposed by Beets (1948). According to
that, the Paleozoic sequence of these boreholes includes a Lower
Ataqa Group of Early Carboniferous age, and an Upper Ataqa Group
of Late Carboniferous - Permian age, also referred to as the "red
shales series".

The Lower Ataqa Group consists of some 200 m of massive
limestone and dolomites (biomicrite, biosparite and dolomicrite),
in places sandy, with recrystallized fossils. The upper part of
this sequence contains sandstones, and shales with coaly layers,
occasionally with thin limestone and silt horizons. Cores taken
from this interval in Ayun Musa 1 and 2 contained plant remains
(*Lepidodendropsis* sp.), together with brachiopods, pelecypods,
bryozoans, crinoids, and trilobite remains which were thought to
point to an Early Carboniferous age.

The Upper Ataqa Group, some 400 m thick, consists of massive
sandstones overlain by dark, coaly and silty shales. Cores from
Ayun Musa boreholes contained *Calamites*, brachiopods and pelecypods
of reputed Late Carboniferous - Permian age. No fossils were
found in the other boreholes, but lithological and electrolog
correlations are good.

The paleontological dating, which in reality indicates
Ordovician to Permian, occasionally Carboniferous - Permian age,
was based to some extent on correlation with the proven Carboniferous
section of the Ras Gharib boreholes, 180 km to the south. On the
other hand, comparison of the palyno-assemblage of Nakhl 1 and Abu
Hamth 1 (defined by A. Horowitz) with similar assemblages in
boreholes of southern Israel indicates possibly younger ages. The
Lower Ataqa Group in the Sinai boreholes was found to contain the
palynomorphs *Aculeisporites variabilis, Falcisporites zapfei,
Grandispora ullrichii, Lueckisporites virkkiae, Klausipollenites
schaubergeri, Veryhachium reductum* and *Veryhachum riburgense*.

The Upper Ataqa Group was found to contain *Sulcatisporites
kraeuseli, Verrucosisporites applanatus, Corollina meyeriana,
Cyathidites minor* and *Concavisporites jurienensis*. Similar
assemblages are found in foraminifera- and conodont-dated intervals
of the Negev boreholes, suggesting consequently an Early-Middle
Triassic age for the Upper Ataqa Group and a Late Permian age for
the Lower Ataqa Group (some Late Triassic palynomorphs, found in
Upper Ataqa samples, are due to caving).

The basal sands which occur at the base of the Permian sequence
in southern Israel (Sa'ad Formation) are not present in Sinai,
possibly due to its nearer proximity to the Arabo-Nubian massif.
Igneous rock underlying the Permian in Abu Hamth 1 was found to be
extrusive dolerite known to occur within the Precambrian Zenifim
arkose.

4. The Permian in Egypt: Permian formations have been
described from outcrops in the Eastern Desert west of the Gulf of
Suez, and from Ataqa 1 borehole, some 50 km to the north (Fig. 1).

According to Abdallah and El Adindani (1965) and Abdallah
et al. (1965), the Permian is represented in two formations: the
Upper Carboniferous-Lower Permian Aheimer Formation, and the
overlying, Permo-Triassic Qiseib Formation.

The Aheimer Formation is exposed on the eastern slopes of the
Northern Galala plateau. Its exposed part is about 250 m thick.
It is believed to overlie conformably the Upper Carboniferous Abu
Darag Formation, although the contact is not exposed. It is
overlain conformably by the Qiseib Formation.

The Aheimer Formation is divided into three units: a lower
unit, consisting mainly of shales interbedded with thin layers of
fossiliferous limestone, a middle unit consisting of sandstones
alternating with sandy limestone and crinoidal limestone layers,
and an upper unit consisting of sandstones, shales and silts.

The fossil-rich limestones of the Aheimer Formation include
spiriferid brachiopods, corals, gastropods and pelecypods which
according to Abdallah and El Adindani indicate Later Carboniferous-
Early Permian age. Said and Eissa (1969), however, examined the
microfauna (foraminifera, conodonts) of these layers, and found
indication for Late Carboniferous only.

The Qiseib Formation crops out above the Aheimer Formation,
reaching 43 m. Its exposures continue southward to Wadi Araba,
increasing in thickness to some 190 m. The formation, which is
variegated and has therehore been known as "red beds", consists of
alternating sandstones, silts and dark shales. A 5-12 m section
of fossiliferous limestone and marl is present in its upper part
in the Abu Darag area.

The Qiseib Formation overlies, in north-south succession, the
Aheimer, the Abu Darag (Upp. Carboniferous) and the Rod el-Hamal
(Upp. Carboniferous) formations, and according to Abdallah et al.
no unconformity is detected anywhere along the contacts. The
Qiseib Formation is overlain unconformably by the Lower Cretaceous
Malha Formation.

A Permo-Triassic age bracket is assigned to the Qiseib Formation,
for the following reasons: a. There is no demonstrable unconformity
between the Qiseib Formation and the underlying Aheimer Formation
whose upper parts are of Early Permian age. b. The fossils found
in the carbonatic upper section in the Abu Darag area may be as
young as the Middle Triassic (Muschelkalk?). These include the
gastropod Naticopsis sp. and the pelecypods Hoernesia sp. and

Pleuromya sp. Silicified tree trunks were found in the clastic section, but were not identified.

Druckman *et al.* (1970), on the basis of other considerations, raised possibilities of doubt as to the Permian age of the Qiseib Formation. They suggested that the Qiseib Formation may be correlative to a series of red beds found in the Um Bogma area, southwestern Sinai, named by them the Budra Formation and argued to be of Triassic age. Weissbrod (in prep.) raises the possibility that both the Qiseib and Budra 'red beds' are equivalent to the Lower-Middle Triassic 'red shales series' or Upper Ataqa Group of the central Sinai boreholes (see above). In that case fossiliferous carbonates found in the upper part of the Qiseib Formation are equivalent to the massive, Middle Triassic carbonatic series that overlie the Upper Ataqa Group in the Sinai boreholes.

The above evidence indicates that the Eastern Desert exposures may possibly not include the Permian at all, although it is doubtlessly represented in the subsurface section of Ataqa 1, which correlates well with the central Sinai boreholes.

5. The Permian in Saudi Arabia: Formations of Permian age are exposed along nearly continuous outcrops in the arcuate cuestas that line the eastern margin of the exposed basement, stretching for 1200 km from the region of Bani Khatmah in the south (lat. 18°00'N) to the Great Nefud in the north (lat. 28°10'N) (Fig. 1). Permian sections were identified also in many boreholes throughout the country.

Steineke and Bramkamp (1952) grouped the Permian sediments into two groups: the Khuff Formation and the overlying Sudair Shale. Their stratigraphy was described in detail by Powers *et al.* (1966).

The Khuff Formation, which is generally 200 to 300 m thick, is composed mainly of carbonates (biomicrite and biosparite, micrite and oomicrite), but in places it contains also shales, sandstones and gypsum. Three members are distinguished in the northern part of the exposure belt: the Khuff Limestone, overlain by the Midhnab Shale, which is followed by the Khartam Limestone. South of lat. 24°00'N this subdivision becomes indistinct due to replacement of the shale by limestone, and extensive dolomitization along the entire section. South of lat. 23°00'N the clastic components become dominant in the section, and in the southernmost outcrops it is completely sandy. In this area the formation is less than 100 m thick, apparently due to stratigraphic truncation.

The Khuff Formation, which was deposited after a period of extensive erosion, overlies older formations with an unconformity that ranges from the Carboniferous in the north to the igneous basement in the south, where erosion cut deepest.

Among the fossils of the formation are poorly-preserved
nautiloids, brachiopods (*Derbyia*), gastropods (*Bellerophon*) and
pelecypods (*Aviculopecten*). Subsurface samples contain foraminifera
(*Colaniella parva?*) and a microfloral assemblage (Hemer, 1965)
that includes *Hoffmeisterites microdens, Fimbriaesporites
fimbriatus* and *Lueckisporites virkkiae*, which indicate Late Permian.

The Khuff Formation is overlain conformably by the Sudair
Shale, which ranges from 185 m in the north to 115 m in the south.
It crops out in patches between lat. 27°10'N and lat. 19°00'N, and
consist of silty shales with interbeddings of sandstone, gypsum
and thin layers of limestone. No fossils were found in the outcrops,
but subsurface samples have yielded Late Permian palynomorphs
(*Hoffmeisterites microdens* and species of *Lueckisporites*) from the
lower part of the section, and Early Triassic striated bisaccate
pollen from the upper section.

THE PERMIAN-TRIASSIC BOUNDARY

Of all the described localities, the section has been examined
in greatest paleontological and stratigraphicl detail in southern
Israel, and only here are enough data available for placing the
Permian-Triassic boundary. Since sedimentation is continuous, the
boundary is not marked in the lithology and can be established
only by biostratigraphic means, for which only microfloral elements
and conodonts are available (the subsurface material having
yielded no macrofossils).

Horowitz (1970, 1973, 1974), on the basis of material from
Zohar 8 and 'Avdat 1, places the boundary rather high in the
Zafir Formation, at the top of the *Klausipollenites schaubergeri*
zone, which he defines as Upper (though not necessarily uppermost)
Permian. The overlying strata belong to the *Taeniaesporites
kraeuseli* zone, which he regards as Middle-Upper Triassic. The
Lower Triassic is thus not represented, implying the possibility
of a biostratigraphic hiatus (although Horowitz, 1974 mentions
a possible Lower Triassic subzone in the upper Zafir Formation
in Zohar 8).

On the other hand Dunay (pers. comm.), as well as Picard and
Flexer (1974) and Druckman (1974), consider not only the Zafir
Formation as Triassic, but also the upper part of the underlying
Yamin Formation as Lower Triassic. They note that many palynomorphs
of the *Klausipollenites schaubergeri* zone continue into the
Triassic and that the top of the Upper Permian, i.e. the Permian-
Triassic boundary, is marked by the upper appearance of the
diagnostic Permian species *Lueckisporites virkkiae*. This floral

boundary is found in the upper part of the Yamin Formation, in which Dunay found also the Upper Permian species *Protohaploxypinus minor*, *P. jacobii*, *Striatoabieites richteri*, *Gardenasporites leonardii*, and *Petricolpipollenites bharadwaji*.

The section between this boundary and the top of the *Klausipollenites schaubergeri* zone contains typical Lower Triassic (Greisbachian-Dienerian) elements, such as *Conaletes* spp., *Densoisporites nejburgii* and *Endosporites papillatus*. The Lower Triassic (late Lower Scythian, or Smithian) age of the Upper Yamin-Lower Zafir interval is also independently indicated by the presence of Pachycladinid and Hadrodontid conodonts (Hirsch and Gerry, 1974).

It seems therefore that even if a biostratigraphic hiatus is present between the Permian and Triassic, it is much smaller than suggested by Horowitz, involving only the basal Triassic. It should be noted that no sedimentary break can be discerned in any of the borehole sequences.

PERMIAN REGIONAL PALEOGEOGRAPHY

Enough data are available for broad lithostratigraphic and biostratigraphic (microfloral) correlations between the above-described regions (Table 1). There remains reasonable doubt, however, concerning the Permian age of the section in the Eastern Desert, Egypt. Table 1 presents the alternative possibility. It should be noted that lithostratigraphic boundaries are diachronous throughout the region, according to the geographic situation in relation to the Arabo-Nubian massif.

The paleogeographic evolution of the Permian may be tentatively summarized as follows: Regional uplifting took place during the Late Carboniferous (possibly contemporaneously with the Hercynian mountain-building elsewhere). Subsequent erosion removed thick Paleozoic and Precambrian deposits, often cutting deep into the igneous basement.

During the Early Permian subsidence began, and fluvial-mixed environment sediments were deposited in the northern parts of the region (Israel, Jordan).

The transgression reached its maximum during the Late Permian, covering all the north-northeastern margins of the massif and possibly parts of its center, setting the transgression pattern for all subsequent Mesozoic transgressions. Sediments include massive shallow-water carbonate series interbedded with shales, and a progressively more clastic and sandy facies toward the massif, with signs of a peritidal environment.

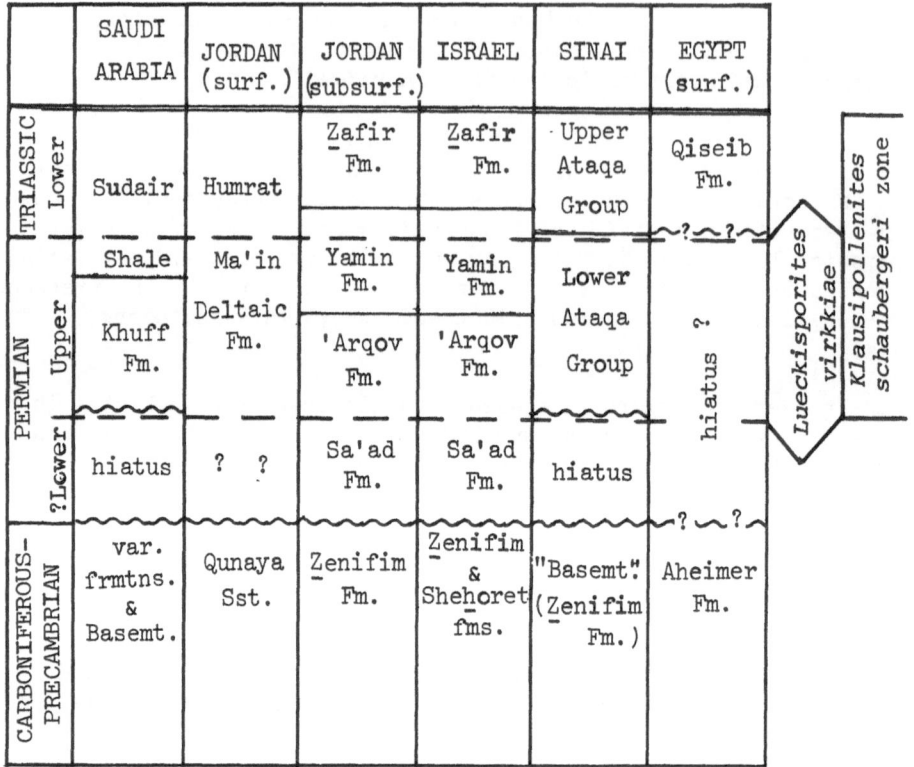

	SAUDI ARABIA	JORDAN (surf.)	JORDAN (subsurf.)	ISRAEL	SINAI	EGYPT (surf.)	
TRIASSIC Lower	Sudair	Humrat	Zafir Fm.	Zafir Fm.	Upper Ataqa Group	Qiseib Fm.	Lueckisporites virkkiae — Klausipollenites schaubergeri zone
						~?~~?~	
PERMIAN Upper	Shale	Ma'in Deltaic Fm.	Yamin Fm.	Yamin Fm.	Lower Ataqa Group	?	
	Khuff Fm.		'Arqov Fm.	'Arqov Fm.		hiatus	
PERMIAN ?Lower	hiatus	? ?	Sa'ad Fm.	Sa'ad Fm.	hiatus		
						~?~~?~	
CARBONIFEROUS– PRECAMBRIAN	var. frmtns. & Basemt.	Qunaya Sst.	Zenifim Fm.	Zenifim & Shehoret fms.	"Basemt." (Zenifim Fm.)	Aheimer Fm.	

Table 1.   Proposed correlation of the Permian-Lower Triassic in
           Near East.

With the beginning of the Triassic the character of the
sedimentation becomes more regressional, and inundations appear
to be less extensive.   Sediments are of a mixed facies (carbonate,
shale, sandstone) in the northern areas, and continental-deltaic
("red bed series") toward the massif, with more silty shales and
sandstones, as in Sinai and Saudi Arabia.   A small biostratigraphic
(microfloral) hiatus marks the Permian-Triassic boundary.

A c k n o w l e d g m e n t s: The author expresses his thanks
to R. Dunay for important unpublished information, and to I. Perath
for help in preparing the manuscript.

REFERENCES

Abdallah, A.M. and El Adindani, A., *Stratigraphy of Upper Paleozoic
    rocks western side of the Gulf of Suez.*   Geol. Surv. Min. Res.
    Dept., Paper 25, Cairo, 1965.
Abdallah, A.M., El Adindani, A. and Fahmy, N., *Stratigraphy of the
    Lower Mesozoic rocks western side of Gulf of Suez, Egypt.*

Geol. Surv. Min. Res. Dept., Paper 27, Cairo, 1965.

Beets, C., *Correlation of the Paleo- and Mesozoic in Egypt.*
A.E.O., Rept. 697, 1948.

Bender, F., *Geologie von Jordanien.* Gebr. Borntraeger. Berlin,
Stuttgart, 1968.

Cox, L.R., Further notes on the Transjordan Trias. *Ann. Mag. Nat.
Hist.,* Ser. 10, 10, 1932.

Derin, B., *Micropaleontological report, Zavoa' 1 well.* Israel Inst.
Petrol. Rept. 3/72, 1972.

Druckman, Y., The stratigraphy of the Triassic sequence in southern
Israel. *Geol. Surv. Israel Bull.,* 64, 1974.

Druckman, Y., Weissbrod, T. and Horowitz, A., *The Budra Formation:
a Triassic continental deposit in southwestern Sinai.* Geol.
Surv. Israel, Rept./OD/3/70, 1970.

Druckman, Y., Hirsch, F., Dunay, R. and Weissbrod, T., *Stratigraphical correlation of Lower Triassic and Paleozoic formations
east and west of the Dead Sea Rift Valley* (in press).

Freund, R., Garfunkel, Z., Zak, I., Goldberg, M., Weissbrod, T.
and Derin, B., The shear along the Dead Sea Rift. *Phil. Trans.
Roy. Soc. London,* A 267, 1970.

Gerry, E., *Paleozoic and Triassic ostracoda from outcrops and
wells in southern Israel.* Israel Inst. Petrol. Rept. 1/67,
1967.

Hamaoui, M., Note on Upper Paleozoic microfossils from southern
Israel. *Geol. Surv. Israel Bull.,* 47, 1969.

Hemer, D.O., *Application of palynology in Saudi Arabia.* Fifth
Arab Petr. Cong., Cairo, 1965.

Hirsch, F. and Gerry, E., Conodont- and Ostracode- biostratigraphy
of the Triassic in Israel. *Schriftenrh. Erdwiss. Komm. Österr,
Akad. Wiss.,* 2, Wien, 1974.

Horowitz, A., *Palynostratigraphy of the Upper Paleozoic - Lower
Mesozoic sequence in Zohar 8 borehole (southern Israel).* Geol.
Surv. Israel, Rept. P/1/70, 1970.

Horowitz, A., Late Permian palynomorphs from southern Israel.
*Pollen et Spores,*XV, 2, 1973.

Horowitz, A., Palynostratigraphy of the subsurface Carboniferous,
Permian and Triassic in southern Israel. *Geoscience and Man.,*
IX, 1974.

Picard, L. and Flexer, A., *Studies on the stratigraphy of Israel:
The Triassic.* Israel Inst. Petrol. Tel Aviv, 1974.

Powers, R.W., Ramirez, L.F., Redmond, C.D. and Elberg, E.L.,
*Geology of the Arabian Peninsula; Sedimentary geology of
Saudi Arabia.* U.S. Geol. Surv. Prof. Papers 560-D, 1966.

Quennell, A.M., The geology and mineral resources of (former)
Trans-Jordan. *Colonial Geol. and Min. Res.,* London, 1951.

Said, R. and Eissa, R.A., *Some microfossils from Upper Paleozoic
rocks of western coastal plain of Gulf of Suez region, Egypt.*
Proc. 3rd Afric. Micropal. Colloq., NIDOC, Cairo, 1969.

Schlatter, L.E., *Geological report on well Abu Hamth-1.* A.E.O.,
Rept. 889, Cairo, 1951.

Steineke, M. and Bramkamp, R.A., Stratigraphical introduction.
    In Arkell, W.J., Jurassic ammonites from Jebel Tuwaiq, central
    Arabia. *Royal Soc. (London) Phil. Trans.,* 236, London, 1952.
Weissbrod, T., The Paleozoic of Israel and adjacent countries,
    part I: The subsurface Paleozoic stratigraphy of southern
    Israel. *Geol. Surv. Israel Bull.,* 47, 1969.
Weissbrod, T., *The Paleozoic of Israel and adjacent countries,
    parts III-V* (in preparation).
Wetzel, R. and Morton, D.M., Contribution à la géologie de la
    Transjordanie. *Notes Mêm. Moyen-Orient,* VII, Paris, 1959.

LATE VARISCAN NONMARINE BASIN DEPOSITS, NORTHWEST AFRICA: RECORD OF HERCYNOTYPE OROGENY*

F.B. Van Houten

Department of Geological and Geophysical Sciences
Princeton University, Princeton, New Jersey 08540

ABSTRACT. The principal Paleozoic deformation, metamorphism, and plutonism in northwest Africa occurred during late Carboniferous phases of the Variscan Orogeny. Fifteen remnants of nonmarine deposits scattered across the Moroccan and Oranian mesetas record predominantly vertical displacement during late Variscan phases. Nine deposits are thick (as much as 2000m), conglomeratic, lithic-rich sequences, commonly with lignitic layers and rarely with coal. Several remnants yield Westphalian C to Autunian plant fossils. A few deposits contain rhyolitic to andesitic lava, two of which yield radiometric ages of 270 ± 17 m.y., and 219 ± 10 m.y. (based on single samples). A similar magmatic-tectonic framework dominated late Carboniferous to middle Permian nonmarine basin deposits in the inner Variscan zones of southwestern Europe.

INTRODUCTION

For many years European geologists have recognized the significant differences between their Alpine and Variscan (Hercynian) domains. The former displays extensive horizontal displacements, low temperature-high pressure metamorphism, and meager intramontane plutonism and volcanism; the latter is characterized by

---

* This investigation was supported by National Science Foundation Grant GF-32410x, comprising a special foreign currency grant from the Office of International Programs and a dollar grant from the Earth Sciences Section of the Division of Environmental Sciences.

vertical displacements, high temperature - low pressure metamorphism, common intramontane plutons, lava, and local basins, and
reactivated basement. Zwart [1] generalized these contrasts as
alpinotype and *hercynotype* deformation. The alpinotype is the
pith of mobilist concepts; the hercynotype has supported fixist
views of a globe whose energy is applied vertically to the lithosphere.

Today geologic speculation is overwhelmed by the mobilist
notions of plate tectonics - large-scale "horizontal thinking"
[2]. Accordingly, orogenies are now attributed to interaction
between lithospheric plates, either one-sided cordilleran convergence of continental and ocean crust or two-sided continental
collision. Gansser [3] has pointed out that in collision mountain
belts, like the Alps and Himalayas, lateral displacement of the
original structural elements prevails, whereas in cordilleran
chains, like the Andes, the major elements remain essentially in
place and vertical displacement predominates.

Alpinotype deformation, with its marked horizontal shortening,
fits rather well into the modern paradigm. But hercynotype orogeny
is more obdurate. In fact, after a very instructive review of
the Variscan elements of central Europe Krebs and Wachendorf [4]
conclude that the Variscan geosyncline and orogenic evolution
cannot be explained by the new global tectonics. But it has been
tried. The several recent attempts to account for hercynotype
deformation by plate tectonics have constructed contrasting
models, and most have ignored the problem of producing the characteristic vertical displacements by horizontal drift. Some
authors suggest that the European Variscides resulted from two-
sided continental collision [5-8], and compare their development
with that of the Himalayan deformation of the Tibetan Plateau [9].
Other authors, impressed by vertical displacement in the Andes,
have invoked one-sided cordilleran convergence [10-11].

Clearly, the distinctive Variscides cannot be explained by a
simple plate tectonics model. Generalizations of this sort have
little more to contribute to the problem. Instead, additional
information about the specific development of hercynotype orogens
is urgently needed. For this reason it is useful to summarize
the record of late Variscan nonmarine deposits and tectonic
activity in northwest Africa, and, in addition, to emphasize the
ways they differ from Triassic counterparts.

GEOLOGIC SETTING

The basement of the Moroccan and Oranian mesetas was fashioned
largely by the late Paleozoic Variscan Orogeny [12]. Locally,
early Carboniferous marine deposits and volcanic rocks (Fig. 1)

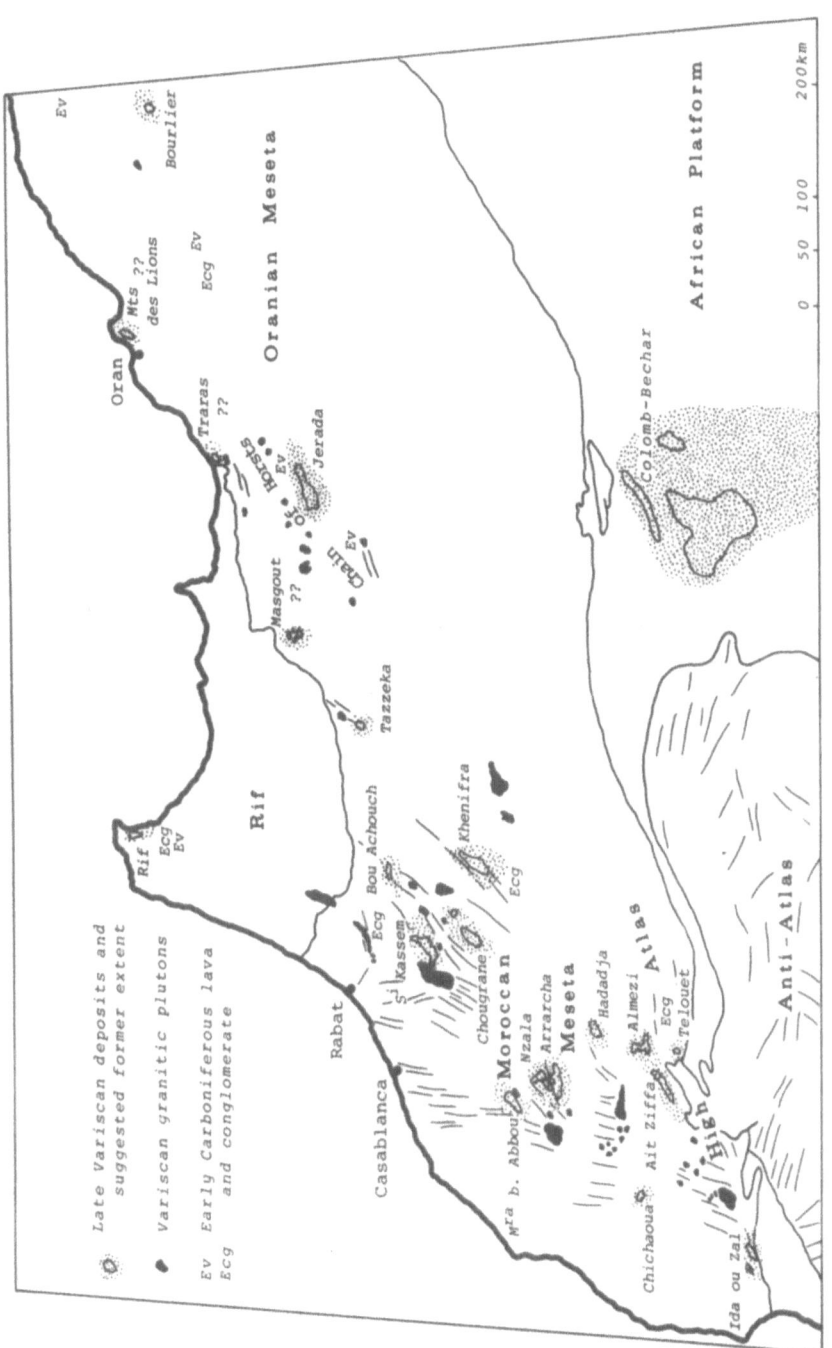

Fig. 1. Late Paleozoic Variscan framework of Morocco and Algeria, with sketched structural trends.

reflect early Variscan activity. But the main phases prevailed
from latest Namurian to the end of Westphalian time [13], producing an emergent terrain of thick continental crust north of the
African Platform.

During late Carboniferous and early Permian time late Variscan
deformation was marked by vertical displacement, intrusion of
scattered granitic plutons, extrusion of acidic to intermediate
lava, and accumulation of local, conglomeratic nonmarine basin
deposits. The broad belt of activity extended 1500 km from the
western High Atlas in the southwest, northeastward across the
central High Atlas and Moroccan Meseta, to the Chain of Horsts
and Oranian Meseta, as far as north-central Algeria. Then a final
phase of deformation destroyed the basins, followed by a long
interval of erosion. Late Permian and early Triassic deposition
was very local or never occurred on the Variscan orogen. Early
Triassic nonmarine sediments associated with extensive middle
Triassic marine deposits were confined to the north flank of the
Moroccan and Oranian mesetas and to adjacent Tunisia, and to the
Saharan Platform to the southeast (Fig. 2). Thick late Triassic
deposits were concentrated in fault basins that crudely outlined
the mesetas in trends of subsidence that became the principal
tracts of early Jurassic sedimentation [14], as well as the locus
of Atlasic deformation during the Alpine Orogeny [15]. Throughout
most of the past 225 m.y. northwest Africa had a broad northeast-
erly paleoslope away from the eastern margin of the rifting North
Atlantic Basin.

LATE VARISCAN DEPOSITS, NORTHWEST AFRICA

Description

Most of the known record of late Variscan events is preserved
in the western half of the orogen [16-18]; additional traces probably are still buried beneath the Mesozoic-Cenozoic mantle on the
eastern half. Fifteen remnants of late Paleozoic nonmarine deposits on the Moroccan and Oranian mesetas (Fig. 1) record sedimentation in intramontane basins and on more stable interbasin areas.
In addition, several other relevant conglomeratic nonmarine deposits crop out in an external foredeep south of the orogen, on the
unstable north flank of the Oranian Meseta, and to the north in
allochthonous Variscan blocks from the Rif to the Petite Kabylie
(Fig. 2).

Fossil plants from eight sequences (Ida ou Zal, Ait Ziffa,
Almezi, Sidi Kassem, Bou Achouch, Khenifra, Jerada, and north-
eastern Rif) indicate ages from Westphalian C to Autunian. Single
whole-rock potassium-argon analyses of associated lavas, made by
Geochron Laboratories, gave 270 ± 17 m.y. (Chougrane) and 219 ± 10

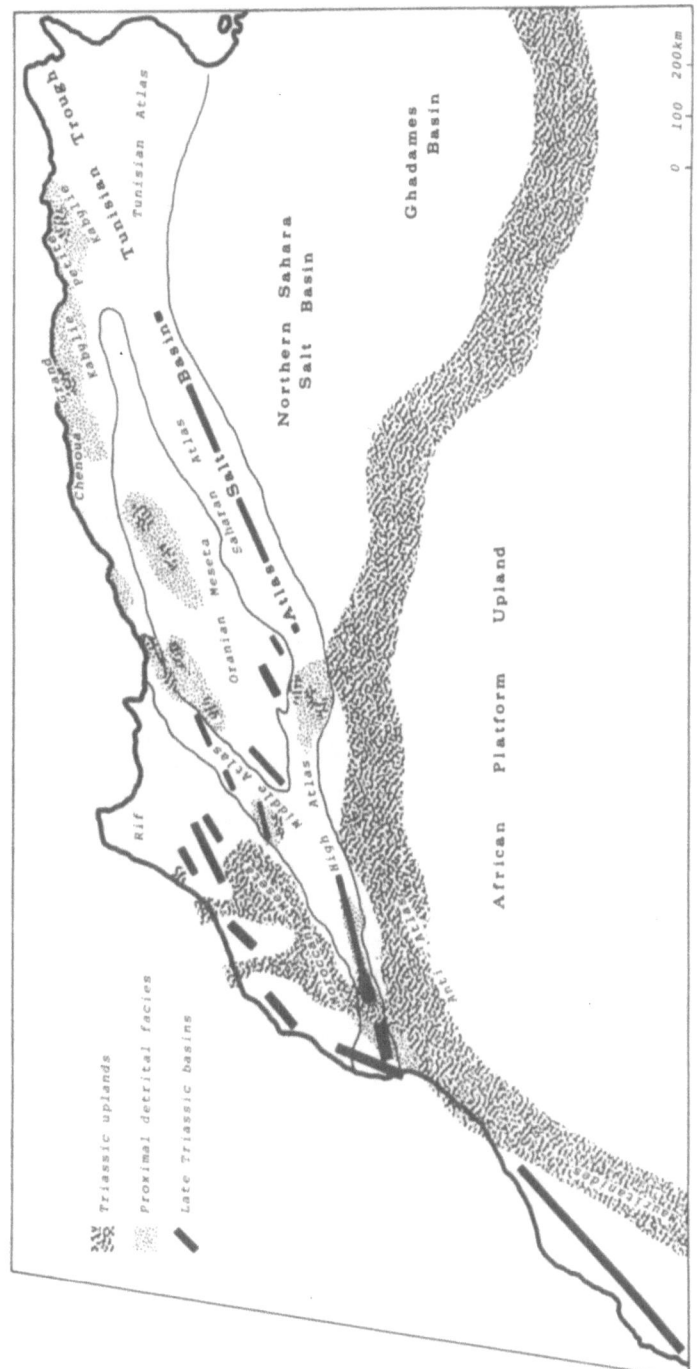

Fig. 2. Triassic basins on Variscan orogen and African Platform, northwest Africa.

m.y. (Chichaoua). Although the other remnants have not been dated
their common stratigraphic position above older Paleozoic rocks
and unconformably below Triassic strata points to a late Paleozoic
age for most of them.

Large intramontane, limnic basins within the orogen (Ida ou
Zal, Almezi, Hadadja, Mechra ben Abbou, Nzala Arrarcha, Sidi Kassem,
Chougrane, and Khenifra) contain detrital deposits as much as 2000m
thick that are nonconformable on metamorphosed Namurian and older
marine strata. Polymictic conglomerates predominate in the lower
10's to 100's of meters. Most of them were eroded from older
Paleozoic quartzite, phyllite, and limestone; a few were derived
in part from nearby late Paleozoic plutonic and volcanic rocks.
Most of the sandstones are lithic-rich and only moderately well
sorted. Lignitic layers are present in several of the sequences
and beds of coal in two (Ida ou Zal and Sidi Kassem). The oldest,
Late Westphalian, sequence contains the most coal. Some part of
each of these deposits is reddish-brown, but the color is rarely
the striking hue of the overlying late Triassic rocks, and beds
of salt are conspicuously absent. The original extent of each
basin fill cannot be reconstructed accurately, but present size
and facies distribution suggest that most of them were at least 50
to 75 km long (Fig. 1). Even though many of the thick conglomerates
indicate a nearby source none has been related to a specific border fault.

A large paralic basin in the northern Variscan domain (Jerada)
was the site of more or less continuous sedimentation from early
Carboniferous marine deposits to a thick, late Carboniferous non-
marine, coal-bearing sequence with conglomerate in its lower part.
Late Namurian to early Westphalian deformation that affected most
of the orogen was relatively minor here. A post-Westphalian phase
destroyed the basin.

Several remnants of sequences less than a few hundred meters
thick (Telouet, Ait Ziffa, Chichaoua, Bou Achouch, Tazzeka, and
Bourlier) are predominantly fine-grained deposits, although some
have a thin basal conglomerate. Two contain thin lenses of lignite
or coal, and one has a few stringers of gypsum. Most of these
thin, undeformed deposits apparently accumulated on more stable
areas of the mesetas.

Acidic to intermediate lava, commonly 10's of meters thick,
is associated with six of the meseta sequences, some with thick
basin deposits (Hadadja, Mechra ben Abbou, Nzala Arrarcha, and
Chougrane), and some with thin interbasin patches (Chichaoua, Bou
Achouch).

On the north flank of the Oranian Meseta, from the Middle
Atlas belt eastward to north-central Algeria, several undated

patches of reddish-brown conglomerate, sandstone, and mudstone, a few hundred meters thick, cover autochthonous Variscan basement (Masgout, Traras, Monts des Lions, Zaccar). North of the mesetas rather similar conglomeratic sequences in nappes of the internal Alpine zone (Rif, Chenoua, Grand Kabylie, Petite Kabylie; Fig. 2) are essentially conformable on early to middle Carboniferous strata, indicating that Variscan deformation was subordinate on these allochtonous blocks [19]. Autunian plants from the deposit in northeastern Rif [20] suggest that some of the undated patches may also be of late Carboniferous or early Permian age; others probably are late Triassic deposits [21].

The broad Colomb-Bechar Basin on the southern foreland east of the Anti-Atlas contains an essentially continuous late Carboniferous succession 2000m thick. Marine deposits and paralic coal measures accumulated in Westphalian C time. Then conglomerates introduced the final, Westphalian D and Stephanian nonmarine sedimentation, with coal in the lower part and red beds and stringers of gypsum in the upper. Only minor post-Stephanian deformation affected the foredeep.

Discussion

Commonly, the discordant late Paleozoic limnic deposits in intramontane basins have been called *molasse* [22], and they belong to that tectofacies to the extent of comprising detritus from uplands produced by late Variscan deformation. Nevertheless, they differ from typical Alpine Molasse [23] in not having accumulated in cratonic foredeeps bordering the orogen, and in commonly containing lava erupted within the mountain belt. If the term, molasse, is applied to the late Variscan deposits it should be modified by an adjective describing the style of the basin. The widely used adjective, *intramontane*, as well as preplatform [22] and successor [24], denote the site within the orogen. These deposits have also been referred to as "post-orogenic". They are not; they accumulated after the main phases of Variscan deformation, as products of final (post-paroxysmal) phases of the orogeny.

Many older, relevant publications state or imply that late Triassic nonmarine basin deposits of northwest Africa, as well as the late Paleozoic ones, record the last phases of Variscan deformation and destruction of the mountain chain. This notion is simply wrong. According to current ideas about continental drift the Triassic tectonic patterns, basins, and associated igneous rocks were produced by a new regime of rifting that was imposed on the cratonized Variscan domain in early Mesozoic time.

The hercynotype framework [25] dominated the well-known late Carboniferous to middle Permian nonmarine sedimentation in central and southwest Europe [26-30], as it had in northwest Africa. The Moroccan and Oranian mesetas are most like the inner zones of the orogen, including the Bohemian Massif, Massif Central, and the Iberian Meseta. They share a common episode of late Westphalian to Autunian, and locally Saxonian, basin filling, followed by later Permian disturbance that disrupted the basins. This late orogenic, intramontane molasse comprises an irregular progression from coal measures to redbeds, thus recording a general trend of increasing aridity in both central and southwest Europe and in northwest Africa. During late Permian and early Triassic post-Variscan stability the interior of the Variscan domains was extensively eroded. Then, in Spain and Portugal, as in western Algeria and Morocco, a new episode of Triassic rifting fractured the flanks of the cratonized mesetas, and thicker Triassic sections accumulated along these marginal trends.

The basic difference between late Variscan intramontane molasse and the succeeding Triassic graben facies is easily obscured because the two shared many sedimentary features. Moreover, where Triassic basin deposits are absent, as in the Massif Central [25], interpretation of the Permian episodes of vertical displacement is ambiguous. Here the middle Permian pattern of deposition differed from the early Permian one, suggesting a change in style of deformation [29, 31]. Nevertheless, these episodes are generally attributed to the close of the Variscan Orogeny. Late Carboniferous to Permian basin deposits in the Bohemian Massif are also commonly considered to be late Variscan intramontane molasse [30]. Jindrich [32] has suggested, however, that they filled post-Variscan grabens of a late Paleozoic rift system; but Havlena [22] insists that they are preplatform molasse that accumulated during platformization of the Variscides, and that the orogen was not completely cratonized until late Permian time.

In Iberia and northwest Africa late Variscan and Triassic basins are readily segregated by their basically different regional distribution relative to the Variscan orogen. This clear contrast in the trends of late orogenic and new extensional basins affords a useful clue for interpreting basin development in hercynotype domains.

REFERENCES

1.   H.J. Zwart, 1967, The duality of orogenic belts: *Geologie en Mijnb.*, v. 46, p. 283-309.
2.   L.P. Zonenshain, 1975, Problems of global tectonics: *Amer. Assoc. Petrol. Geologists*, v. 59, p. 124-133.
3.   A. Gansser, 1973, Orogene entwichlung in den Andean, im

Himalaja, und den Alpen: *Eclogae Geol. Helv.*, v. 66, p. 23-40.

4. W. Krebs and H. Wachendorf, 1973, Proterozoic-Paleozoic geosynclinal and orogenic evolution of Central Europe: *Geol. Soc. America Bull.*, v. 84, p. 2611-2630.

5. C.F. Burrett, 1972, Plate tectonics and the Hercynian Orogeny: *Nature*, v. 239, p. 155-157.

6. R. Laurent, 1972, The Hercynides of south Europe, a model: *24th Intern. Geol. Cong., Rept.*, Sec. 3, p. 363-370.

7. R.V. Burne, 1973, Paleogeography of southwest England and Hercynian continental collision: *Nature (Phys. Sci.)*, v. 241, p. 129-131.

8. G.A.L. Johnson, 1974, Crustal margins and plate tectonics during the Carboniferous: *7th Cong. Intern. Stratigr. Geol. Carbonifere, C.R.*, Bd. III, p. 261-265.

9. J.F. Dewey and K. Burke, 1973, Tibetan, Variscan, and Precambrian basement reactivation, products of continental collision: *Jour. Geology*, v. 81, p. 683-692.

10. A. Nicolas, 1972, Was the Hercynian orogenic belt of Europe of the Andean type? *Nature*, v. 236, p. 221-223.

11. R. Riding, 1974, Model of the Hercynian foldbelt: *Earth Planet. Sci. Ltrs.*, v. 24, p. 125-135.

12. A. Michard and J. Sougy, 1975, L'orogenese hercynienne a la lisiere nord-ouest de l'Afrique: *C.N.R.S., France, Colloq. Intern.*, 53p. (in press).

13. G. Choubert and A. Faure-Muret, 1971, The Hercynian period: p. 353-359 in G. Choubert, ed., *Tectonics of Africa*; Unesco, Paris, 602p.

14. G. Choubert and A. Faure-Muret, 1962, Evolution du domain Atlasique Marocain depuis les temps Paleozoiques: p. 447-527 in M. Durand-Delga, ed., *Livre a la memoire du Professor Paul Fallot*; Mem. Soc. Geol. France, v. 1, 656p.

15. J.F. Dewey, W.C. Pitman, W.B.F. Ryan, and J. Bonnin, 1973, Plate tectonics and the evolution of the Alpine system: *Geol. Soc. America Bull.*, v. 84, p. 3137-3180.

16. E. Roch, 1950, Histoire stratigraphique du Maroc: *Serv. Geol. Maroc, Notes et Mem.*, N. 80, 435p.

17. G. Choubert, 1956, Maroc: *20th Cong. Geol. Intern., Lex. Stratig. Intern.*, v. 4, f. 1a, 165p.

18. M. Durand-Delga, 1962, Algerie: *21st Cong. Geol. Intern., Lex. Stratig. Intern.*, v. 4, f. 1b, p. 1-132.

19. M. Durand-Delga, 1967, Structure and geology of the northeast Atlas Mountains: *Petrol. Explor. Soc. Libya, 9th Ann. Field Conf.*, p. 59-83.

20. Y. Milliard, 1959, Sur l'existence du Permien dans le massif paleozoique interne du Rif (Maroc): *C.R. Acad. Sci., Paris*, T. 249, p. 1051-1052.

21. M. Durand-Delga, 1955, Etude geologique de l'ouest de la Chaine Numidique: *Bull. Serv. Carte Geol. de l'Algerie, 2nd Ser. Stratig.*, N. 24, 533p.

22. V. Havlena, 1974, Einteilung der kohlen führenden variszichen

Ablagerungsräume und deren Auffüllungen: *7th Cong. Intern. Stratig. Geol. Carbonifere, C.R., Bd. III*, p. 169-188.

23. F.B. Van Houten, 1974, Northern Alpine Molasse and similar Cenozoic sequences in southern Europe: *Soc. Econ. Paleon. Mineralogists,* Sp. Publ. 19, p. 260-273.

24. G.H. Eisbacher and H. Gabrielse, 1975, The molasse facies of the Columbian orogen, Canadian Cordillera: *Geol. Rundschau,* Bd. 64, p. 85-100.

25. J. Debelmas and F. Ellenberger, 1974, Le bati hercynien et ses noyaux anciens: p. 9-14 in J. Debelmas, ed., *Geologie de la France,* v. 1; Doin, Paris, 296p.

26. J. Aubouin, 1965, *Geosynclines:* Elsevier, Amsterdam, 335p.

27. J.P. Bard, R. Capdevila, P. Matte, and A. Ribeiro, 1973, Geotectonic origin for the Iberian Variscan orogen: *Nature (Phys. Sci.),* v. 241, p. 50-52.

28. H. Falke, 1972, The paleogeography of the continental Permian in central-, west-, and in part of south Europe: p. 281-299 in H. Falke, ed., *Rotliegend*; E.J. Brill, Leiden, 299p.

29. R. Feys and C. Greber, 1972, L'Autunien et le Saxonien en France: p. 114-136 in H. Falke, ed., *Rotliegend*; E.J. Brill, Leiden, 299p.

30. V. Holub, 1972, Permian of the Bohemian Massif: p. 137-188 in H. Falke, ed., *Rotliegend*; E.J. Brill, Leiden, 299p.

31. P. Vetter, 1973, Le Carbonifere superieur et le Permian du Massif Central: p. 169-213 in C.E. Wegmann, ed., *Geologie, geomorphologie, et structure profonde du Massif Central, France;* Plein Air Serv., Clermont-Ferrand, 607p.

32. V. Jindrich, 1971, New views in tectonic significance of platform sediments in the Bohemian Massif, Czechoslovakia: *Geol. Soc. America Bull.,* v. 82, p. 763-768.

GENERAL GEOLOGY AND VERTEBRATE BIOSTRATIGRAPHY OF THE DUNKARD BASIN

Richard Lund

Department of Biology, Adelphi University
Garden City, New York, U.S.A.

ABSTRACT

Vertebrate faunal successions in the Appalachian basin can be used to delimit biostratigraphic zones correlable with both Western United States, Nova Scotian and European sections. These zones are correlable with the Westphalian, Stephanian and Autunian of Europe, and the Missourian, Virgilian and Wolfcampian Lower Clear Fork beds of Texas. Although precise placement of the Carboniferous-Permian boundary is not possible, there is general acceptance of the Stephanian-Autunian (Virgilian-Wolfcampian) boundary for the systematic change as well. As good lower Wolfcampian vertebrate faunas are established in the Appalachian basin by the middle of the Monongahela group, Pittsburgh formation, a plausible place to draw this boundary would seem to be at the base of the Benwood carbonate. Pittsburgh formation.

INTRODUCTION

Sedimentary rocks of the Dunkard group of northern West Virginia, southwestern Pennsylvania and eastern Ohio represent the youngest preserved Paleozoic sediments in eastern North America. The Dunkard group is underlain successively by the Monongahela, Conemaugh and Allegheny groups. Dating of the Dunkard group as Wolfcampian (Autunian) Lower Permian, and of the non-marine Monongahela and upper Conemaugh groups as Virgilian (Stephanian) Upper Pennsylvanian has been done

H. Falke (ed.), The Continental Permian in Central, West, and South Europe, 225-239. All Rights Reserved.
Copyright © 1976 by D. Reidel Publishing Company, Dordrecht-Holland.

on the basis of first occurences of plants in the absence of marine fossil invertebrates [1-3]. Correlation
of the middle Conemaugh group marine Ames limestone as
Virgilian (Stephanian) has been done on the basis of
conodonts and fusulinids [4.5]. Several fossil vertebrate horizons in the interval from the Allegheny,
through the Dunkard have been closely correlable with
vertebrate faunas in Illinois, Texas, Nova Scotia and
Europe [6-8]. Information from the fossil plants has
corresponded closely with the vertebrate story.

This paper incorporates information gleaned from
highly fragmentary vertebrate remains extracted and
concentrated from over 50 localities. These localities
span the interval between the lower Kittaning coal,
Allegheny group, and the top of the Greene formation,
the uppermost unit of the Dunkard group. The primary
aim of this work is to outline a refined biostratigraphic zonation of the non-marine upper Appalachian
basin sediments based upon vertebrate fossils, with
correlations to other areas. The work is by no means
complete, either in the field or the laboratory. At
the very least, more adequate environmental representation is necessary, as is, of course, better quality
material.

ACKNOWLEGEMENTS

This work would not have been possible without the
financial support of the Pittsburgh Foundation, as well
as the encouragement of Carnegie Museum. To Dr. James
A. Barlow, of the West Virginia Geologic Survey, whose
keen interest and sharp eye has resulted in a wealth of
Greene formation material, my thanks as well.

GEOLOGICAL SETTING

Pennsylvanian strata of West Virginia and adjacent
states were deposited in two successive basins, bordered
on the east by the unstable folded belt area of the
Appalachian Mountains. The earlier southern basin
thickens to the southwest from central West Virginia,
with the basinal axis near the border with the unstable
area. This basin contains Mississippian, and lowest
Pennsylvanian sediments and lower Pennsylvanian coals,
topped by the Charleston group. The later northern
basin contains the basal Pottsville and the Allegheny
groups, which thicken to the north, overlain successive-

ly by the Conemaugh, Monongahela and Dunkard groups, which thicken to the east and southeast from eastern Ohio through southwest Pennsylvania and northern and central West Virginia. The Charleston group, slightly older than the Pottsville of the northern basin, marks the end of significant deposition southeast of the hinge line between the two basins [9].

The northern Pennsylvanian basin was located on the western continental margin of the tectonically active Appalachian fold belt and received most of its sediments from this source. The northeastern Pennsylvanian anthracite coal field evidently constitutes Pottsville, Allegheny and possibly Conemaugh equivalents involved in immediately subsequent deformation. The northern basin was bordered on the west by the Ohio-Indiana platform of the Cincinnati Arch province, a persistent highland which affected marine communications with the midcontinent seaway further west. Widespread transgressive and regressive episodes from the west-southwest upon the coastal plain punctuated the Allegheny and the lower half of the Conemaugh groups, while deltaic deposition predominated between marine events [10-12]. A cessation of marine influence after the middle of the Conemaugh is accompanied by a shift to a broadening coastal plain regime characterized by river and lake deposition [9, 13-16]. A limited brackish water incursion at the level of the Washington Coal in Ohio marks the last known Paleozoic marine influence in the eastern United States [17].

Virtually every lithologic unit in the Dunkard basin has a name, and the names of units differ from state to state. Names of the major coal units, important marker beds, remain constant, however.

The base of the Allegheny group is drawn at the bottom of the Clarion coal. Lower, middle and upper Kittaning coals, and the lower and upper Freeport coal are contained within it, as well as several marine intervals in the lower half of the unit. The top of the Upper Freeport coal is considered the upper boundary of the group. The Conemaugh group extends from the top of the upper Freeport coal to the bottom of the Pittsburgh coal, and contains a number of small coals of no economic importance. Several marine incursions are present in the lower half of the unit.

The Monongahela group includes the Pittsburgh coal as its base, the Redstone, Uniontown and Waynesburg

coals as well as other locally mineable coals, and the
thick, extensive Benwood limestone sequence in the middle.
The top of the Waynesburg coal, the bottom of the
Cassville shale, is considered the upper boundary of the
Monongahela group.

The Dunkard group is divided into a lower unit, the
Washington formation, and an upper unit, the Greene
formation. The Washington formation extends from the
top of the Waynesburg coal to the top of the upper
Washington limestone, and includes several locally
prominent coals such as the Waynesburg "A", the Washington coal and the Hundred coal, as well as the
locally well developed lower and upper Washington
limestones.

The Greene formation contains no major marker beds
that can be traced with confidence from one hill to the
next, a problem compounded by steep, heavily vegetated
topography and very prominent thinning of beds between
the Washington coal and the limestones in the middle of
the Greene formation. Middle and high Greene formation
limestones are quite fossiliferous. The top of the
Greene formation is covered only with soil; there is no
evidence of further deposition in the area [9,13,18].

Arkle [9,13,18] has extensively documented
stratigraphic and facies relationships in the Monongahela and Dunkard groups. He describes three predominent
facies, grey, transitional and red which are found in
northwest to southeast progression. The grey facies
containing thin bedded shales, sandstones, clays, coals
and fresh water limestones, is characteristic of delta
plain and lower coastal plain environments. Clays
representing soil zones underly the coals and may contain caliches grading into bedded freshwater limestones
of variable extent. Coals may be succeeded by thin
black shales or channel sandstones.

Southward and Southeastward the grey shales and
clays give way to variegated red and yellow shales;
carbonate sequences thin to nodular zones or iron-rich
zones and disappear, and massive channel sands become
more common. Coals become thinner and discontinuous.
This is the transitional facies which marks an environment bordering on an upper coastal plain environment,
the red facies.

The red facies consists chiefly of red and yellow
shales, occasionally containing zones of high-iron

hematite and limonite, and massive to lenticular sandstones. Coals, limestones and grey shales are absent.

Fish, amphibian and reptile remains, usually
dissociated, are common in the fresh water limestones
and carbonaceous shales and clays, as well as in the
black roofing shales of coals. They are far less common
in sand distributary channels and usually absent in the
grey, micaceous flood-plain shales. All vertebrate
material is rare in the red facies.

A progressive, although unsteady spread of the red
facies to the north and northwest takes place through
the upper Conemaugh, Monongahela and Dunkard groups.
The spread of this piedmont environment is interrupted
by several major lake and coal swamp episodes, resulting in the Pittsburgh-Redstone coals, the Benwood
limestone, Uniontown coal, the Washington coal and
limestones, and the Rockport limestones. Climatic
change as seen in the facies change becomes progressively more upland, and consequently somewhat drier, but
no evidence of marked aridity aside from strong
seasonality is present in the sequence.

The vertebrate fossils extracted and analysed from
the Dunkard Basin are almost exclusively from the "grey
facies" or "transitional facies" of Arkle [9] , representing a coastal plain environment, albeit a coastal
plain environment being greatly modified through time
by regression of the sea southwestward. The biostratigraphic zones indicated by the fossil vertebrates do
not correspond to stratigraphic boundaries. They do
correspond roughly to major changes in sedimentary
regimen indicating significant environmental change.

Fossil vertebrates from the fresh water limestones
were extracted by Amine-220 in 10% Formic Acid, sieving
through 60-mesh screen, concentrating and cleaning with
Amine-220 in Methyl Amyl Alcohol, and resieving through
60-mesh prior to picking [16]. All collections have
been deposited with Carnegie Museum, Pittsburgh,
Pennsylvania.

BIOSTRATIGRAPHY

Lower Kittaning Shale, Allegheny Group, through the
Mason Shale, Conemaugh Group, (Missourian, Westphalian)
     Two famous cannel shale vertebrate faunas have

been known from the Appalachian basin Carboniferous for
a century, the Darlington Cannel and the Linton Cannel.
The Darlington Cannel is approximately at the level of
the Lower Kittaning Coal, and is roughly equivalent in
age to the Mazon Creek faunas of southwest Illinois,
while the far richer deposit at Linton, Ohio is directly
overlain by the Upper Freeport Coal and is comparable
in age to Nyrany, Bohemia [8, 19]. Locally, several
minor localities in black shales and one in a freshwater
limestone reinforce the faunal unity of this zone, while
comparable faunas in Illinois, Nova Scotia and Europe
provide an evolutionary scale [8]. All known faunas in
this sequence are eudeltaic; both the Darlington Cannel
and Linton faunas are found in oxbow lakes.

The extensive Linton fauna contains articulated
fish, amphibians and early reptiles, with affinities
to not only members of other Pennsylvanian faunas but
to descent Permian forms. Of these animals, several
are useful in considerations of the Biostratigraphy of
the Dunkard Basin. The acanthodian Acanthodes cf A.
marshi is ubiquitous in both fresh water and marine
deposits from the Allegheny group to the top of the
Greene formation. Acanthodes is known to occur as high
as the base of the Cimarronian (top of the Leonardian)
of Oklahoma [20]. Several pleuracanth sharks were
named by Newberry [6], principally from isolated teeth
supplemented by jaw fragments. These tooth types,
Orthacanthus (diplodus) compressus, O. (diplodus) latus
and Xenacanthus (diplodus gracilis, are distinguishable
and common although no claim can be made for the taxo-
nomic integrity of the two Orthacanthus until Newberry's
material is reexamined. Teeth of an unnamed helodontid
[21] occur in the upper Freeport limestone at the Linton
site, but not from the cannel itself ("Orodus" of
Simpson, [9]). Teeth of this fish also persist throughout the Dunkard basin section, are unknown in local
marine units and are unequivocally fresh water. The
ichthyodurulite Euctenius [22, fig. 207] believed to
be associated with claspers of some chondrichyhyans
[21] has also been reported from the Linton cannel.

Two dipnoans are known from the Linton cannel,
Sagenodus serratus on the basis of jaws (associated
jaws and a skull roof are known from Illinois) and
S. ohioensis from a single skull roof [23]. The
earliest record of the Monongahela-Gnathorhiza group
of dipnoans is a single pterygoid element from the
Upper Freeport limestone at Linton [15, 16]. Lungfish
are unfortunately very rare from either horizon.

Three paleoniscoid fish are found in the faunas
of this zone, Haplolepis, Pyritocephalus, and
Elonichthys peltigerus [19, 16]. Only Haplolepis is
locally common, all three provide links with the Mazon
Creek fishes, and Haplolepis and Pyritocephalus are
among the very rare fish useful in intercontinental
correlation [8], both forms being limited to the
Westphalian of Europe. Haplolepis is found in various
black shales between the roof shale of the Lower
Kittaning Coal and the Mason Shale (=Gallitzen Shale)
below the marine Brush Creek event of the Lower
Conemaugh, as well as in fresh-water Upper Freeport
Limestone, usually accompanied by Rabdoderma cf. R.
elegans, a coelancanth also very commonly found at
Linton.

The nectridean amphibian Diceratosaurus, one of
three common nectridean genera from Linton, is also
known from the European Westphalian. Diceratosaurus
is the apparent ancestor of the horned nectridean
Diploceraspis of younger rocks in this area.

Ewing Limestone, Conemaugh Group, through the Fishpot
Shale, Monongahela Group (Virgilian, Stephanian)

Published fresh water and terrestrial vertebrate
fossils from this interval have been from four
principal horizons, the Round Knob red beds [24], the
Grafton Sandstones [24, 25], the Duquesne Limestone
[15], and the lower Pittsburgh limestone [24]. In
addition, a number of unpublished Carnegie Museum
localities collected by the author, including a
paleonscoid fauna from the Birmingham Shale, give
some representation through this interval.

Several brief marine transgressions punctuate the
lower half of the Conemaugh group. The Ames limestone,
the uppermost correlable bed, has been dated as
Virgilian on the basis of fusulinids and conodonts [4,5].

Strong fluctuations of sea level predominated
through the Morgantown interval, coupled with contemporaneous tectonic activity (Khoury, personal communication) and delta building. Development of a pond and
stream coastal plain topography was a feature of the
post-Morgantown interval, culminating in the vast lakes
of the Pittsburgh Formation. Most of the localities
in the local Virgilian are pond and lake deposits,
except for the Round Knob and the Grafton, which are
riverine deposits.

A thin fresh water shale below the marine Woods
Run limestone, the Ewing limestone and the Rock Riffle
limestone [26], all older than the Ames marine interval,
have produced useful vertebrate material. The Ewing
limestone of Ohio has frequently been confused with the
Rock Riffle, which lies in the Round Knob Red Beds,
(Percy Raymond's Bone Bed), [24], and occurs only as a
caliche through eastern Ohio and southwestern Pennsylvania. Higher on the delta plain, however, the Rock
Riffle occurs as well bedded limestones near Morgantown,
West Virginia. Chondrichthyans from this interval include teeth referable to Xenacanthus gracilis,
Orthacanthus latus, O. compressus, and O. aff.
platypternus. The latter teeth differ from true O.
platypternus in having oval, coarsely fluted cusps.
Hybodus allegheniensis is known from the Mason Shale
upward [15], as is the helodontid. Osteichthyan
remains include Elonichthys aff. E. peltigerus and a
variety of paleoniscoid scales unknown in lower horizons.
Haplolepis and Rhabdoderma are absent.

Terrestrial faunal elements are known, again from
fragmentary material, from the Round Knob red beds at
Adderly, in Turtle Creek Valley, and from Grafton sands
north of Pittsburgh and west of Steubenville, Ohio
[24,25,27]. Large Carboniferous embolomerous amphibia,
Eryops, Astreporhachis ohioensis and Neopteroplax
conemaughensis occur together with the early diadectid
seymouriamorph Desmatodon hollandi and the very tiny
and very early fin-backed pelycosaur Edaphosaurus
raymondi. This fauna, typically eudeltaic [28], is
also typically Stephanian in its association of well
established Carboniferous forms with early representatives of prominent upland Permian tetrapods [29,30].

Lacustrine faunas from the Duquesne Limestones [15]
and Birmingham shale, Clarksburgh limestone, lower
Pittsburgh limestones [24,31] and the roofing shales
of the Pittsburgh Coal, as well as a poor and enigmatic
oxbow lake fauna from the Fishpot shale, give a fairly
good representation of the upper portion of the Virgilian. The chondrichthyans, including Euctenius,
continue without appreciable morphologic change through
the interval, although Hybodus allegheniensis becomes
more common in the upper beds. Sagenodus cf. S.
serratus (not S. periprion [15,24]) is present,
although relatively rare. Monongahela stenodonta
appears to diverge into several polymorphs towards the
top of the Conemaugh group. The most common paleoniscoid is Elonichthys aff. E. peltigerus, which differs

from the Linton form in having unornmented dorsal ridge
scales. The highly ornate, thin and pectinate flank
scales seem undistinguishable from those of the Allegheny
Group fishes. Several other peculiar paleoniscoids
occur, one of these, with cycloid scales and an unornmented skull, having an affinity with Sphaerolepis
of the upper Westphalian and lower Stephanian of
Czechoslovakia [32]. The nectridean Diploceraspis
conemaughensis is known from the Duquesne Limestone and
lower Pittsburgh limestone. The lacustrine faunas
change gradually through to the Fishpot Shale, with the
addition of a large paleoniscoid with highly ornamented
rhombic scales at around the Clarksburgh horizon, the
apparent disappearance of Elonichthys above the roof of
the Pittsburgh Coal, and the establishment of a small
paleoniscoid with thick scales ornamented with few
ridges of enamel.

The basal Virgilian in the Appalachian basin is
marked by an almost complete discontinuity among the
paleoniscoid fish and coelacanths. Elements seem to
be added, presumably by immigration, throughout the
interval. Among the amphibia there is no faunal discontinuity, but pond and stream dwellers evolve from
established Lower Carboniferous groups while a more
terrestrial component emerges which presages Wolfcampian and Leonardian upland forms.

Benwood Carbonate, Monongahela Group, through the Greene
Formation, Dunkard Group (Wolfcampian, Autunian)

Virtually the only faunas known from the Monongahela group are those of mine, and several of these have
yet to be adequately exploited and analyzed. Dunkard
group fossil vertebrates, while quite common [24,33-35]
are as yet based on fragmentary remains and are chiefly
from the Washington Formation.

Sedimentary regimes fluctuate drastically during
this interval, but generally represent coastal plain
environments. Broad, shallow lakes predominated in the
Pittsburgh formation, with no coarse detrital input.
Strong river and pond topography succeeded it in
Uniontown and Waynesburgh formations. The Washington
contained large lakes and rivers, while the Greene
represented predominantly a time of ponds, small lakes
and rivers, small   The last evidence of marine conditions in the region occur at the level of the Washington Coal, Washington Formation. Upland red beds

become prominent near the source area to the southeast,
but there is no evidence for a major climate shift
towards aridity as is seen near the Texas Leonardian-
Guadalupian boundary.

Faunas from above the Fishpot shale are qualita-
tively different from those below, as typical Stephanian
representatives such as Elonichthys aff. E. peltigerus,
Desmatodon, Edaphosaurus raymondi and Diploceraspis
conemaughensis give rise to typical lower Wolfcampian
forms. While there is as yet only one articulated bony
fish from the entire interval, and thus no paleoniscoid
identifications are possible, elasmobranch, dipnoan and
tetrapid evidence is common and qualitative paleniscoid
evidence is obvious.

The earliest terrestrial fauna clearly representative of the Wolfcampian is from the Monongahela "B"
horizon at Elm Grove, West Virginia. The tetrapods
include Diploceraspis burkei, Edaphosaurus boanerges,
a large rhachitome tentatively identified as Edops,
Zatrachys serratus (Norton, inms.) and Lysorophus
dunkardensis. D. burkei is the most common vertebrate
of the Greene formation [33], except perhaps for
Orthacanthus, and Edaphosaurus boanerges, the most
typical, or most easily recognizable pelycosaur of the
Washington and lower Greene formations as well as the
Wichita group of Texas [30]. Zatrachys is confined to
Wichita and lower Clear Fork of New Mexico [36].

Recent additions to our knowledge of the tetrapods
has been among upland forms from the Washington and
lower Greene formations. Two finds have been particularly noteable, that of typical Diadectes [35] and
Zatrachys serratus from several horizons in the Washington and Greene formations (Norton, in ms.). All tetrapod information reinforces the unity of the upper
Pittsburgh, Washington and lower Greene and its similarity to the Wichita and lower Clear Fork Beds of Texas.

Chondrichthyan teeth above the Benwood carbonate
member of the Pittsburgh formation undergo a number of
changes. Xenacanthus cf. X. leudersensis [37] replaces
X. gracilis. Teeth of Orthacanthus latus and O.
compressus shape are succeeded by teeth of the O.
texensis type [38] in what seems to be gradual phyletic
change. Teeth of O. aff. platypternus are succeeded by
teeth closer to O. platypternus. They differ from O.
aff. platypternus in that the cusps are smooth, slightly
compressed and bladed, and there is less size difference

between major cusps. They differ from typical O.
platypternus [38] in having a prominent basal tubercle
and in cusps being somewhat apically divergent. There
seems to be some ecological segregation of O. texensis
and O. cf. platypternus in beds above the Ninevah
limestone, mid-Greene formation, O. texensis being
found more commonly in deeper water or fluviatile
sediments. The pleuracanth sharks of the Dunkard group
thus resemble those of the Wichita and basal Clear Fork
of the southwestern United States. The helodontid
remains persist through the top of the Dunkard. Hybodus
becomes very rare around the middle of the Greene.

The genus Sagenodus is quite common in the Wolfcampian of the Appalachian basin. As in the Wichita
beds of Texas, Sagenodus seems divisible into at least
three species, S. dialophus, S. periprion and S.
porrectus on the basis of tooth plates [23].

Populations of Monongahela stenodonta from the
upper Conemaugh as well as the midcontinental upper
Virgilian of Peru, Nebraska consist of several polymorphs presaging the origins of M. dunkardensis [16],
Gnathorhiza serrata [39]and G. dikeloda [40], as well
as at least two other lepidosirenid taxa commonly found
in the Greene formation. Gnathorhiza serrata does not
persist in the Dunkard Basin, and G. dikeloda is known
from one fragment from the red facies of Belpre, Ohio;
both are commonly known from the Leonardian of Texas
and Oklahoma [40]. Progressive change in characters
of M. dunkardensis toward smaller size and shorter jaws
occurs throughout the Dunkard group. Another diminu-
tive offshoot of the Monongahela stock develops single
bladed upper and lower jaws and evidently a long, narrow
snout; this form is increasingly common toward the
middle of the Greene formation. A large lepidosirenid
divergent from M. stenodonta in the loss of homeostatic
control limiting it to a four ridged upper and three
ridged lower jaw, increases in size through the Washington and Greene formations. While ecological segregation,
and hence lepidosirenid species differentiation of the
various lungfish can be demonstrated by the lower
Dunkard Colvin Run Limestone, both divergence and abundance increase to a zenith in the higher beds of the
Greene formation.

The highest beds of the Greene formation, those
around the Windy Gap limestone, outcrop over a very
limited area and are difficult to trace. The few
localities sampled have yielded fossils consistent

with lower high Greene formation faunas. An isolated
report of <u>Edaphosaurus</u> <u>cruciger</u> [24] from the Greene
formation is regarded as dubious upon reexamination of
the material by Berman (pers. comm.).

The vertebrate faunas of the Dunkard group are
thus consistent from the middle of the Pittsburg formation, Monongahela group, to the uppermost fossiliferous
units of the Greene formation. Most faunal elements,
and evolutionary changes within taxa, where they can
be demonstrated, indicate a strong correlation to
Wichita and lower Clear Fork (Wolfcampian and lower
Leonardian) faunas of Texas and Oklahoma. These in
turn have been amply compared and correlated with
Autunian faunas elsewhere in Eurasia [7,24,30].

REFERENCES

1. W.C. Darrah, American Carboniferous Floras. Second
    Heelen (Holland) Carboniferous Congress Summary;
    Issued from the Botanical Museum, Harvard University,
    1935.
2. W.M. Fontaine and I.C. White, The Permian or Upper
    Carboniferous Flora of West Virginia and Southwestern
    Pennsylvania. 2nd Geol. Survey of Pennsylvania, PP;
    143 pg 1880.
3. I.C. White, Stratigraphy of the bituminous coal
    field of Pennsylvania, Ohio and West Virginia. U.S.
    Geol. Survey Bull 65, 212 p. 1891.
4. G.K. Merril, Geol. Soc. Amer. Spec. Paper 10., 1966.
5. N.G. Lane et al. Geol. Soc. Amer. Mem. 127, 1970.
6. J.S. Newberry. Descriptions of Fossil Fishes. Geol.
    Survey of Ohio, Paleontology, 1(2):247-358, 1873.
7. A.S. Romer. Texas Permian redbeds and their verte-
    brate faunas, in studies on fossil vertebrates,
    T.S. Westoll, ed. London, 1958.
8. D. Baird, A Haplolepid fish fauna in the early
    Pennsylvanian of Nova Scotia. Paleontology,
    5(1):22-29, 1962.
9. T. Arkle, Jr. The configuration of the Pennsylva-
    nian and Dunkard (Permian?) Strata in West
    Virginia: a challenge to classical concepts, in
    some Appalachian coals and carbonates; models of
    ancient shallow-water deposition, Geol. Soc.
    America preconvention field trip guidebook,
    p. 55-88, 1969
10. E.G. Williams, et al. Cyclic sedimentation in the
    Carboniferous of Western Pennsylvania. 29th Field
    Conf. of Pennsylvania Geologists, Oct. 1964.
11. J. Donahue and H. Rollins. Paleoecological Anatomy
    of a Conemaugh (Pennsylvanian) marine event. Geol.
    Soc. Amer. Spec. Paper 148, 1973.
12. A.C. Donaldson ed. Anatomy of an ancient shallow
    water delta, its rocks, fossils, dpositional
    environments and practical applications. Field
    Trip Guidebook for Interstate 79 between Morgantown
    and Fairmont, West Virginia. W. Va. Univ.,
    Morgantown, 1971.
13. T. Arkle, Jr. Monongahela Series, Pennsylvanian
    System, and Washington and Greene Series, Permian
    system of the Appalachian Basin, in Guidebook for
    field trips, Pittsburgh meeting, Geol. Soc. Amer.
    115-141, 1959.
14. A.C. Donaldson, ed. Some Appalachian coals and
    carbonates, models of ancient shallow water de-

position. Geol. Soc. Amer. Preconvention field trip
Guidebook, 348 pp., 1969.

15. R. Lund, Fossil Fishes from Southwestern Pennsylvania,
    I: Fishes from the Duquesne Limestones (Conemaugh
    Pennsylvanian). Ann. Carnegie Museum 41(8):231-
    261, 1970.

16. R. Lund, Fossil Fishes from Southwestern Pennsyl-
    vania, II: Monongohela dunkardensis, new species
    (Dipnoi,Lepidosirenidae) from the Dunkard Group.
    Ann. Carnegie Museum 44(7):71-101, 1973.

17. H.L. Berryhill, Jr. Geology and Coal resources of
    Belmont County, Ohio, U.S. Geol. Survey Prof. Paper
    380:113 pp. 1963.

18. T. Arkle, Jr., I.C. White Memorial symposium Field
    Trip, Log of Trip, Sept. 27-29, 1972. West Virginia
    Geol. Survey, Morgantown, W. Va. 1972.

19. T.S. Westoll, The Haplolepididae, a new family of
    Late Carboniferous bony fishes. Bull. Amer. Nat.
    Hist. 83(1):1-122, 1944.

20. L.C. Simpson, Paleoecology of the East Manitou site
    Southwestern Oklahoma. Oklahoma Geol. Notes 34 (1):
    15-27, 1974.

21. R. Lund. Vertebrate Fossil Zonation and correlation
    of the Dunkard Basin. in I. C. White Symposium: The
    Age of the Dunkard. West Virginia Geol. Survey,
    Morgantown, W. Va. 1972 (in press).

22. G.R. Case. Fossil sharks, a pictorial review.
    Pioneer Litho., N.Y. 1973.

23. A.S. Romer and H.J. Smith, American Carboniferous
    Dipnoans Jour. Geol. 42(7):700-719, 1934.

24. A.S. Romer. Late Pennsylvanian and early Permian
    Vertebrates of the Pittsburgh-West Virginia region.
    Ann Carnegie Mus. 33:47-112, 1952.

25. P.P. Vaugn, A. Platyhystrix- like amphibian with
    fused vertebrae, from the upper Pennsylvanian of
    Ohio. Jour Paleo. 45(3):464-469, 1971.

26. D.D. Condit. The Conemaugh formation in Ohio. Geol.
    Survey Ohio 4th Ser. Bull. 17:5-260, 1912.

27. A.S. Romer. The larger embolomerous Amphibians of the
    American Carboniferous. Bull. Mus. Comp. Zool.,
    Harvard Univ. 128(9):415-454, 1963.

28. P.P. Vaughn, Early Permian Vertebrates from Southern
    New Mexico and their paleozoogeographic significance.
    Contrib. Los Angeles Co. Mus. 166:1-22, 1969.

29. A.S. Romer, The late Carboniferous Vertebrate fauna
    of Kounova (Bohemia) compared with that of the
    Texas Red Beds. Amer. Jour. Sci., 243:417-442, 1945.

30. E.C. Olson, Parallelism in the evolution of the
    Permian Reptilian faunas of the old and new worlds.
    Fieldiana: Zool- 37:385-401, 1955.

31. W.E. Moran, Location and stratigraphy of known occurences of fossil tetrapods in the Upper Pennsylvanian and Permian of Pennsylvania, West Virginia and Ohio. Ann. Carnegie Mus. 33(1):1-45, 1952.

32. B.G. Gardiner, Further notes on paleoniscoid ifshes with a classification of the Chondrostei. Bull. Brit. Mus. Nat. Hist. Geol. 14(5):143-206, 1967.

33. J.R. Beerbower, Morphology, paleoecology and phylogeny of the Permo-Pennsylvanian Amphibian Diploceraspis. Bull. Mus. Comp. Zool. 130(2): 31-108, 1963.

34. E.C. Olson, Trematops Stonei, sp. nov (Temnospondyli, Amphibia) from the Washington formation, Dunkard Group, Ohio. KIrtlandia, 8:1-12, 1970.

35. D.S. Berman, A small skull of the lower Permian reptile Diadectes from the Washington formation, Dunkard Group, West Virginia. Ann Carnegie Mus. 43(3):33-46, 1971.

36. W. Langston, Permian Amphibians from New Mexico. Univ. Calif. Publ. Geol. Sci. 29(7):349-416, 1953.

37. D.S. Berman, Vertebrate Fossils from the Leuders formation, lower Permian of North-central Texas. U-Calif. Pub. Geol. Sci. 86, 1970.

38. N. Hotton, III, Jaws and Teeth of American Xenacanth Sharks, Jour. Paleo 26(3):489-500, 1952.

39. K.J. Carlson, The skull morphology and estivation burrows of the Permian lungfish Gnathorhiza serrata. Jour. Geol. 76:641-663, 1952.

40. E.C. Olson and E. Daly, Notes on Gnathorhiza (Osteichthyes, Dipnoi). Jour. Paleo. 46(3):371-376, 1972.

DIAGENETIC ORIGIN OF CONTINENTAL RED BEDS*

Theodore R. Walker

Department of Geological Sciences, University of
Colorado

ABSTRACT. Convincing counterparts of classic ancient continental
red beds, such as those of Permo-Triassic age in Europe and North
Africa, and those of Pennsylvanian to Triassic age in western interior North America, are forming diagenetically today in deposits
of Cenozoic age in the arid and semi-arid regions of southwestern
United States and northwestern Mexico. Sequences of sediments that
range from Holocene to mid-Tertiary age show critical early stages
of reddening, and reveal that the pigment (hematite) originates
from intrastratal alteration of iron-bearing minerals. Any type
of iron-bearing mineral provides a potential source of iron for
the pigment, but the most commonly occurring parent minerals are
detrital ferromagnesian silicates such as olivine, augite, hornblende, biotite, etc., and detrital and authigenic iron-bearing
clay minerals.
    Owing to the regional aridity, the unstable ferromagnesian
silicates are not destroyed by surface weathering and thus they
become important constituents of the basin sediments. Those minerals which form equidimensional grains, e.g., olivine, augite,
hornblende, etc., occur as accessory minerals throughout the
deposits; micaceous minerals, on the other hand, e.g. biotite and
clay minerals, having low settling velocities, characteristically
are more highly concentrated in fine-grained, low-energy deposits.
In addition, detrital clay minerals collect in the intersticies of
coarser-grained, higher energy deposits, where they are mechani-

* This research has been supported by the National Science
  Foundation (Grant No. 33560) and the University of Colorado
  Council on Research and Creative Work.

H. Falke (ed.), The Continental Permian in Central, West, and South Europe, 240-282. All Rights Reserved.
Copyright © 1976 by D. Reidel Publishing Company, Dordrecht-Holland.

cally infiltrated with influent surface water. Wherever any of
these minerals occur, they provide sources of iron for pigment if
the chemistry of the interstitial water is favorable for the formation and preservation of ferric oxides.

The sediments are not red when deposited, but they redden
with time. The iron-bearing minerals are not in equilibrium with
the migrating ground waters and hence they alter diagenetically
(mainly by hydrolysis) and release iron which, owing to the oxidi-
zing, alkaline nature of the interstitial water, precipitates as
hematite, or a precursor oxide which converts to hematite upon
aging. The mechanically infiltrated clay characteristically
reddens first (within thousands to tens of thousands of years),
probably because it occurs in highly permeable sediments which
frequently are recharged by influent, oxygen-bearing surface water.
Subsequent alteration of the framework silicates releases additional iron which results in more extensive hematite formation and more
intensive reddening. Thin sections and SEM studies reveal two
types of alteration of framework silicates: dissolution, and replacement by clay. Both alterations release iron to the interstitial water. X-ray analyses show that the authigenic daughter
clay is mixed-layer illite-montmorillonite with 0 to 20 percent
non-expandable layers, microprobe analyses show that it is iron-
bearing, and thin sections show that it reddens in situ. Similar
clay also commonly is precipitated interstitially. These authigenic clays together with the reddened, mechanically infiltrated clay,
form bright red matrix analogous to that in ancient red beds.
Other authigenic minerals also commonly form at this stage of diagenesis, such as quartz, adularia, and zeolites. Of these, at
least quartz and adularia commonly are stained red and they contribute to the formation of the red matrix.

After prolonged intrastratal alteration (many tens of millions of years) the more unstable species of ferromagnesian silicates (e.g., olivine, augite, and hornblende) are destroyed completely, leaving no direct evidence that they originally were
present. The authigenic hematite, however, unless dissolved by
subsequent chemical reduction, persists indefinitely and its
euhedral crystalline nature, which is best observed at high-
magnification on polished surfaces or using electron microscopy, confirms its diagenetic origin.

INTRODUCTION

Most major sequences of red beds throughout the geologic
record are continental deposits, and most of them have strikingly
similar characteristics which suggest that they have had a common
origin. For example, they typically are orogenic deposits and they
commonly are highly feldspathic, indicating that source area highlands commonly contain crystalline rocks. They usually consist

mainly of non-marine sediments that occur in sequences which are
hundreds to thousands of meters thick and which, although interrupted by paleo-highlands, have an areal extent covering tens to
hundreds of thousands of square kilometers. They characteristically consist of marginal fanglomerates which grade basinward into
fluvial sandstones, and finally into finer grained sediments that
reflect playa-lacustrine and/or nearshore marine environments.
Most so-called classic red bed sequences contain or are regionally
associated with thick, widespread evaporite deposits. Red beds
typically contain associated drab and variegated sediments, but
wherever red pigment occurs it characteristically is concentrated
in fine-grained deposits (e.g., shale and siltstone) and in the
clayey matrix of medium and coarse-grained deposits (e.g., sandstone and conglomerate).

Considerable controversy has existed concerning the nature
of the conditions that formed the red beds and the controversy
has centered on the origin of the hematite that forms the red
pigment. This paper will develop the hypothesis that the pigment
in red beds having the above-described characteristics, regardless
of the age or location of the deposits, is a product of post-
depositional (i.e., diagenetic) alterations of iron-bearing detrital grains including clay minerals. The evidence presented will
show that these alterations can be documented, step by step, in
analogous red beds which are forming today in the arid regions
of the southern Basin and Range Province of southwestern United
States and northwestern Mexico. The evidence also will show that
the alterations unquestionably continue for many tens of millions
of years after the sediments are deposited, and that they probably
continue indefinitely.

The examples of Cenozoic red beds illustrated herein come
mainly from two locations (Fig. 1): 1) A canyon which the writer
refers to as Cañon Rojo (Fig. 2), located on the west side of the
Cocopah Mountains in northeastern Baja California, Mexico, about
30 kilometers (18 miles) south of the United States border (see
Larson and Walker, 1975, Fig. 1, for a more detailed map). About
350 meters (1000 feet) of unnamed red beds of late Tertiary age are
exposed here. 2) Outcrops on the flank of San Diego Mountain
(Fig. 3) located about 42 kilometers (25 miles) north of Las Cruces,
New Mexico. About 1500 meters (4500 feet) of dominantly red beds
are exposed here. Near the base and near the top of this sequence
are lavas which have been dated at 26 and 9 million years b.p.
respectively, indicating that these red beds are dominantly of
Miocene age. The sequence is part of the Sante Fe Group and has
been named the Hayner Ranch Formation by Seager and others (1971).
In this paper, these deposits are referred to as the Hayner Ranch
red beds.

Fig. 1.   Location map.

Fig. 2.   West flank of Cocopah Mountains, showing location of
Cañon Rojo.   View is looking southeast about 30 kilometers south
of the United States border.

Fig. 3. Type area of Hayner Ranch Formation. View is looking
southwest from south flank of San Diego Mountain, about 42
kilometers north of Las Cruces, New Mexico.

These two locations have been selected for emphasis because
they are particularly informative. Analogous examples, however,
occur at many other places in the southern Basin and Range Province
where erosion has exposed the Cenozoic basin-fill deposits.

Before proceeding with a discussion of these deposits and
the evidence supporting the above-mentioned hypothesis, a few
basic principles need emphasis.

BASIC CONCEPTS

Mineral Stability Law

A fundamental geologic principle which commonly seems to be
ignored, overlooked, or underrated by geologists is the so-called
law of stability of minerals (Keller, 1969), which states that
minerals are stable (i.e., in equilibrium with their surroundings)
only in the environment in which they form. Or, stated in another
way, if changes occur in the environment in which a mineral has
formed, or if the mineral is transported to a different environment, it is not likely to be in equilibrium with its new surroundings and it will tend to change to a new form which is stable
under the new conditions. This principle is accepted universally
to explain processes of surface weathering, but many geologists
seem to overlook the fact that the principle also applies intra-

stratally wherever ground water migration occurs. Indeed, it
seems axiomatic that grains of silicate minerals such as feldspars, ferromagnesian silicates, etc. which form under conditions of high temperature, high pressure, and a deficiency of
oxygen, carbon dioxide, and water, are not in equilibrium with
migrating subsurface water, either above or below the water table
where temperatures and pressures are low, and oxygen, carbon
dioxide, and water commonly are abundant. In this paper it will
be shown that intrastratal alteration of these minerals is commonplace, and that the alterations play an important role in the
formation of the pigment in red beds.

Hydrolysis of Silicate Minerals

The process by which silicate minerals alter intrastratally
is primarily hydrolysis, and water is the only reagent needed to
accomplish the alteration (see Krauskopf, 1967, p. 113-116 for a
concise discussion of the process). The alterations occur
wherever silicate minerals are in contact with migrating water
with which they are not in equilibrium, and assuming that the
soluble products of the alterations are removed either in solution
or by precipitation of authigenic minerals, the alterations
should continue as long as remanents of the minerals remain in
the sediments. The most common unstable minerals are ferromagnesian silicates such as augite, hornblende, biotite, etc., which
are rich in iron, and therefore as these minerals alter much
iron is set free and is thus available for forming hematite
pigment wherever the interstitial chemical environment is favorable for precipitation of ferric oxides.

The rate at which these minerals alter is slow even in
geologic terms. The writer's studies of Cenozoic deposits indicate that although the alterations begin immediately following
deposition, complete removal requires many tens of millions of
years, and some minerals, such as biotite, may persist indefinitely. The potential for the formation of hematite pigment in
sediments that contain iron-bearing silicates at the time of
deposition therefore persists for a geologically long time.

A major point to be emphasized here is that conditions po-
tentially are favorable for the formation of red beds at any time
that iron is set free by the alteration of any type of iron-
bearing mineral. Whether or not red beds develop, however,
depends upon whether or not the interstitial chemical environment
is favorable for the formation and preservation of hematite, as
described in the following discussion.

Hematite Stability

The pigment in red beds consists predominantly of hematite, although many red beds, particularly those that are in the early stages of development, also contain ferric hydrate precursors of hematite which convert to hematite upon aging (Walker, 1967; Berner, 1969; Langmuir, 1971).   It follows, therefore, that the pigment in red beds reflects interstitial conditions that are in equilibrium with hematite.   At our present state of knowledge, we may not know all of the factors involved in creating favorable conditions, but we do know that Eh and pH are critical, and they may be the controlling factors.

A diagram showing the stability of hematite in relation to Eh and pH is shown in Fig. 4.   When applied to geologic situations, this diagram shows that iron released by intrastratal hydrolysis can remain in solution as ferrous iron or precipitate as ferric oxide, depending on the Eh and pH of the water.   If the interstitial environment lies in the stability field of ferrous ions, the iron either remains in solution and migrates with the interstitial water, or it precipitates as a ferrous iron-bearing authigenic mineral (e.g., pyrite, siderite, many clay minerals, etc.).   Sediments with such interstitial environments will be gray or drab colored and they may become partly depleted in iron.   If, on the other hand, the interstitial environment is

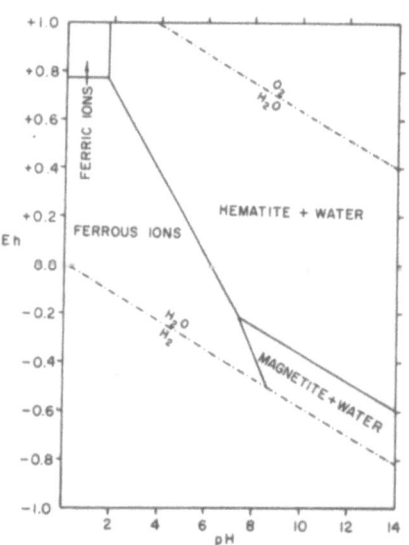

Fig. 4.   Stability fields of hematite and magnetite in water. After Garrels and Christ, 1965.

such that the Eh and pH of the water lies in the stability field
of hematite, the iron will precipitate interstitially as hematite,
or as a precursor oxide that ultimately ages to hematite.   Such
sediments are destined to redden with time.   The chemistry of
interstitial water does not remain constant indefinitely, and
hence changes in Eh and pH may occur at any time during the later
history of the sediment.   Such changes may cause later bleaching
of sediments that formerly were red, or later reddening of
sediments that formerly were drab colored.   For example, a
decrease in either Eh or pH of sufficient magnitude to shift the
interstitial environment from the stability field of hematite into the stability field of ferrous ions (Fig. 4), would cause any
iron released by subsequent intrastratal alteration to remain
in solution, and also cause previously precipitated authigenic
hematite to dissolve.   Under these conditions, further reddening
would be halted and the previously reddened sediments would begin
to lose their color.   Conversely, an increase in Eh and pH of
sufficient magnitude to shift the environment from the ferrous
ion field into the hematite field (Fig. 4), would have the opposite effect; that is, iron released by subsequent alteration, as
well as that already in solution, would precipitate as hematite
(or its precursor oxide) and the previously drab colored sediments would begin to redden.

Significance of Common Association of Red Beds and Evaporite
Deposits

It is well known that throughout the world most major sequences of red beds are associated with extensive evaporite
deposits.   Outstanding examples are red beds of Permo-Triassic
age in Europe and North Africa and those of Pennsylvanian to
Triassic age in western interior North America--and there are
many others.   The evaporite deposits have important implications
that must be considered in interpreting the origin of the associated red beds because they clearly indicate that regionally
arid climates prevailed at the time of deposition.   The association does not prove that all red beds form in deserts (see
Walker, 1974), but it does convincingly indicate that arid
climates are particularly favorable for their formation.   It
follows therefore that modern deserts should contain evidence of
the early stages of formation of red beds that are counterparts
of these classic ancient examples.   It is important therefore to
focus attention on geologic processes that are active in modern
deserts.

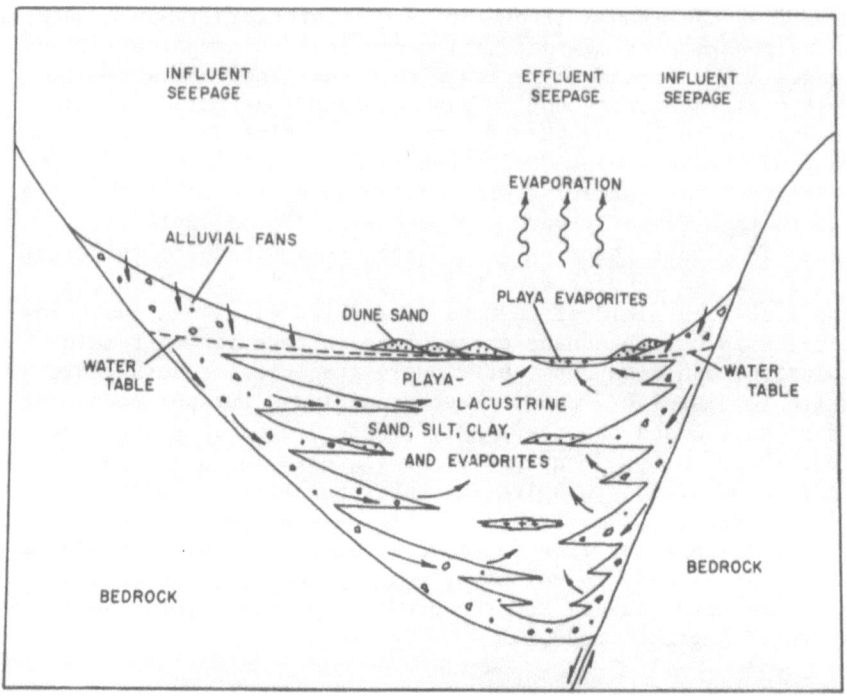

Fig. 5. Schematic diagram of desert basin showing sedimentary
facies and water circulation. After Davis and DeWiest, 1966.

Sedimentology and Hydrology of Desert Basins

       The sedimentary facies and paths of ground water circulation
that typically occur in desert basins are portrayed schematically
in Fig. 5. No dimensions are implied in the diagram; the basin
may be a narrow trough or hundreds of kilometers across, the
marginal highlands may be high mountains or low-lying hills, and
the sediments may be a thin veneer or thousands of meters thick.
Regardless of the dimensions, the sedimentary facies and the paths
of ground water circulation will conform by and large to predict-
able patterns, as explained subsequently.

       The margins of desert basins characteristically are underlain
by coarse-grained alluvial deposits (fanglomerates) which are
composed of detritus derived locally from the adjacent highlands.
These deposits typically grade basinward into finer grained alluvial deposits and finally into playa-lacustrine sediments that
commonly include bedded evaporites. Micas and clay minerals,
both of which may be important sources or iron, typically are concentrated in the low-energy basin-center sediments. In addition,

sands may be locally concentrated into dunes. Owing to the regional aridity, very little chemical weathering occurs in the source
area highlands and consequently, wherever the provenance lithologies contain unstable silicate minerals, these minerals persist in
the sediments that are transported into the basin. Moreover, these
unstable minerals are not likely to be destroyed by surface weathering within the basin because on a geologic time scale deposition
in the basin is essentially continuous, and consequently there is
inadequate time for much weathering to occur at the surface of the
deposits before they are buried by younger sediments. Desert
basins therefore are ideally suited for the preservation of unstable minerals as the sediments initially accumulate.

Owing to the high permeability of the alluvial fans, water
table gradients on the margins of the basins characteristically
are low and the depth to the water table, particularly in the
upper reaches of the fans, may be many tens of meters below the
ground surface. Surface drainage in these areas is ephemeral
and influent and therefore whenever runoff occurs water flows
into the alluvium. Effluent seepage, on the other hand, occurs
closer to the basin center where the saturated zone lies nearer
to the surface and ground water is lost to the atmosphere by
evaporation and evapo-transpiration. These conditions of influent seepage on the basin margins and effluent seepage nearer
the basin center results in the migration of ground water from
the margins of the basin toward the center along the flow lines
shown by arrows in Fig. 5.

Under these geologic conditions two diagenetic processes
occur that play important roles leading to the ultimate reddening
of desert sediments. 1) Influent surface water carries suspended clay into the permeable alluvium, and the clay, which
characteristically is not red when infiltrated, reacts with
oxygenated, alkaline subsurface water and becomes reddened post-
depositionally. 2) The prolonged contact of unstable iron-
bearing framework silicate grains with migrating ground water
causes these framework minerals to alter intrastratally and release iron which in oxygenated, alkaline interstitial environments
precipitates as hematite (or a precursor ferric hydrate), causing further reddening. For convenience, in the following discussion reddening associated mainly with alteration of the infiltrated clay is designated as "early diagenetic" if it occurs without
pronounced alteration of the framework grains, and reddening
associated with extensive alteration of the framework grains is
designated as "late diagenetic." Admittedly, the distinction is
arbitrary and subjective, because the processes involved in the
reddening associated with both types of alteration are similar
and gradational.

MECHANICAL INFILTRATION OF CLAY AND THE FORMATION OF EARLY
DIAGENETIC PIGMENT

Mechanical infiltration of clay into permeable sediments
occurs wherever influent seepage accompanies alluviation. It is
a particularly important mechanism for producing secondary matrix
in desert alluvium because water tables in deserts characteris-
ticallly are low and thus whenever runoff occurs streams characteristically are influent. The interstitial pore spaces through
which the infiltrating water passes are many times larger than
the clay particles and therefore suspended clay is not filtered
from the water. If ponding occurs, infiltration into the alluvium may be inhibited locally and temporarily by accumulation of
silt and clay on the bottom, but such natural filters have little
significance because they occur only locally and normally they
are removed during the initial stages of subsequent runoff.

Field studies coupled with electron microscopy and petrographic studies by the writer and Anthony J. Crone, show that
mechanically infiltrated clay is an almost ubiquitous constituent of coarse alluvium (i.e., sand, sandy gravel, gravelly
sand, etc.) of Quaternary and late Tertiary age throughout the
arid regions of North America. However, clay normally is not
present in the intersticies of analogous freshly deposited
alluvium, because owing to its much lower settling velocity the
clay stays in suspension and moves with the surface water as the
heavier coarser particles lag behind. Examination of samples
collected from pits dug into Holocene alluvium, however, reveals
that infiltrated clay commonly occurs within a few feet below
the surface and it commonly is concentrated on the upper surfaces
of buried pebbles where it accumulates by gravity-settling and
produces geopedal fabric (Crone, 1975). Such clay represents
early stages of accumulation resulting from modern infiltration
that inevitably occurs during each period of surface runoff.
Clay of similar origin normally is much more abundant in analogous
older alluvium (i.e., Pleistocene and late Tertiary age), such
as that exposed in deep excavations in alluvial fans or in the
walls of incised arroyos. In such deposits the clay commonly
forms well developed coatings on framework grains and the coatings
can be readily seen under a ten-power hand lens.

Clay which has been mechanically infiltrated has a clastic
texture which is readily recognized when samples are examined
using a scanning electron microscope (SEM). The clay is composed
of mixed sizes of platelets which in early stages of accumulation
are concentrated on framework grains either as sinuous ridges
and bridges which form at the menisci between films of pellicular
water (Fig. 6), or as upper-surface coatings which form geopedal
fabric (Fig. 7). Textures and structures similar to these have

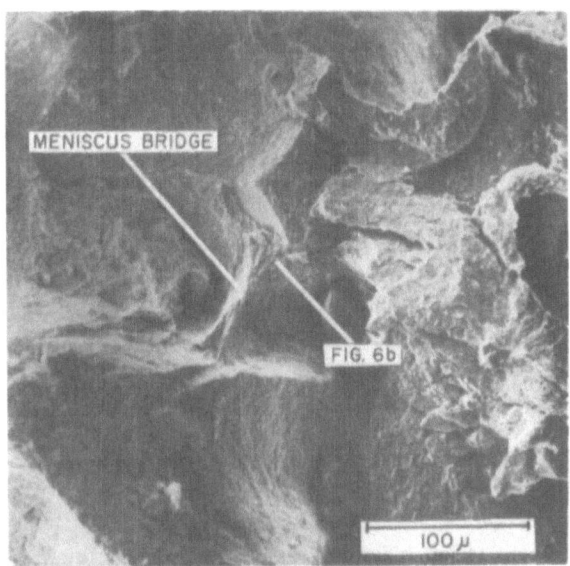

Fig. 6a. Mechanically infiltrated clay forming meniscus bridge and grain coatings. Pleistocene alluvium in dissected terrace, Cañon Rojo. Sample BE-33b. SEM photomicrograph.

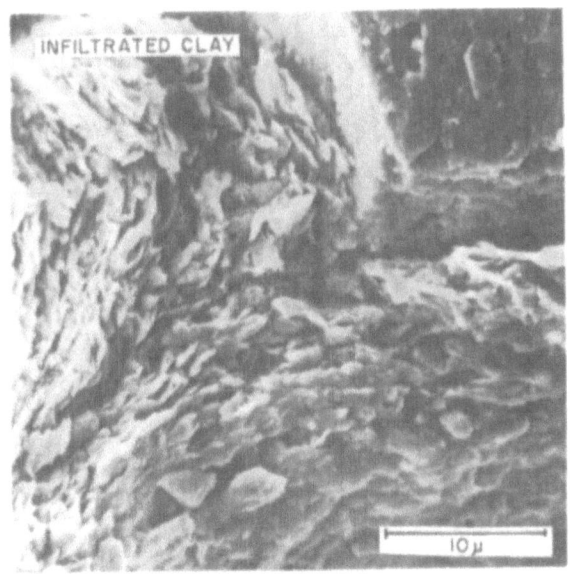

Fig. 6b. Close-up of Fig. 6a, showing clastic texture of clay. Color of this clay is 7.5YR 6/4 to 7/4 (pink to light brown).

Fig. 7. Early stage of accumulation of mechanically infiltrated
clay (dark coatings) showing geopedal fabric and meniscus bridges
between framework grains. Note general absence of clay on
bottom surfaces of grains. Pleistocene alluvium in dissected
terrace, Cañon Rojo. Sample BE-33b. Thin section photomicrograph.

been produced artificially by infiltrating clay-rich water
through sand in laboratory experiments, and the similarities
provide reassuring evidence that the clay in the natural deposits
has originated by infiltration (Crone, 1974; 1975). In advanced
stages of accumulation, the clay surrounds the grains in continuous coatings composed of aggregates of descrete platelets
oriented generally parallel to grain surfaces (Fig. 8). In
time, presumably owing to repeated wetting and drying, the
density of the clay increases but its original clastic texture,
although obscured, is still recognizable (Fig. 8). Existing
data are inadequate to determine how long the clastic texture
persists, but it must persist for many millions of years because
it is clearly detectable in red beds of Miocene age in the Hayner
Ranch Formation.

    X-ray analyses show that the infiltrated clay is composed of
a mixture of clay minerals that includes the same types as those
occurring in the source areas. For example, the interstitial
clay typically has the same mineralogy as clay collected from
sun-cracked mud on the surface of modern alluvium derived from
the same source areas.

    The infiltrated clay is irregularly distributed in the sedi-

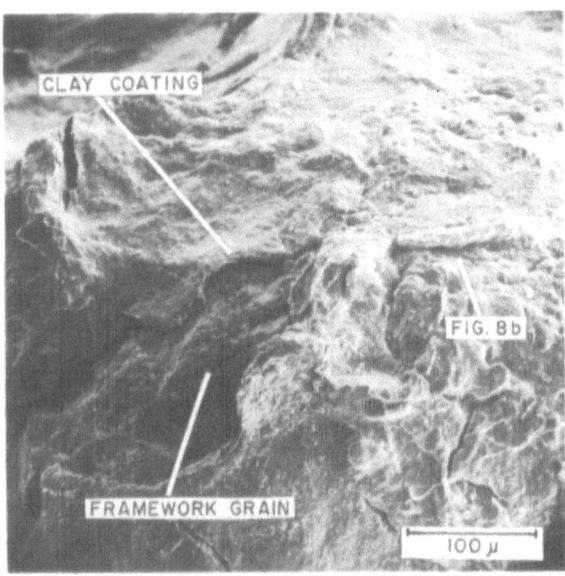

Fig. 8a.   Coating of infiltrated clay on framework grains.
Pleistocene alluvium in dissected terrace, Cañon Rojo,   Sample
BE-33a.   SEM photomicrograph.

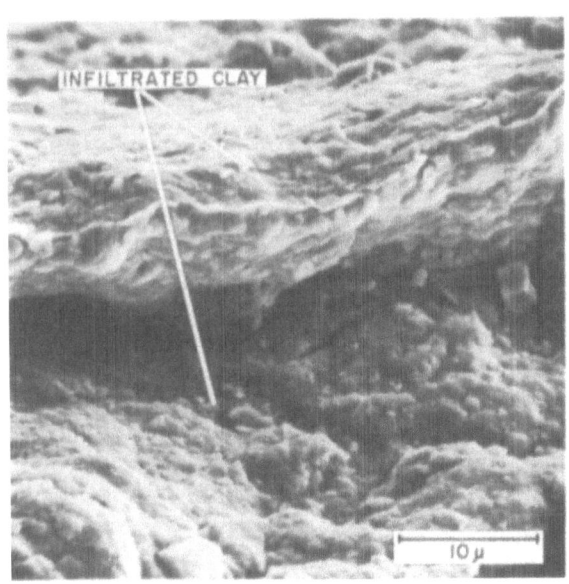

Fig. 8b.   Close-up of Fig. 8a, showing dense clastic texture.
Compare with Fig. 6b.   Color of this clay is 5YR 6/4 (light
reddish brown).

ments because of the nonuniform nature of influent seepage.
The clay accumulates whever interstitial conditions allow it
to settle from suspension. The amount that accumulates depends
in part on the nature and persistence of these conditions, in
part on the amount of clay suspended in the infiltrating water,
and in part on the frequency of episodes of influent seepage.
Following are four inferred mechanisms that explain patterns of
distribution of infiltrated clay that commonly occur in Cenozoic
deposits in the arid regions of North America.

     Concentrations in vadose zone. If influent surface water
enters sediments in which films of pellicular water are not
fully developed on the framework grains, the available water will
be consumed in replenishing these films and it will not reach
the water table (Fig. 9a). Under these conditions the suspended
clay remains with the pellicular water and accumulates on the
surfaces of the grains in coatings composed of aggregates of
clay platelets oriented parallel to the grain surfaces and in
ridges and bridges that traverse the grains along the menisci
between adjacent films of pellicular water (see Figs. 6 and 8).
This process is important wherever and whenever influent seepage
of surface water is inadequate to reach to the water table. The
infiltrated clay accumulates from the surface downward and the
thickness of the interval of accumulation depends on the depth of
penetration of the influent seepage, which of course varies with
each period of runoff. If the sediments near the surface periodically are reworked, as would be true in drainageways and in areas
between drainageways affected by sheetflow, the clay will be removed from the upper sediments by each successive flow and the
upper part of the deposit will be essentially clay-free. Under
these conditions, accumulation of infiltrated clay begins below
this active zone.

     Concentrations near the water table. Outcrops of the
Cenozoic deposits in places display concentrations of infiltrated
clay in nearly horizontal layers which cut across primary bedding
structures, suggesting that they reflect former positions of the
water table. The concentrations can be explained by differences
in the velocity of water migration above and below the water
table. For example, gravity water (i.e., water in excess of
that required for pellicular films) moves relatively rapidly in
the vadose zone because it moves under the influence of a vertical gradient. Once the pellicular films of water are replenished,
the excess water, along with the suspended clay contained in it,
migrates rapidly downward until it reaches the saturated zone.
Within the saturated zone, however, the migration rate is much
slower because here ground water moves under the influence of
the nearly flat water table gradient. The infiltrated clay
therefore, once it reaches the water table, has time to settle
from suspension and it tends to become concentrated in the upper

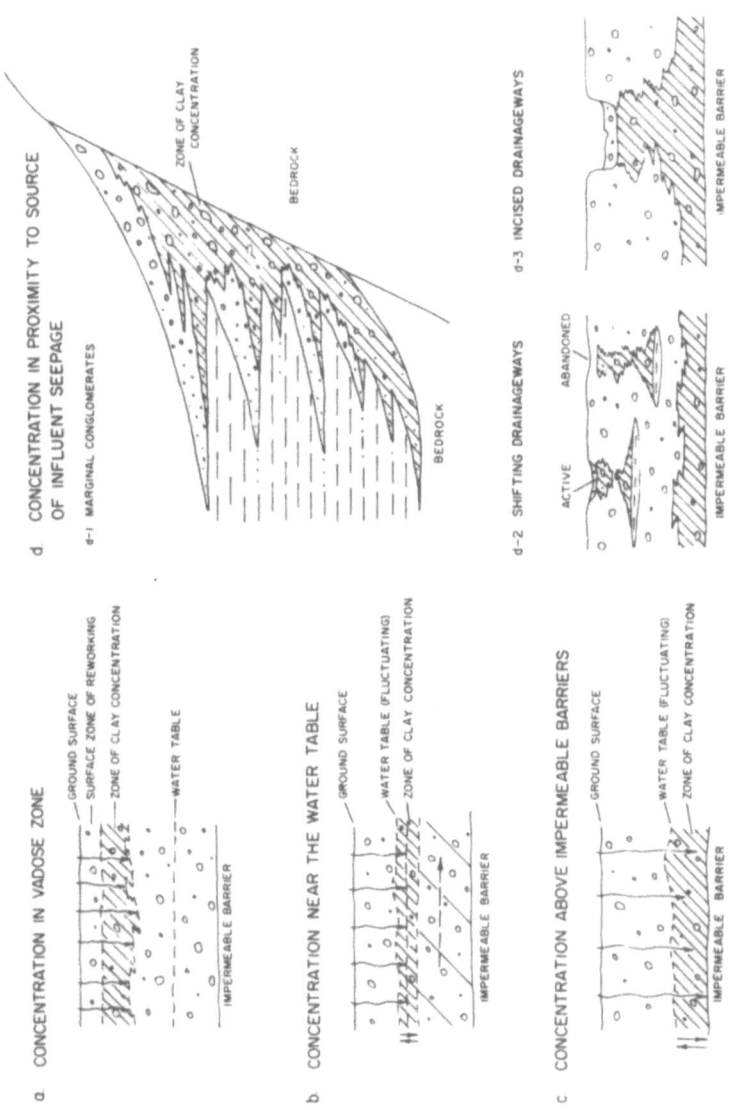

Fig. 9. Schematic portrayal of various conditions leading to concentration of mechanically infiltrated clay.

saturated zone where the decrease in velocity first occurs
(Fig. 9b). Prolonged maintenance of the water table in nearly
the same position (i.e., without major seasonal fluctuations)
should enhance such concentrations of clay within a relatively
narrow vertical range.

Concentrations above impermeable barriers. Wherever influent seepage occurs in coarse-grained alluvium, infiltrated
clay characteristically is concentrated above layers that serve
as barriers to the downward migration of the influent surface
water (Fig. 9c). These barriers may be the bedrock below the
alluvium, beds of fine-grained sediments within coarser, more
permeable alluvium, cemented zones within the alluvium, buried
soils, or any other type of material that would tend to retard
the downward movement of the water. During periods of intermittent surface runoff, perched water tables tend to develop
above these barriers, creating conditions similar to those just
described that allow the infiltrated clay to settle from suspension. During dry intervals, on the other hand, influent
seepage temporarily ceases, the perched water tables subside (the
saturated zones above the barriers may drain completely), and the
remaining suspended clay becomes concentrated in pellicular
films and forms clay coatings, bridges, etc., as in the vadose
zone described heretofore. As these events are repeated again
and again during alternating periods of runoff and drought, a
great deal of infiltrated clay accumulates in the intermittently
saturated zone and ultimately it may fill or nearly fill interstitial voids in the alluvium above the barriers.

Concentrations in proximity to sources of influent seepage.
Inasmuch as infiltration of clay occurs whever influent seepage occurs, the amount of clay infiltrated should generally increase in proximity to principal recharge areas. For example,
coarse-grained marginal conglomerates are likely to contain
more infiltrated clay than finer-grained distal sands because
nearly every period of runoff recharges the marginal conglomerates, whereas only major storms provide enough runoff to extend
to the distal areas (Fig. 9d-1). Similarly, more clay is infiltrated into alluvium located beneath ephemeral drainageways than
into alluvium located beneath divide areas between the drainageways (Fig. 9d-2). As channels shift position during alluviation,
the location of influent seepage similarly shifts, causing infiltration to essentially cease in one place and begin anew in another (Fig. 9d-2). Where channels are incised, however, they
tend to remain in essentially the same position and thus they
recharge the same area for long periods. This condition permits
the underlying alluvium to accumulate larger amounts of infiltrated clay than can accumulate in areas beyond the influence
of the incised channels (Fig. 9d-3).

All of the above-listed conditions probably occur repeatedly during the deposition of thick sequences of desert alluvium. Hence the distribution of infiltrated clay within such deposits is likely to be highly irregular.

The infiltrated clay is not red when deposited, but it reddens with time. Within a geologically short period of time this clay begins to give the deposits a reddish-yellow color which in early stages of development usually conforms to one or more of the patterns just described. It is this type of early diagenetic reddening that is mainly responsible for the reddish-yellow (about 5YR 6/6 on the Munsell Soil Color Chart) pigment that characterizes so many of the younger (i.e., Plio-Pleistocene age) sediments in the Santa Fe and Gila Groups throughout New Mexico and Arizona. Excellent examples are exposed in the valley of the Rio Grande and its tributaries between Taos and Santa Fe, New Mexico (Fig. 1), as well as almost all other places where sediments of this age are exposed in southwestern United States.

The following conditions are inferred to explain the reddening. The clay platelets are more or less continuously bathed by subsurface water which intermittently is replenished by renewed infiltration. The water is oxygenated when it enters the sediment because there is little or no decaying organic matter to consume oxygen either at the surface or within the sediments. Thus the water provides a medium for post-depositional alteration of the clay and a source of oxygen for the formation of iron oxides. Moreover, for reasons explained heretofore, the clay commonly is concentrated in areas where sediments are alternately saturated and aerated, and during the periods of aeration the availability of oxygen is increased. Conditions such as these, in which supplies of both water and oxygen are constantly being renewed, are ideally suited to cause any iron that may be associated in any way with the clay to ultimately form iron oxide and cause the clay to redden.

A reasonable estimate of the nature of occurrence and amount of iron associated with the clay at the time of initial infiltration can be obtained from studies of clay in associated modern alluvium. Previous work (Walker and Honea, 1969) has shown that clay in the modern alluvium throughout the desert contains iron, and that the iron occurs in part as non-red, limonitic coatings (average amount is less than 1.0 percent), and in part as a constituent of the clay mineral lattice (average amount is about 4.5 percent). Both types of occurrences provide important sources of iron for the red pigment.

The earliest detectable reddening occurs in deposits of probable late Pleistocene age (e.g., those shown in Fig. 6) and it may simply reflect aging of the limonitic coatings (Walker, 1967; Berner, 1969; Langmuir, 1971), but in time, as the clay platelets are repeatedly, or in some instances continuously, bathed by the migrating subsurface waters, iron from within the clay mineral lattice probably is released and contributes to further reddening. That iron released by alterations of the clay minerals plays an important role in much of the reddening is indicated by the fact that infiltrated clays are reddest in locations where independent evidence indicates that they have been in prolonged contact with ground water. For example, in deposits showing early stages of reddening, the interstitial infiltrated clay is the first constituent of the sediment that becomes reddened, and invariably it is much redder above imper-meable barriers where perched water tables are likely to have occurred, than it is in the overlying vadose zone (the clay shown in Fig. 8 is an example).

The mineralogy of the earliest formed red iron oxide pig-ment is uncertain. X-ray traces of concentrates of the reddened clay typically show no evidence of hematite, and examination of the clay with a SEM at 50,000x magnification shows no evidence of hematite crystals on the surface of the clay. The writer concludes that the pigment at this stage is either hematite which is present in amounts that are too small to be detected by X-ray analysis and too finely crystalline to be detected at even these high magnifications, or more likely, it is an amor-phous reddish precursor of hematite. Later in the diagenetic history of these red beds, however, crystalline hematite is commonplace, and it can be identified both by X-ray analyses and by SEM studies, as will be shown in the subsequent discussion.

INTRASTRATAL ALTERATION OF FRAMEWORK SILICATE MINERALS AND THE FORMATION OF LATE DIAGENETIC PIGMENT

Owing to the essentially continuous circulation of ground water from the margins to the central parts of the basins (Fig. 5), the detrital silicate minerals that comprise the framework sand grains continuously are bathed by water with which they are not in equilibrium. Consequently, these unstable minerals undergo alterations and form new minerals which are in equilibrium with the interstitial water. Studies of first-cycle arkoses collected from throughout southwestern United States and northwestern Mexico, indicate that the minerals gener-ally alter instrastratally in the order of Goldich's Weathering Series (Goldich, 1938), the most unstable minerals altering first. For example, olivine, which is a common original con-stituent of sediments derived from basaltic source rocks, alters first, followed by augite, hornblende, and calcic plagioclase, all of which seem to alter at about equal rates. Sodic plagio-

clase, K-feldspars, and of course quartz, persist longest.
Biotite has variable stability; it commonly shows evidence of
initial stages of alteration where hornblende, for example,
shows advanced alteration, but biotite seems to be able to
persist indefinitely.

Significantly, the most unstable grains are ferromagnesian
silicates. As these minerals alter they release important
quantities of iron, and wherever the interstitial environment
lies in the stability field of hematite (Fig. 4), the iron
precipitates as hematite or as a ferric hydrate precursor of
hematite. The red authigenic ferric oxide commonly forms
aureoles around the altered grains (Fig. 10). The ferromagnesian
silicates continue to provide sources of iron for pigment for
as long as relicts of any of them remain in the sediments.
It is certain that the alterations continue for at least tens of
millions of years after deposition, because relicts of partly
dissolved augite (Fig. 11) and hornblende (Fig. 12 and 13) are
common in red beds in the Hayner Ranch Formation which, on the
basis of associated dated basalts, are known to have been deposited during the interval between 26 and 9 million years b.p.
It is likely, however, that the alterations continue for much
longer intervals, because much augite and hornblende still remain in the dated Miocene deposits and because it is well known
that biotite, which is an important intrastratal source of iron,
persists in various stages of alteration in red beds of all ages.

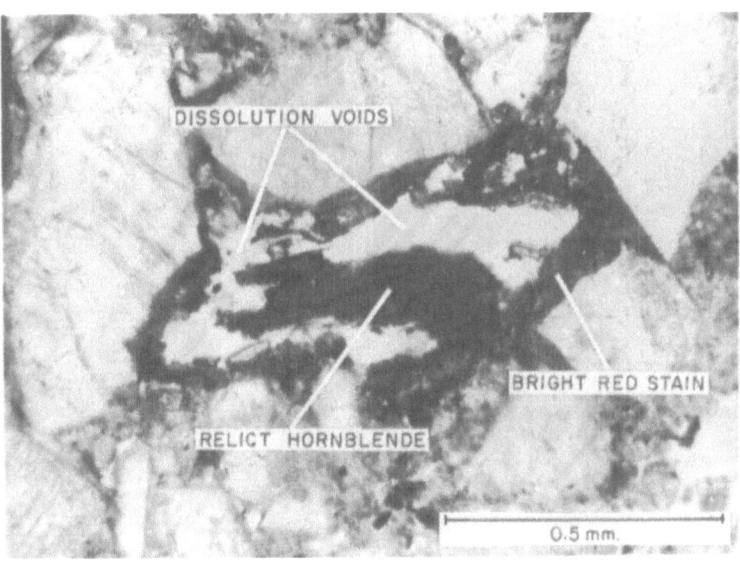

Fig. 10. Relict of dissolved hornblende grain with aureole of
bright red stain. Cañon Rojo red beds. Color of aureole is
9R 3/6 (dark red). Sample WFRB-5-5. Thin section photomicrograph.

Fig. 11. Relict of partly dissolved augite grain. Hayner Ranch
red beds. Color of matrix is 9R 4/4 (weak red) to 9R 4/6 (red).
Sample BC-23a. SEM photomicrograph.

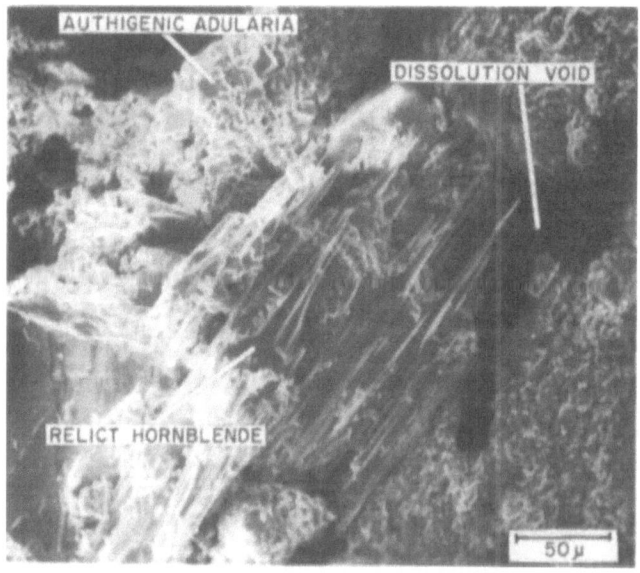

Fig. 12. Relict of dissolved hornblende as seen under SEM. Matrix
contains authigenic adularia. Color of matrix is 10R 5/4 (weak
red). Hayner Ranch red beds. Sample BC-27b.

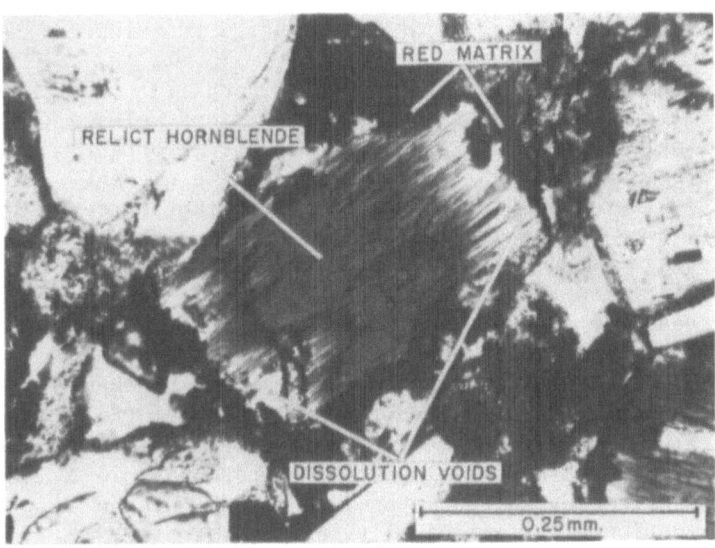

Fig. 13.   Relict of dissolved hornblende as seen in thin section.
Hayner Ranch red beds.   Color of matrix is 8R 4/6 (red).
Sample BC-22.

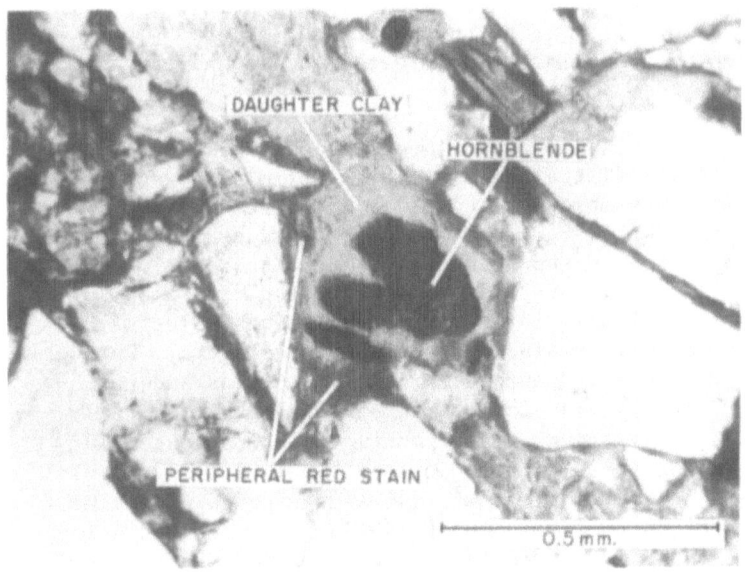

Fig. 14.   Hornblende showing replacement by clay in Cañon Rojo
red beds.   Color of peripheral stain and matrix is 9R 5/6 (red).
Sample BE-32s.   Thin section photomicrograph with reflected light.

Studies by the writer and B. Waugh (1973), show that detrital silicate minerals, including feldspars, undergo two kinds of intrastratal alteration: dissolution (Figs. 10, 11, 12, 13, 18, and 22a), and replacement by clay (Figs. 14, 15, and 17). Where ferromagnesian silicates are present, both alterations release iron to the interstitial waters, and hence both are important mechanisms for providing sources of iron for hematite pigment. These commonplace alterations also have other important effects on the characteristics of the sediments. For example, they decrease the percentage of unstable silicates in the framework fraction and thereby increase the quartz/feldspar ratios. Thus the alterations increase the mineralogical maturity of the sediments diagenetically. Moreover, to the extent that the silicates are replaced by clay, the alterations increase the amount of matrix in the sediments. Thus, the so-called textural maturity (Folk, 1951; Blatt, et al., 1972) of the sediments is decreased diagenetically. In this regard it is important to note that in many places authigenic clay in the Cenozoic deposits, either alone or in combination with mechanically infiltrated clay, produces more than 15 percent interstitial matrix, giving the deposits a graywacky-like texture. It should be emphasized that the high matrix content reflects post-depositional alterations, not original deposition by debris flows. These deposits therefore, although they have a different geologic setting, support the contention of Cummins (1962), that the matrix-rich texture of many ancient graywackes is a diagenetic fabric, not an original fabric. Similar post-depositional mineralogical and textural changes should be anticipated in all ancient first-cycle non-marine red beds.

The authigenic clay in the Cenozoic deposits is readily distinguished from infiltrated clay by its striking boxwork appearance when dry samples are examined with the SEM (compare Fig. 15 with Figs. 6b and 8b). Moreover, the mineralogy of the authigenic clay reflects the chemistry of the interstitial water, and it normally is strikingly different from the mineralogy of the detrital, mechanically infiltrated clay which reflects the types of clay minerals available from the source areas. Throughout the desert regions of southwestern United States and northwestern Mexico, authigenic clay characteristically is randomly interstratified, mixed-layer illite-montmorillonite containing only a small amount (0-20%) of non-expandable layers.* A typical X-ray trace of clay obtained from a sample containing almost exclusively authigenic clay is shown in Fig. 16a. As would logically be expected, many diagenetically altered sediments contain both infiltrated and authigenic clay, and if both are abundant the presence of each can be verified in these geologically young sediments by X-ray analysis (Fig. 16b), and/or by SEM studies (Fig. 17).

---

*Determined by X-ray methods outlined by Reynolds and Hower (1970).

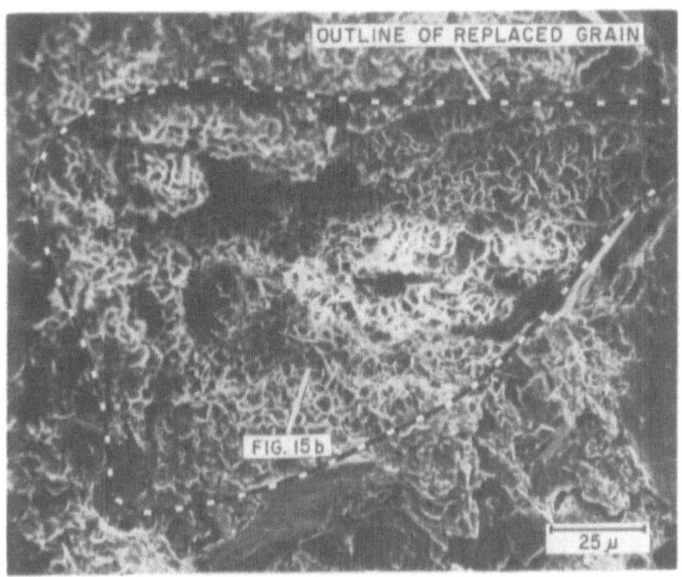

Fig. 15a. Unstained authigenic IM clay with less than 20 percent
non-expandable layers. Cañon Rojo red beds.   Sample BE-32s.
SEM photomicrograph.   Compare with Fig. 14.

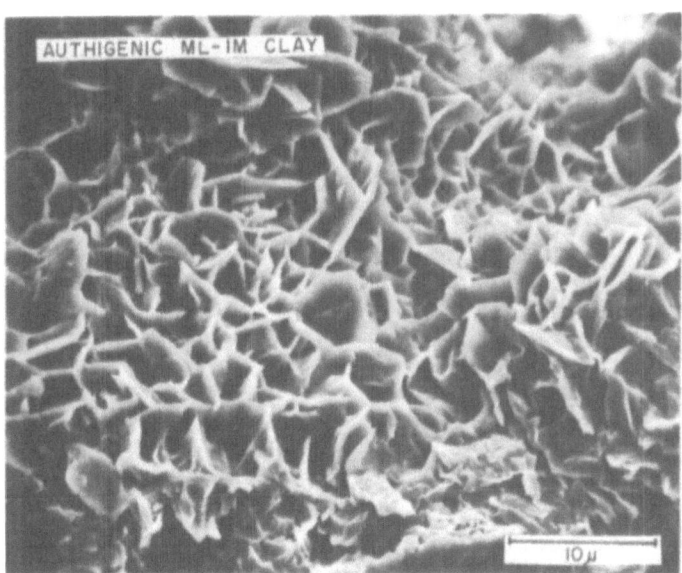

Fig. 15b. Close-up of Fig. 15a, showing boxwork texture of
authigenic IM clay.

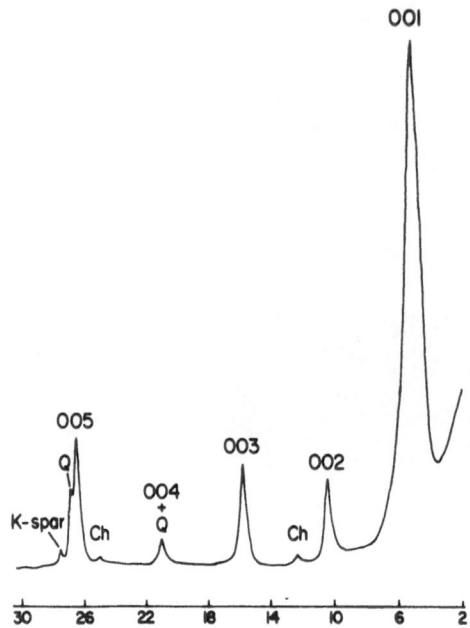

Fig. 16a. X-ray diffraction trace after glycolation of minus-two micron fraction of sample containing almost exclusively replacement IM clay in Cañon Rojo red beds. Sample BE-32s. See Fig. 15.

IM = MIXED-LAYER ILLITE-MONTMORILLONITE
Ch = CHLORITE
 I = ILLITE
 Q = QUARTZ
 G = GOETHITE

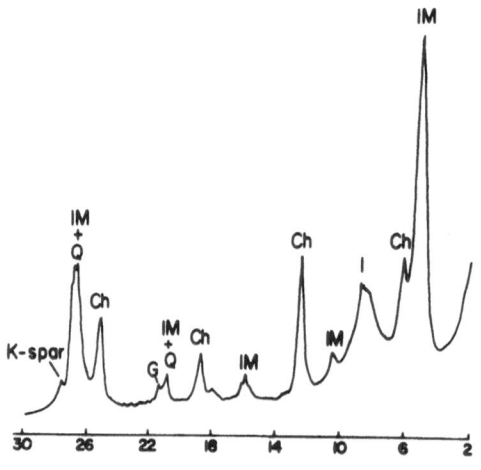

Fig. 16b. X-ray diffraction trace after glycolation of minus-two micron fraction of sample containing mixture of authigenic IM clay and infiltrated clay in Cañon Rojo red beds. See Figs. 17c and 17d.

The authigenic clay, like the mechanically infiltrated clay
described earlier, reddens with time. Thin sections and SEM
studies of samples containing hornblende in advanced stages of
replacement by clay, commonly show that the authigenic clay is
stained bright red. The staining of the clay characteristically
begins on the periphery of the replaced grains and progresses
inward toward the center (Fig. 14), and/or it begins along
fractures or cleavage planes which guided the earlier replacement, and diffuses laterally into the clay. In advanced stages
of staining the daughter clay becomes uniformly stained.

Earlier studies (Walker, Ribbe, and Honea, 1967) have shown
that the daughter clay which replaces hornblende is iron-bearing,
and thus the authigenic clay, like the infiltrated clay described
earlier, can provide its own source of iron for the stain. In
addition, most of the iron originally present in the hornblende
is released as the hornblende alters to clay (Walker, et al.,
1967), thus providing another important local source of iron. The
only other agent needed is oxygen, and apparently, in view of the
widespread occurrence of stained authigenic clay, the interstitial
water provides an adequate source of oxygen if given enough time.

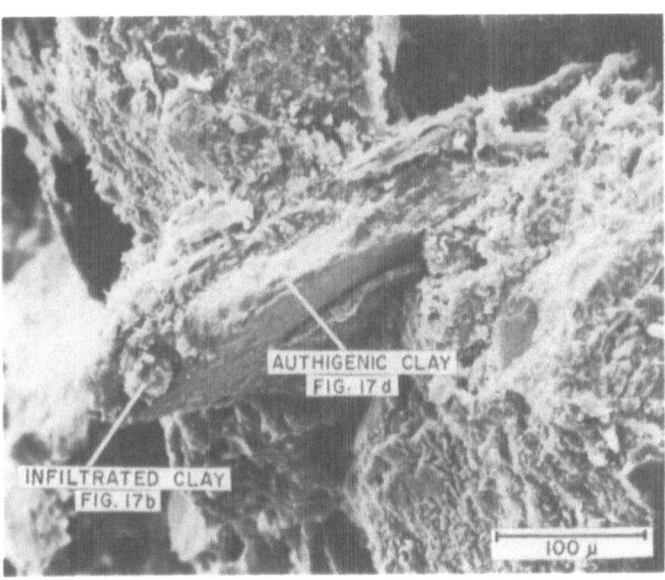

Fig. 17a. Red interstitial matrix containing mixture of authigenic
and infiltrated clay in Cañon Rojo red beds. Sample V-CR-Misc-1.
SEM photomicrograph. See enlargements.

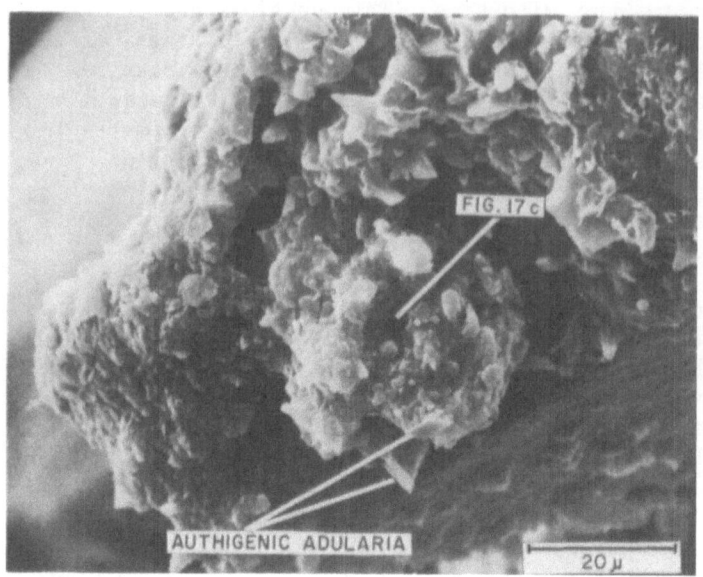

Fig. 17b. Close-up of Fig. 17a, showing concentrate of infiltrated
clay.   Color of this clay is 8R 4/6 (red).

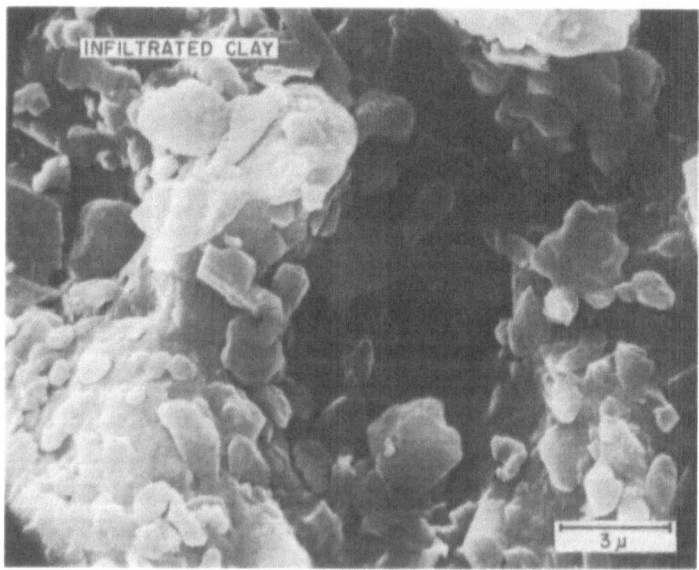

Fig. 17c. Close-up of Fig. 17b, showing clastic texture of
mechanically infiltrated clay.

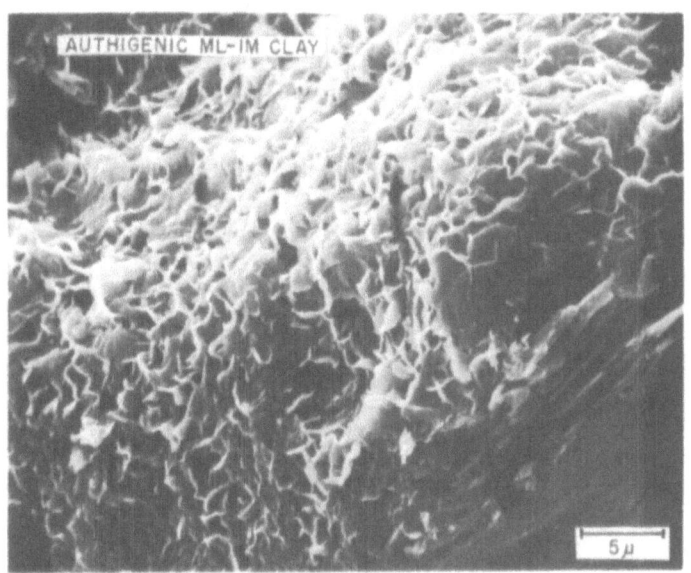

Fig. 17d. Close-up of Fig. 17a, showing authigenic clay. Color
of this clay is 8R 4/6 (red).

At this stage of diagenetic development, bright red hematite crystals commonly are formed which are large enough to be
detected by X-ray analysis and with the SEM. The crystals seem
to form earliest in voids left by the dissolution of the iron-
bearing silicate grains (Fig. 18), and between cleavage lamellae
in biotite (Fig. 19). Energy dispersive X-ray analyses using
the SEM confirm that the crystals shown in Figs. 18 and 19 are
iron oxide, and X-ray diffraction analyses of the minus-two
micron fraction confirm the presence of hematite in both samples.

Also at this stage of diagenetic development, the red beds in
some places contain abundant interstitially precipitated authigenic clay (Fig. 20). X-ray analyses show that this clay, like
that which replaces hornblende described previously, is composed
of randomly interstratified mixed-layer illite-montmorillonite
with a very small percentage of non-expandable layers. Moreover,
energy dispersive X-ray analyses show that the clay is iron-bearing,
and thin sections show that it commonly is stained bright red.
Wherever it is present and stained, the precipitated clay comprises an important component of the interstitial red matrix.
It is important to emphasize therefore that three types of clay
of post-depositional origin are known to occur in the red interstitial matrix in the Cenozoic red beds: 1) detrital clay which
has been mechanically infiltrated into the sediments, 2) authigenic
clay which has replaced detrital silicates, and 3) authigenic clay

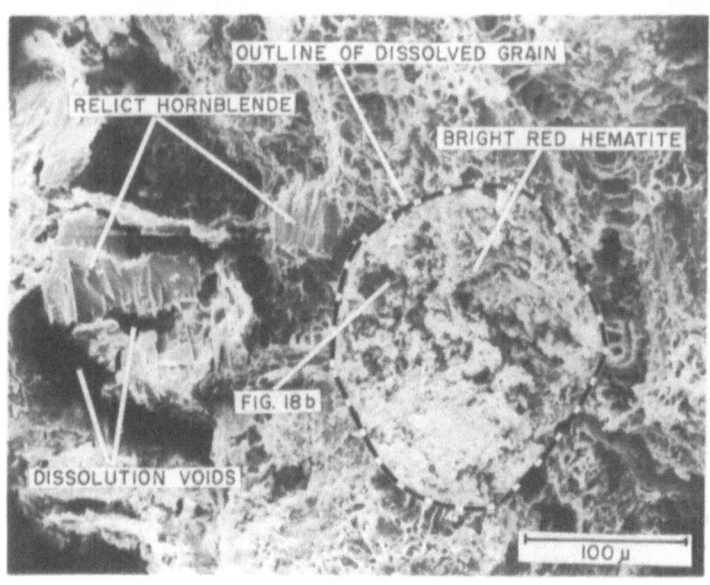

18a. Intrastratally dissolved ferromagnesian silicates in
Hayner Ranch red beds. Color of matrix is 9R 4/4 (weak red) to
9R 4/6 (red). Color of area containing hematite crystals (Fig.
18b) is 7.5R 4/8 (red). Sample BC-23a. SEM photomicrograph.

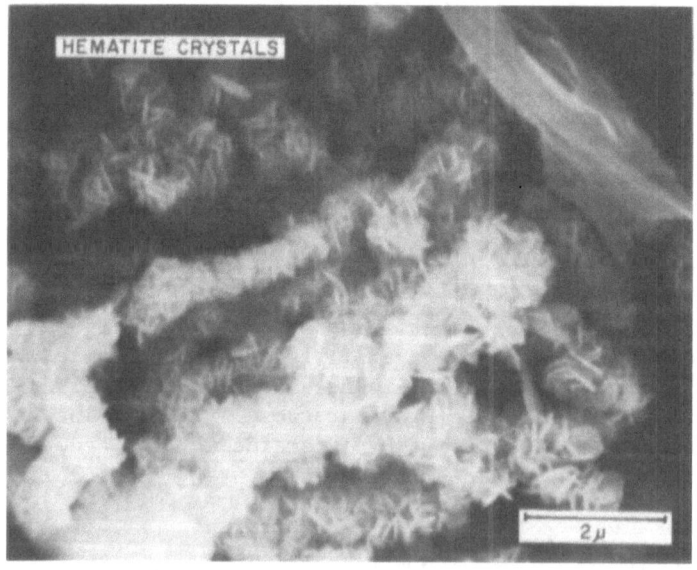

Fig. 18b. Close-up of Fig. 18a, showing authigenic hematite
crystals on wall of void left after dissolution of silicate grain.

Fig. 19a. Authigenic hematite between biotite cleavage lamellae in Cañon Rojo red beds. Color of crystalline hematite is 7.5R 4/8 (red). Sample WFRB-1. SEM photomicrograph.

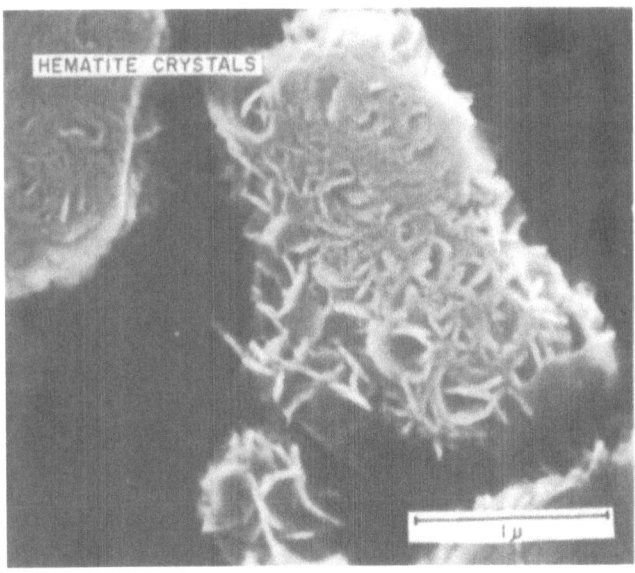

Fig. 19b. Close-up of Fig. 19a, showing finely crystalline authigenic hematite.

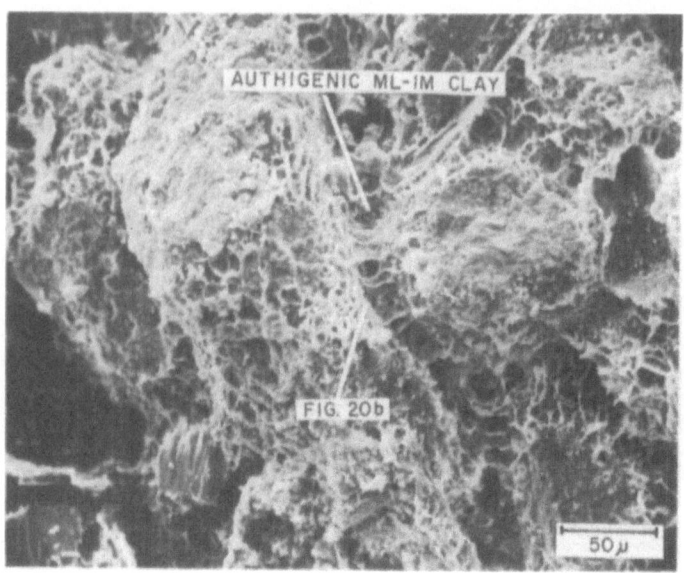

Fig. 20a.   Interstitially precipitated authigenic iron-bearing
IM clay in Hayner Ranch red beds.   Color of matrix is 9R 4/4
(weak red) to 9R 4/6 (red).   Sample BC-23a.   SEM photomicrograph.

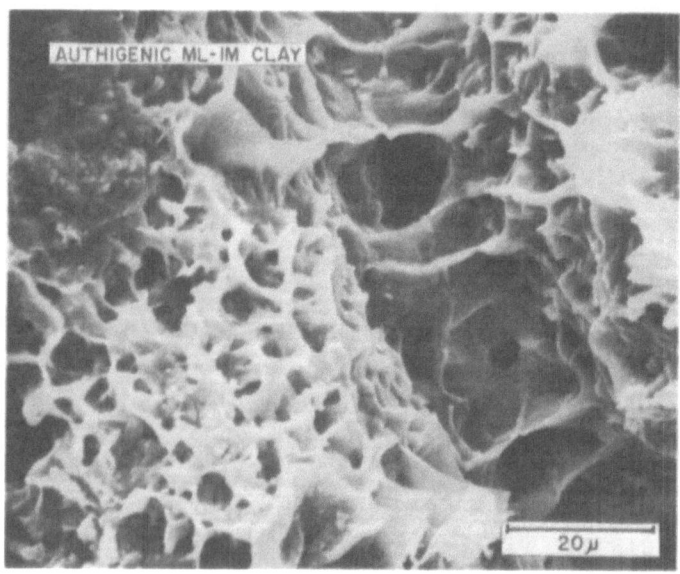

Fig. 20b.   Close-up of Fig. 20a, showing box-work texture of
authigenic IM clay.

which has been precipitated in interstitial voids. Presumably, each type of clay may occur alone or each may occur in any combination with the others. Each of these types of clay characteristically is not red initially, but each reddens with time if the interstitial environment is favorable for the formation of ferric oxides.

It also is important to point out that in addition to providing ions for the formation of authigenic hematite and authigenic clay, the extensive intrastratal alterations that have occurred in these deposits have provided a source of ions for the formation of other authigenic minerals, such as zeolites (Fig. 21), adularia (Fig. 22), and quartz (Fig. 22). These minerals are not present everywhere, but each of them is an important constituent of the red beds in some places. The zeolites most commonly occur as clear crystalline cement, but studies of concentrates of the adularia and quartz show that they commonly are stained red, and thus along with the three types of post-depositional clay just mentioned, they can be important constituents of the red interstitial matrix (Fig. 22b).

NATURE OF EVIDENCE PRESERVED IN ANCIENT DIAGENETIC RED BEDS

The Cenozoic red beds are young enough that they still contain the critical early stages of alteration and staining that proves that the pigment forms post-depositionally. However, with increasing age, increased depth of burial, and continued intrastratal alteration, the evidence is obscured. For example, the more unstable species of ferromagnesian silicates (e.g., olivine, augite, and hornblende) in time are destroyed completely, leaving no direct evidence that they originally were present. In addition, the infiltrated and authigenic clays become distorted and squeezed between the remaining stronger framework grains, and hence the original textural characteristics which in younger deposits confirm their diagenetic origin, are lost. Moreover, owing to higher temperatures, higher pressures, and changes in the chemical composition of the ground water, the infiltrated and earlier-formed authigenic clays are not in equilibrium with the new interstitial water and therefore they should be expected to convert to new clay minerals such as mixed-layer illite-montmorillonite containing higher percentages of illite than before, discrete illite, chlorite, etc. Similarly, the authigenic zeolites and adularia are likely to convert to new silicates, presumably some other type of feldspar.

Despite such modifications, many ancient red beds have characteristics which convincingly show that the hematite pigment is authigenic and that it is the product of diagenetic

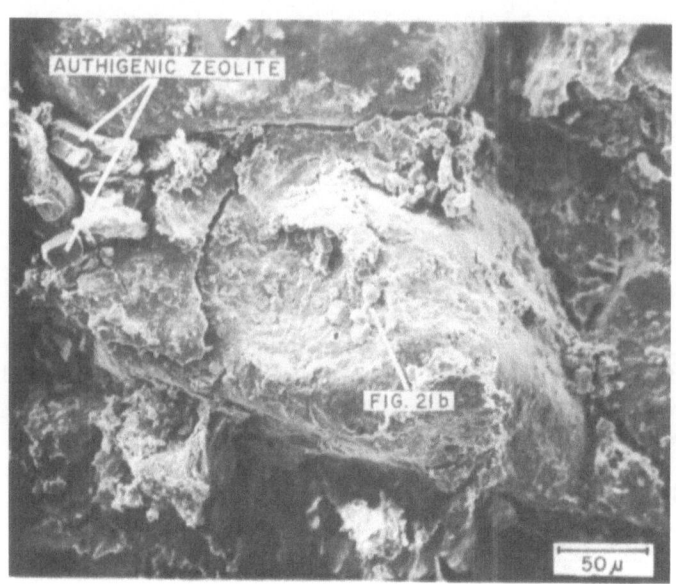

Fig. 21a. Authigenic zeolite (clinoptilolite) in interstitial
matrix. Tesuque Formation (Pliocene) 42 kilometers (25 miles)
southwest of Taos, New Mexico, on U.S. highway 64. Color of
matrix is 6YR 7/3 (pink). Sample Penr. 46. SEM photomicrograph.

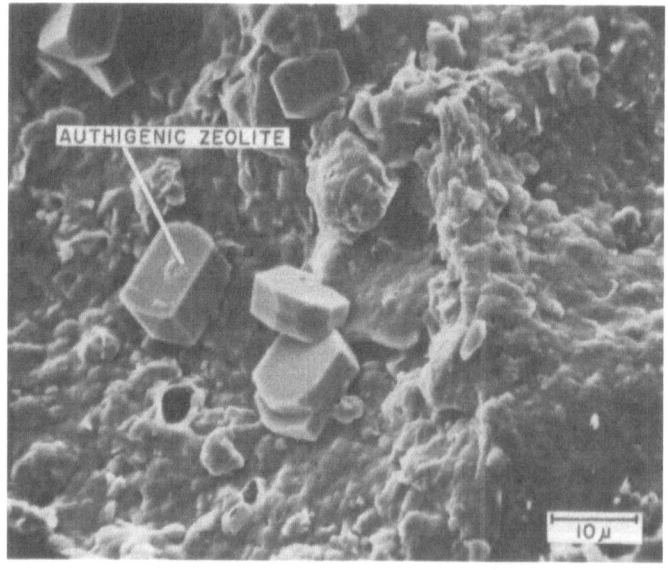

Fig. 21b. Close-up of Fig. 21a, showing clinoptilolite crystals
on infiltrated clay matrix.

Fig. 22a.   Red matrix containing authigenic adularia and quartz in
Cañon Rojo red beds.   Color of matrix is 10R 5/8 (red).   Sample
WFRB-5.   SEM photomicrograph.

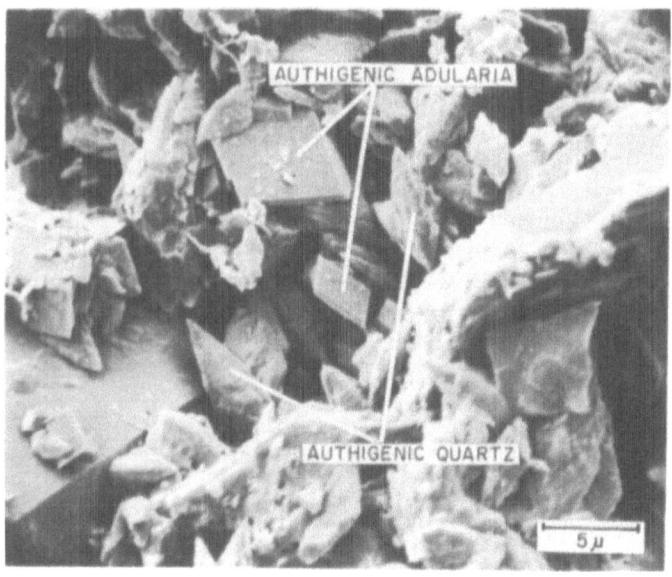

Fig. 22b.   Close-up of Fig. 22a, showing crystals of red authigenic
adularia and quartz.

alterations which are analogous to those in the Cenozoic deposits just described. Following are several characteristics
which the writer has found to be helpful in recognizing ancient
analogues of the Cenozoic red beds. Commonly all of them occur
in the same sample. The first four characteristics usually can
be observed with careful examination using a 10-power hand lens,
but none of these four characteristics proves that the pigment
is authigenic; each merely indicates that the rock has been
altered intrastratally and thus each merely suggests the
possibility of diagenetically-formed pigment. The remaining
four characteristics, however, provide direct evidence of
authigenic hematite, and with the exception of the last one,
they usually can be observed in thin section or by using polished
surfaces and techniques of ore microscopy. The most conclusive
evidence of authigenic hematite, of course, is provided by SEM
studies when the composition of the crystals are simultaneously
determined by energy dispersive X-ray analysis, and the
mineralogy is confirmed by X-ray diffraction analysis

Absence of common ferromagnesian silicates such as augite
and hornblende in red beds derived from source rocks containing
these minerals.

Heavy mineral studies of modern alluvium (Walker, 1974)
from a variety of climatic regions show that regardless of
climate or intensity of source area weathering, if unstable
ferromagnesian minerals are present in the rocks in the provenance area, they will occur as important accessory minerals in
the sediments derived from those sources. Consequently, ancient
red beds which were derived from source rocks containing unstable
iron-bearing minerals almost certainly contained these minerals
when the sediments originally were deposited. If they do not contain these minerals now, the minerals probably have been destroyed
by intrastratal alteration and their destruction would account for
an ample supply of iron for staining. In the writer's opinion,
geologists traditionally have overemphasized the importance of
source area weathering and overlooked the importance of intrastratal alteration to explain the absence of unstable minerals
in sedimentary rocks.

Sand-size concentrations of clay squeezed between stable
framework grains.

This common characteristic of ancient red beds suggests
that unstable silicate minerals have been replaced by clay and
the clay has been squeezed between the remaining more stable
framework grains by the weight of the overburden. If the clay

masses are heavily stained with hematite, they may represent replaced ferromagnesian silicates.

Incongruent bimodal texture

Sandstones and conglomeratic sandstones in ancient red beds commonly consist of framework grains and interstitial clay without intermediate silt sizes. Inasmuch as primary depositional processes are not known to produce similar bimodality in modern sediments, this type of texture suggests that the clay is post-depositional, and thus the clay probably is a product of infiltration and/or authigenesis.

Relicts of altered feldspar

Relicts of altered feldspars which show evidences of dissolution and/or replacement by clay similar to those displayed by the hornblende grains in the Cenozoic deposits described heretofore, are common in many ancient red beds. Occurrences of such relicts indicate that intrastratal alterations have proceeded to the stage where they have attacked the more stable silicates. Such advanced stages of alteration suggest that less stable ferromagnesian silicates probably were present originally but have been destroyed completely.

Hematite aureoles surrounding altered iron-bearing detrital grains

Aureoles of hematite surrounding altered iron-bearing detrital grains provide excellent petrographic evidence that the pigment is diagenetic. In young (i.e. Cenozoic age) red beds these aureoles are likely to occur around all types of partly altered ferromagnesian silicate minerals, but the more unstable silicates normally do not persist in older red beds, and therefore in these rocks the aureoles typically are best developed around biotite grains (Fig. 23). The aureoles are best observed on polished surfaces using reflected light and techniques of ore microscopy, but they commonly are discernible under reflected light in standard thin sections.

Blotchy concentrations of red pigment

When examined under reflected light in petrographic thin section, the pigment in ancient red beds commonly is concentrated in irregular, nonuniformly-stained blotches which commonly contain irregular blebs of dark red to black material (Fig. 24).

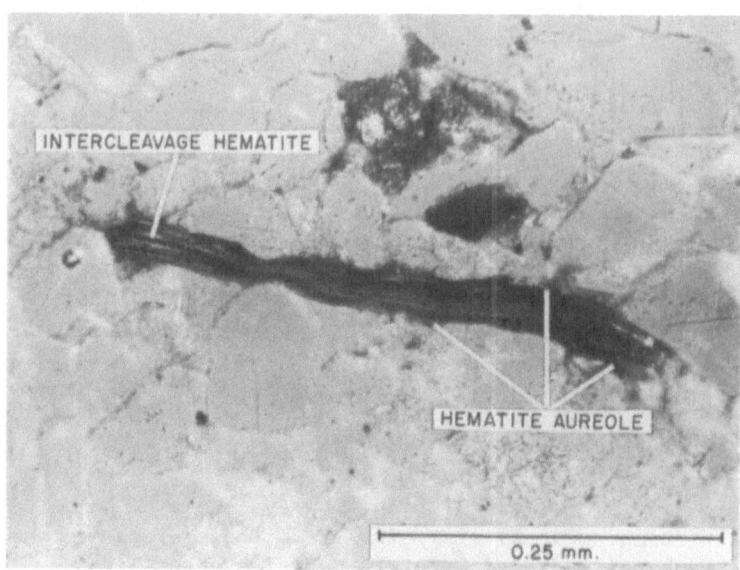

Fig. 23. Authigenic hematite associated with altered biotite.
Maroon Formation (Permian) along Colorado River between Burns
and Dotsero, Colorado. Sample CR-2-359. Thin section photomicrograph. Reflected light.

Fig. 24. Blotchy concentrations of authigenic hematite in Moenkopi
Formation near Cameron, Arizona. Sample GM-S-4-B. Thin section
photomicrograph. Reflected light. Compare with Fig. 26.

The blotches are much too irregular to be explained by depositional processes. SEM studies (see following discussion)
reveal that the blotches represent centers of hematite precipitation and that the variations in redness and associated
irregular dark blebs reflect variations in crystal size (see
Figs. 25b and 26). The precipitation centers in some instances
seem to be distributed randomly through the interstitial
matrix; in other instances they seem to grow outward from
relicts of highly altered detrital iron-bearing grains which
have been extensively replaced by hematite (e.g. biotite,
hornblende, magnetite, etc.).

Euhedral hematite crystals

     The diagenetic origin of hematite pigment in red beds is
most convincingly documented by the occurrence of euhedral
crystals of hematite. In some ancient red beds these crystals
in places are coarse enough to be detected in petrographic
thin section (for an example, see Walker, 1967, Plate 2, Fig.
5). In most instances, however, the crystals are too small
to be resolved with certainty except at very high magnification using the SEM. Studies by this writer, E. E. Larson, and
R. P. Hoblitt, using a combination of ore microscopy techniques,
SEM studies, energy dispersive X-ray analysis, and X-ray
diffraction analysis, have confirmed that authigenic crystalline hematite is commonplace in ancient red beds. The hematite occurs as extremely fine-grained platy crystals (Fig. 25b)
which commonly form rosette-like clusters, and as coarser
platy hexagonal crystals (Fig. 26). The differences in crystal
size probably reflect different ages of crystallization of the
hematite. That is, the fine-grained crystals represent the
most recently precipitated hematite, whereas the coarser crystals represent earlier precipitated hematite which has recrystallized. The platy hexagonal shape of the coarser
crystals (Fig. 26) is strikingly similar to that of authigenic
kaolinite, but the two minerals are readily distinguished
under the SEM by the use of energy dispersive X-ray analyzer
to determine the elemental composition. The examples illustrated in Figs. 24, 25, and 26 have been analyzed in this
manner. In addition, the mineralogy has been further confirmed
by X-ray diffraction analyses of concentrates of the matrix
material.

     Increase in grain size seems to darken the color of red
beds. For example, our studies show that blotchy concentrations of red pigment such as those previously described (Fig. 24),
occurring in the Moenkopi Formation (Triassic age) in north-
eastern Arizona, consist of various sizes of euhedral crystals
of hematite. The brightest red areas are colored by finely

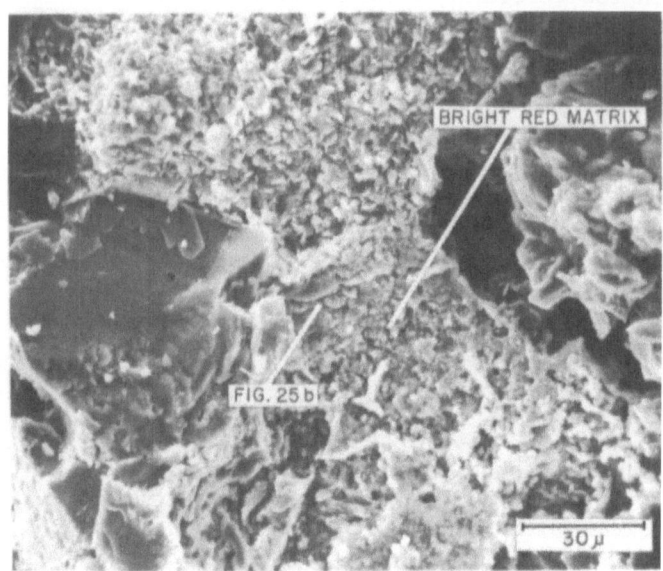

Fig. 25a. Bright red interstitial matrix in Moenkopi Formation
near Cameron, Arizona. Color of finely crystalline hematite is
R 4/8 (red). Sample GM-9. SEM photomicrograph

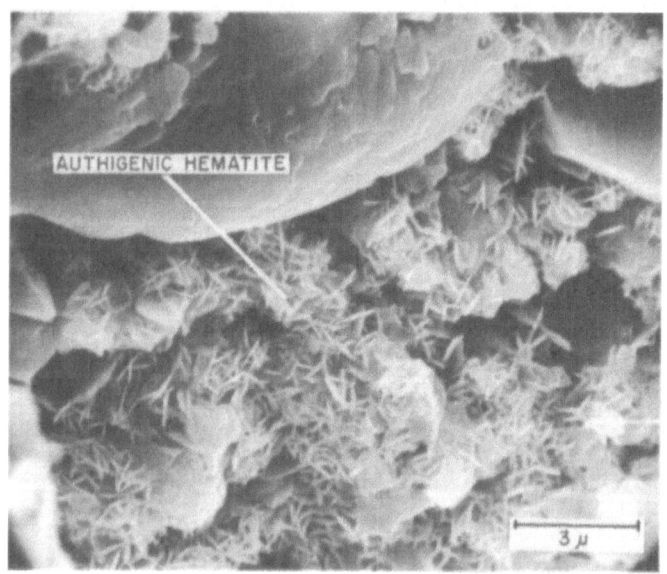

Fig. 25b.   Close-up of Fig. 25a, showing finely crystalline red
authigenic hematite.

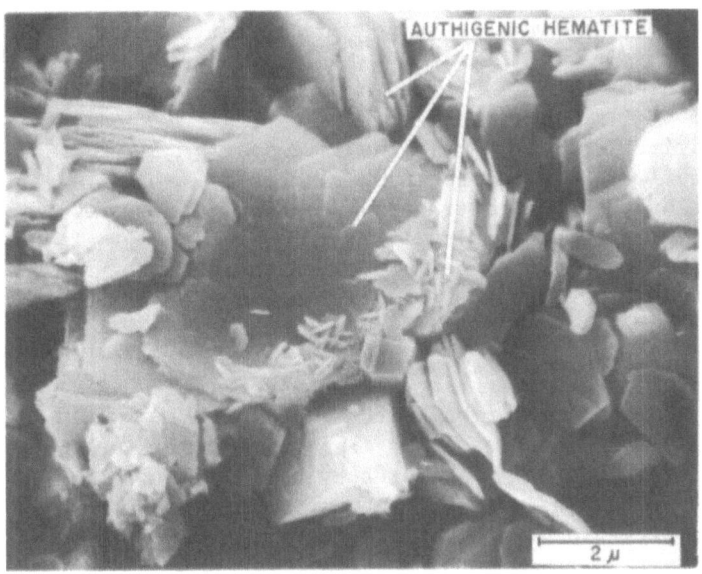

Fig. 26. Finely crystalline and coarsely crystalline authigenic hematite in Moenkopi Formation near Cameron, Arizona. As viewed in thin section under reflected light, color varies from 8R 4/8 (red) in fine crystals, to 8R 3/8 (dark red) to essentially black, in coarser crystals.

crystalline hematite (Fig. 25b) whereas the blebs of dark red to black material (Fig. 24) consist of coarsely crystalline hematite (Fig. 26).

SUMMARY AND CONCLUSIONS

There is no mystery about the manner in which red beds form diagenetically in deserts, because each step in the process can be documented in Cenozoic deposits in southwestern United States and northwestern Mexico. These deposits, which are convincing analogues of many classic ancient red beds, show that the red pigment forms in response to intrastratal alterations of iron-bearing detrital grains under conditions in which the chemistry of the interstitial water favors the formation and preservation of hematite.

The sediments are not red when deposited, but the reddening processes begin soon after deposition, wherever iron-bearing detrital minerals come in contact with migrating subsurface water with which they are not in equilibrium, assuming conditions exist in which the water lies in the stability field of hematite. Inter-

stitial iron-bearing clay minerals, which consist dominantly of
detrital platelets of clay which have been mechanically infiltrated
into the sediments by influent surface water, redden first, probably because they are repeatedly bathed by oxygenated water while
the sediments are still very permeable. The earliest stage of this
reddening probably reflects aging of limonitic coatings on the
clay platelets, whereas later stages probably reflect precipitation of iron released by intrastratal alteration of the clay.
Such reddening is common in deposits of Pleistocene age,
suggesting that this type of reddening begins within thousands
or tens of thousands of years after deposition.

    Prolonged contact with migrating subsurface water, however,
both above and below the water table, ultimately results in
clay replacement and/or dissolution of framework silicates.
Wherever ferromagnesian silicates occur, these alterations
yield iron-bearing daughter clay and/or they release iron to
the interstitial water. Wherever the interstitial water lies
in the Eh-pH stability field of hematite, the released iron
precipitates as hematite or as a red precursor ferric oxide
that converts to hematite upon aging. The daughter clay which
has replaced ferromagnesian silicates reddens simultaneously.
Thus, with time the reddening intensifies.

    The other ions which are released by the alterations
either remain in solution and migrate with the ground water,
or they precipitate as authigenic minerals which are in equilibrium with the interstitial water. Authigenic minerals
which are most commonly formed are adularia, quartz, zeolites,
and clay minerals. These minerals in places comprise part or all
of the interstitial matrix and they commonly are red, and
hence their formation in the presence of precipitating iron
oxide contributes to further reddening of the sediments.

    The formation of hematite from iron released by intra-
stratal alteration of iron-bearing detrital grains, including
clay minerals, can occur at anytime during the post-depositional
history of the deposits, and its formation continues, assuming
that the Eh and pH of the interstitial water remains in the
hematite stability field, until all of the unstable iron-
bearing minerals are completely altered, or until the alteration is halted by cementation of the sediment. Even cementa-
taion, however, may only delay the alterations until subsequent
changes in the chemistry of the interstitial water allow the
cement to be dissolved. The fact that relicts of partly
altered augite and hornblende are still common constituents
in red beds of known Miocene age indicates that the alterations
leading to the formation of hematite pigment continue for at
least tens of millions of years after deposition.

These studies of successively older red beds show a progression in the development of hematite pigment, and successive
stages in the progression are reflected in a general way by
the redness of the sediments. In the earliest stages of
development, such as those occurring in the Pleistocene and late
Tertiary deposits, the pigment seems to consist of amorphous red
ferric oxide which cannot be identified by either X-ray analysis
or SEM studies. At this stage, deposits typically are reddish
yellow (a commonly occurring color is 5YR 6/6). In the intermediate stages of development, such as those occurring in the Miocene deposits, very finely crystalline hematite begins to form,
the identification of which can be confirmed both by X-ray analysis and SEM studies at very high magnification using an energy
dispersive X-ray analyzer. The brightest red colors (e.g. 8R to
9R 4/6 to 5/6) seem to develop at this stage. In advanced stages
of development, such as those occurring in the Triassic deposits,
recrystallization converts the fine-grained crystals into coarser
crystalline hematite, and the sediments become darker red to reddish brown (e.g., 10R 4/6 to 2.5R 4/6). Meaningful ages cannot be
assigned to the boundaries between the various stages because too
many variables exist which influence the alterations, such as differences in the chemistry of the interstitial water, mineralogy
and texture of the sediments, nature of ground water circulation,
and time, nature, and intensity of cementation. Consequently,
alterations leading to the development of the pigment do not begin
everywhere at the same time. Moreover, conditions during the
course of development of the pigment vary from formation to
formation and even from place to place and from time to time
within a single formation.

Finally, it should be emphasized that at least insofar
as this writer is aware, all direct evidence available at the
present time indicates that classic red beds characteristically
are products of post-depositional alterations which are analogous to those occurring today in the Cenozoic deposits described
herein. Exceptions have been claimed, but they are supported by
indirect, inconclusive, and in this writer's opinion, unconvinc-
ing evidence. If readers know of verifiable exceptions, they
would contribute greatly to our understanding of red beds by
presenting direct evidence of another type of origin.

REFERENCES

Berner, R. A., 1969, Geothite Stability and the Origin of Red
        Beds: Geochim et Cosmochim Acta, v. 33, no. 2, p. 267-273.
Blatt, H., Middleton, G., and Murray, R., 1972, Origin of
        Sedimentary Rocks, Prentice-Hall, Englewood Cliffs, New
        Jersey, 634 p.

Crone, A. J., 1974, Experimental Studies of Mechanically-Infiltrated Clay Matrix in Sand: Geol. Soc. America Abstracts with Programs, v. 6, no. 7, p. 701.

Crone, A. J., 1975, Laboratory and Field Studies of Mechanically-Infiltrated Matrix Clay in Arid Fluvial Sediments: Unpubl. Ph.D. thesis, Univ. of Colorado.

Cummins, W. A., 1962, The Greywacke Problem: Liverpool and Manchester Geol. Jour., v. 3, p. 51-72.

Davis, S. N. and DeWiest, R. J. M., 1966, Hydrogeology, John Wiley, New York, 463 p.

Folk, R. L., 1951, Stages of Textural Maturity in Sedimentary Rocks: Jour. Sed. Pet., v. 21, no. 3, p. 127-130.

Garrels, R. M., and Christ, C. L., 1965, Solutions, Minerals, and Equilibria: Harper and Row, New York, 450 p.

Goldich, S. S., 1938, A Study in Rock Weathering: Jour. Geology, v. 46, p. 17-58.

Keller, W. D., 1969, Chemistry in Introductory Geology: 4th edition, Lucas Bros. Pub., Columbia, Missouri, 108 p.

Krauskopf, K. B., 1967, Introduction to Geochemistry, McGraw-Hill, New York, 721 p.

Langmuir, Donald, 1971, Particle-Size Effect on the Reaction Goethite-Hematite Plus Water: Amer. Jour. Sci., v. 271, p. 147-156.

Larson, E. E., and Walker, T. R., 1975, Development of Chemical Remanent Magnetization During Early Stages of Red Bed Formation in Late Cenozoic Sediments, Baja California: Geol. Soc. America Bull., v. 86, p. 639-650.

Reynolds, R. C., and Hower, John, 1970, The Nature of Inter-layering in Mixed Layer Illite-Montmorillonite: Clays and Clay Minerals, v. 18, p. 25-36.

Seager, W. R., Hawley, J. W., and Clemons, R. E., 1971, Geology of San Diego Mountain Area, Dona Ana County, New Mexico: New Mex. State Bureau of Mines and Min. Res., Bull. 97, 38 p.

Walker, T. R., 1967, Formation of Red Beds in Modern and Ancient Deserts: Geol. Soc. America Bull., v. 78, p. 353-368.

_____, 1974, Formation of Red Beds in Moist Tropical Climates: A Hypothesis: Geol. Soc. America, v. 85, p. 633-638.

Walker, T. R., Ribbe, P. H., and Honea, R. M., 1967, Geochemistry of Hornblende Alteration in Pliocene Red Beds, Baja California, Mexico: Geol. Soc. America Bull., v. 78, p. 1055-1060.

Walker, T. R., and Honea, R. M., 1969, Iron Content of Modern Deposits in the Sonoran Desert: A Contribution to the Origin of Red Beds: Geol. Soc. America Bull., v. 80, p. 535-544.

Walker, T. R., and Waugh, B., 1973, Intrastratal Alteration of Silicate Minerals in Late Tertiary Fluvial Arkose, Baja California, Mexico: Geol. Soc. America Abstracts with Programs, v. 7, no. 7, p. 853.

PERMIAN PALEOGEOGRAPHY AND CLIMATOLOGY

A.E.M. Nairn and M.E. Smithwick

Department of Geology, University of South Carolina,
Columbia, South Carolina

ABSTRACT. An empirical attempt has been made to generate summer
and winter surface wind patterns for the Lower and Upper Permian
using as a base paleogeographic maps derived from paleomagnetic
and regional geological considerations. Although problems remain, the maps so produced show general good agreement.

1. INTRODUCTION

Over the years there have been many attempts to evaluate the
climate of the Permian. Most of these attempts were based upon
the interpretation of geological data. On occasion wide-ranging
criteria were used (1,2) and at other times a single criterion
treated in depth formed the basis (3). The success or failure
of the attempts resided largely in the skill of the interpreter
to extract a consistent result from data characteristically imprecise. While at a local or even a regional level the climatic
descriptions may be satisfactory, extrapolation to other regions
is seldom successful. The reason lies partly within the nature
of the data used, but the more important part lies outside of
them. Any extrapolation basically assumes a known geography,
that is latitude and longitude, and if large scale displacements
of unknown amount occur they render such extrapolation invalid.
Thus Brooks (4) found himself in a meteorological quandry if not
an absurdity in his interpretation of Permian climate which Koppen and Wegener (5) escaped by assuming a different geography.

The undeniable conclusion is that while climatic data may
suggest that geographic changes have occurred, seldom is the evidence conclusive, rarely is the interpretation unique, and never

H. Falke (ed.), The Continental Permian in Central, West, and South Europe, 283-312. All Rights Reserved.
Copyright © 1976 by D. Reidel Publishing Company, Dordrecht-Holland.

can it specify the displacement save in the most general terms.
Even further, since climate is a global phenomenon, the changes
recorded in one region may be the result of events affecting another region a hemisphere away. This can be illustrated by brief
reference to the Cenozoic climate of Western Europe which ranged
from sub-tropical to glacial during a time in which only small
latitudinal displacements of Europe were predicted. Some authors,
including one of the present authors, have considered this as
providing a valuable limit in climatic interpretation by indicating the range of climate possible in cool temperate latitudes
during the course of a geological period. While such a conserva-
tive approach is logically impregnable and may even prevent gross
errors, it fails to reflect the possibility that the changes may
have been brought about largely by motions affecting the Antarctic
plate. If true, the cause of the climatic change in Europe lay
entirely outside that continent and could not be predicted by
climatic data from within it.

If then global climate cannot be successfully deduced, induc-
tive or model techniques must be used. Herein lies one insoluble
question, for since the energy drive of climate is solar and variation in this energy flux is always possible, an indeterminacy
will always exist. For the purpose of this paper we intend to
assume that the solar constant was at its present level throughout the Permian and only if no satisfactory model can be made
from a reasonable extrapolation of terrestrial criteria will re-
source be made to an extra-terrestrial cause, a circumstance well
into the future. Taking the first order climatic control as constant, the main thrust in this paper will be to establish the
second order control, the land-sea distribution, together with
the much less extensive and less precise information on topography
which forms the third order control. For this purpose, the basic
idea of the occurrence of large scale horizontal displacements
indicated by paleomagnetism and which are fundamental in "plate
tectonics" will here be assumed. The attempt will be made to
derive an atmospheric and oceanic circulation pattern. Only in
this way for example, can we hope to discern whether or not there
was a Permian monsoon system, and the region affected. The pre-
dictions of this model will then be compared with the inferences
of climate possible from the use of geological criteria.

There is yet one more important consideration, that of correlation. Since the atmosphere and oceans are fluid bodies, the
effects of any important change are rapidly felt on a global
scale; and just as this suggests that climate may be an important
means of monitoring such change it also carries with it the im-
perative need for accurate time correlation if climate is to be
clearly understood. The problems of accurate correlation between
continental and marine sequences and of intercontinental correlation are well enough understood to make it unnecessary to labour

the point. Yet it is manifestly absurd to treat the Permian as
a whole. This for example would combine both glacial and non-
glacial conditions in Australia. We have chosen therefore, to
consider Upper and Lower Permian separately, as initially the
most practical division, in the hope that at some future time it
may be possible to further refine this; possible only through
very detailed regional geological studies with the attendant
correlation problems. The problem of correlation is compounded
by the loss of material due to erosion, so that distributions
are minimal distributions. In all probability vastly larger areas
originally carried a Permian cover, as clearly was the case in
Africa. Even when preserved basin margins retreat so that marginal facies with the information they could have provided may
be lost. Losses thus tend to concentrate in the critical facies.
What is here presented, then is the nature of a report of progress with many gaps, and no doubt errors. The authors welcome
and actively solicit help in refining their model.

2. THE PALEOGEOGRAPHIC MODEL

2.1. The Reconstruction of the Continents

The first stage in deriving a paleogeographic model is to
generate a continental reconstruction, that is to define the
distribution of the lithospheric plates of which the continents
are the visible and most durable portion. Since there is apparently no extant oceanic crust of Permian age, the principal
techniques must involve some physiographic matching, use of paleomagnetic data, or some hybrid. There are several maps available
but though more accurate cartographically, they basically resemble
the classical Wegener map. Most work is based upon the computer-
ized least squares topographic fit of Bullard et al (6) of the
Atlantic region. This suffers from an absence of absolute lati-
tude/longitude constraints unless some additional criterion is
injected as for example by Dietz and Holden (7). Briden et al (8)
used the Bullard et al (6) Atlantic fit combined with the Gondwana
assembly of Smith and Hallam (8) as a basins for the examination
of Permian paleomagnetic data and demonstrated that the latter
data was consistent, to a first approximation, with a dipolar
field. We therefore adopt this as our base map, and use the paleomagnetic data to provide latitudinal control. Reference should
be made to the maps of Van der Voo and French (10) which follow a
similar pattern. The weakness of the reconstruction lies in the
restricted data from large regions of Asia. In order not to pre-
judice the model we have made no use in the model of attempts to
define latitude whether obtained by a variety of geological criteria (11) or from a single criterion. These methods have usually
considered data with an implicit geographical assumption with results not always uncontested (3, 12, 13).

2.2 Land-Sea Distribution

The second step, one rarely developed on the geophysically based models, is to attempt to establish land-sea relationships together with what data may be assembled on topography from regional geological considerations. This approach is not new, for there are world maps (14, 15, 16), and regional studies. Although we have gone to original sources rather than reviews, it is nonetheless clear that we have benefited enormously from these works. In particular we cite for Europe: Glennie (17) Brinkmann (14) and Burek (18); Australia: Brown et al (19); Antarctica: Stump (20); South Africa: Frakes and Crowell (21); South America: Harrington (22) and Bigarella (23); North America: McKee (24); Asia: Nalivkin (25).

Our methods of handling this data are at present empirical, and indeed we are unaware of successful quantitative methods. Prime importance is given to completeness of the geological record, without regard to thickness. This condition is imposed by the impossibility, or rather impracticability, of studying a sufficiently narrow time interval. The thickness and litho-and biofacies are considered together. Clearly in some regions over a time as long as the Lower Permian there may be appreciable facies changes and particularly important in this respect is an alternation of continental and marine conditions. We have assigned the condition which appears to predominate, and where both are of equal duration the relative influence of each will be considered in the discussion.

The thickness of the succession cannot be considered out of the facies context. Where the facies is essentially constant then the thickness reflects the tectonic environment. We hope in the future to demonstrate that it may be possible to derive important presumptive regional numerical information. At the present time one of the more puzzling problems engendered by the current optimism of global tectonics is how established sedimentation patterns can persist through several geological periods in a so-called "stable" craton while it is supposedly moving at a rate of centimeters per year.

We therefore conclude that our data treatment is excessively simple since:
  1) the state of the geological record is notoriously incomplete;
  2) we are compelled to use a long time interval and so suffer from the generalizing effects of time;
  3) we do not yet have sophisticated numerical techniques to handle geological data; and,
  4) the base model may be incorrect.

While these arguments may reduce the value of the conclusions we
reach, we believe that the pattern emerging is sufficiently se-
ducing to warrant further pursuit of its sophistication. The
present article is thus a progress report, and its findings will
be subjected to progressive refinement.

The principal results are summarized in a series of maps.
For each continental unit one map gives the location and approximate extent of and nature of the geologic record. This data is
then transferred to a global reconstruction and forms the basis
for the surface wind and oceanic current map. We report here only
the broad outlines of the regional geology.

2.3.   Northern America (including Greenland)

Outcrops of Permian rocks are largely restricted to the
western half of the continent. In the east, the Dunkard sequence
in Pennsylvania and West Virginia is assigned to the Permian, although its exact stratigraphical position has been contested (26).
Permian rocks also appear along the eastern coast of Greenland
between 71 and 74°N (27), and in the Canadian Arctic Islands (28).

In the Western United States well defined epicontinental
marine basins were bordered by flats upon which developed extensive red beds. Periodically conditions favored the production of
evaporites, and desert sandstones developed near the margins of
the sea. Further to the west the succession passes through miogeosynclinal into a geosynclinal type. This geosyncline stretched
from the Western United States (and presumably) from Mexico towards Alaska.

The general paleogeographic pattern seems to be a region of
general emergence in the eastern part of the country, with the
emergent zone for the most part forming a peneplain for it does
not here seem to have contributed an appreciable sedimentary load
to the west, south or north. As is pointed out in most elementary
texts the final consolidation of the Appalachian-Allegheny mountains occurred during Permian times, the final burst ending in
mid Permian (e.g., Seyfert and Sirken (29). The active erosion
associated with the elevation of a mountain range is seen in eastern Greenland with five to six kilometers of Lower Permian coarse
continental sandstones and arkoses (i.e. molasse). On the western
side of the mountain belt only the sediments of the Dunkard basin
are preserved; the facies present, including coal, suggest that
they were removed from the source area and the lacustrine-fluvia-
tile conditions presumable gave way southwestwards to marine conditions, though this question is not entirely clear.

LOWER PERMIAN

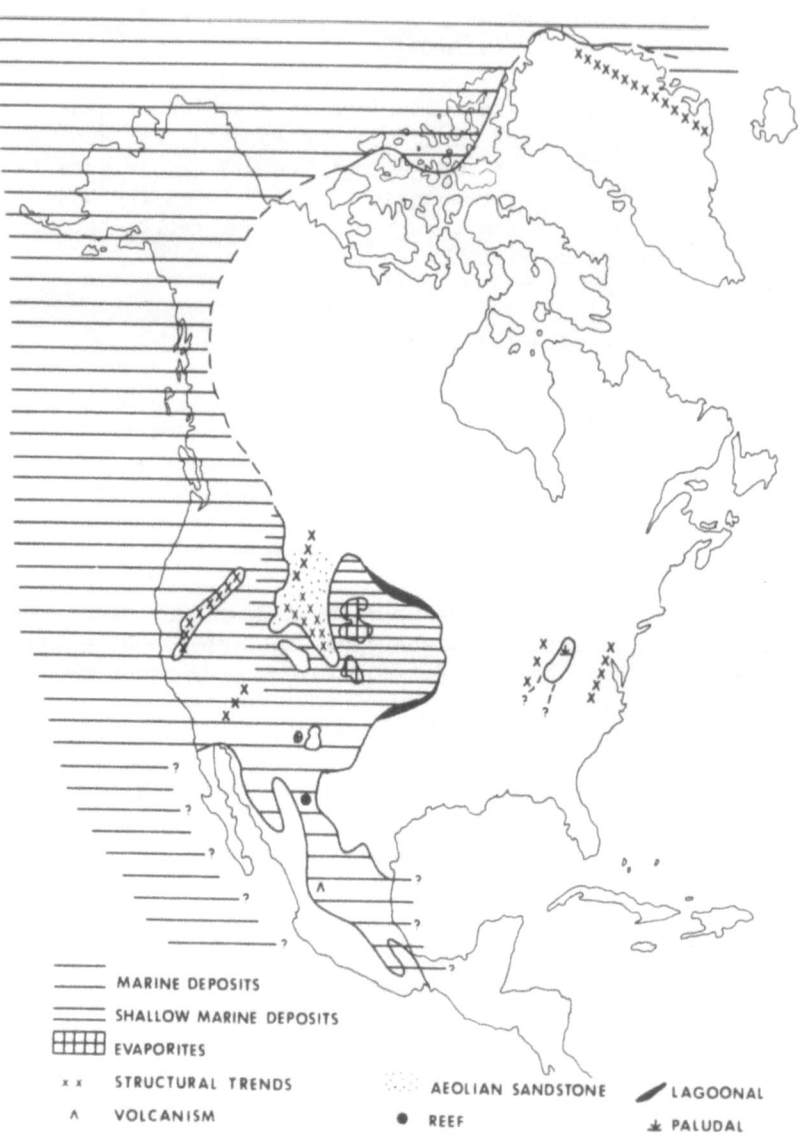

MARINE DEPOSITS

SHALLOW MARINE DEPOSITS

EVAPORITES

x x   STRUCTURAL TRENDS   AEOLIAN SANDSTONE   LAGOONAL

∧   VOLCANISM   ● REEF   ⊾ PALUDAL

Fig. 1. NORTH AMERICA - Paleogeography of North America
        during Lower Permian showing land/sea distri-
        bution, major structural trends and some environ-
        mental indicators.

UPPER PERMIAN

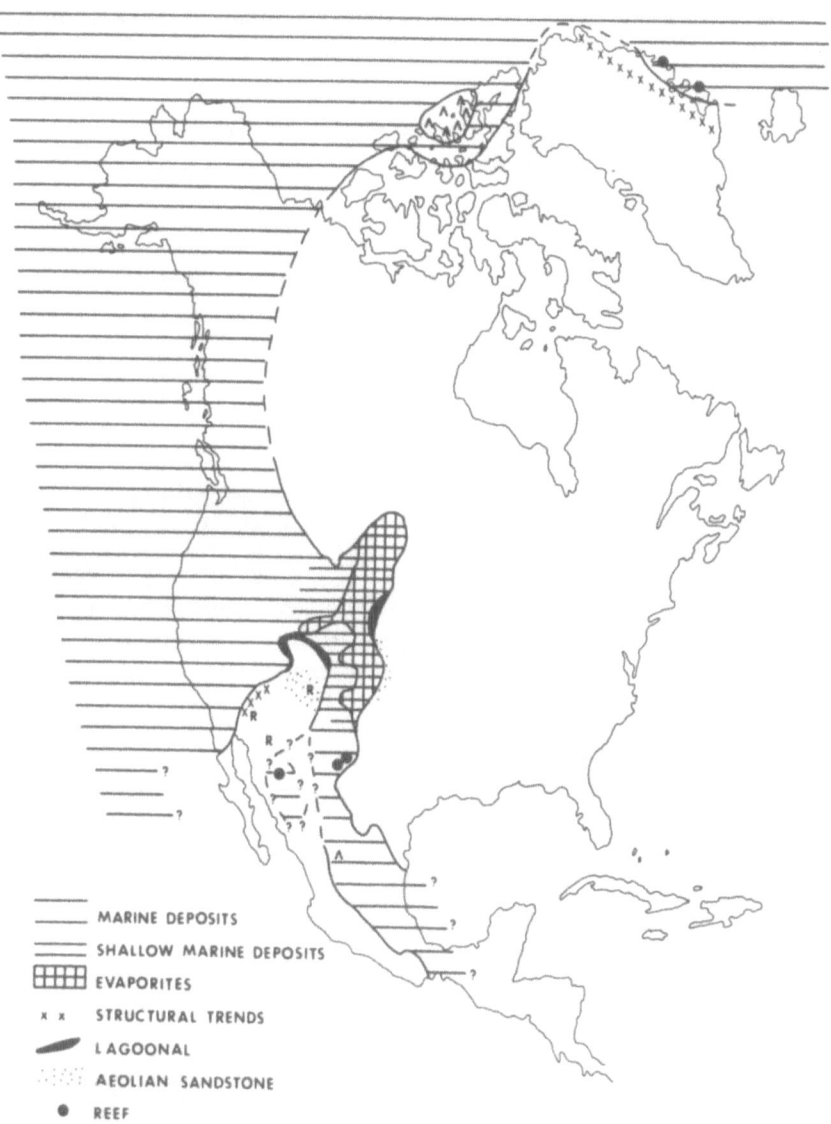

MARINE DEPOSITS

SHALLOW MARINE DEPOSITS

EVAPORITES

x x   STRUCTURAL TRENDS

LAGOONAL

AEOLIAN SANDSTONE

●   REEF

Fig. 2. NORTH AMERICA - Paleogeography of North America
during Upper Permian showing land/sea distribution, major structural trends and some environmental indicators.

The Upper Permian of eastern Greenland, with diverse, near-shore marine and neritic environments well represented, with the occurrence of reefs, dolomites and massive typsum, and only a relatively thin basal conglomerate, implies that the former mountainous region was no longer providing a significant sedimentary supply.

Over the western portion of the United States a shallow epicontinental sea covered a wide area. Its form was affected by a number of positive elements trending either parallel to the Appalachian trend, as for example the Central Nevada Ridge which separates the eugeosyncline from the miogeosyncline, or normal to the trend such as the Uncomphagre Uplift. Paleogeographically in the Lower Permian the Northern Rocky Mountain uplift (see Figure 1) played an important role for it underlies a broad continental salient projecting from the north. The quasi-isolation of the basin to the east was presumably responsible for the formation there of evaporitic sediments.

During the Upper Permian its importance diminished, but its southerly continuation, the Western Arizona Platform seems to have formed a broad positive element carrying redbed lithologies, with reefs and evaporitic basins along its eastern side. This positive zone can be traced south into Mexico although exposures are poor, and at Las Delicias (Durango) highly altered volcanics are found in a detrital sequence which contains limestone (as well as reef) horizons. This Upper Permian includes the Guadalupe basin, classic through the detailed description of P. B. King (30). In the continental area, dune bedded sandstones have provided wind directions which are relatively consistant from Wyoming to New Mexico (31).

In Mexico the Early to Middle Permian is the best represented, occurring in the geosyncline which lies roughly between Placer de Guadalupe and Mexico. This was a relatively narrow trough apparently bordered to the east and west by emergent zones. In the northwest the sea spread over the broad area of Sonora now represented by limestone deposits (32).

The eugeosynclinal deposits now incorporated in the fold belts can be traced from California to Alaska, and is a typical sequence in which rocks of volcanic origin play a large role. The miogeosynclinal belt of carbonates and clastics is well known in the United States but is less familiar in Canada. In the Arctic Islands Thorsteinsson and Tozer (28) suggest that a shoreline lay east of the areas of outcrop on Western Ellesmere Island. Its position was not fixed but migrated back and forth during the Permian for interbedded in the marine sequence on Ellesmere and Cameron Islands and on the Grinnell Peninsula are freshwater limestones and conglomerates. The marine deposits all appear to be

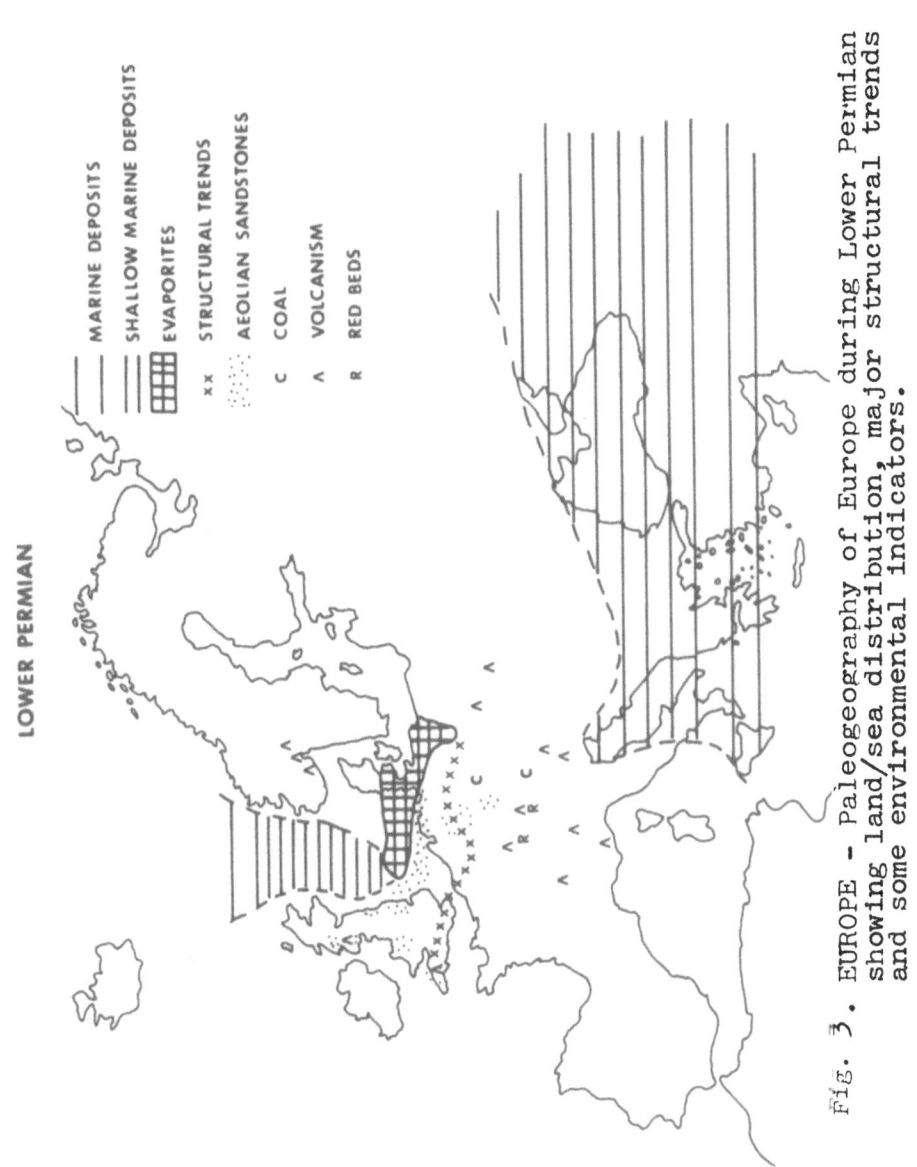

Fig. 3. EUROPE – Paleogeography of Europe during Lower Permian showing land/sea distribution, major structural trends and some environmental indicators.

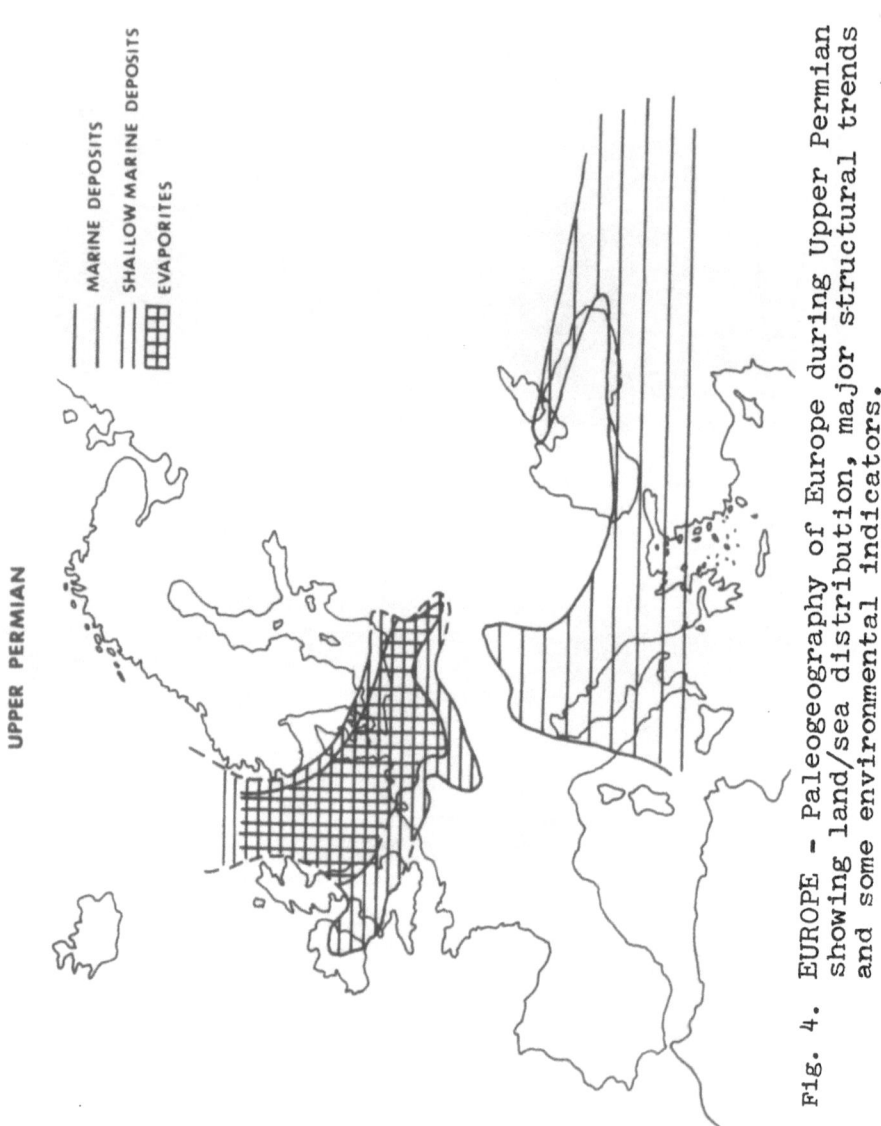

Fig. 4. EUROPE - Paleogeography of Europe during Upper Permian showing land/sea distribution, major structural trends and some environmental indicators.

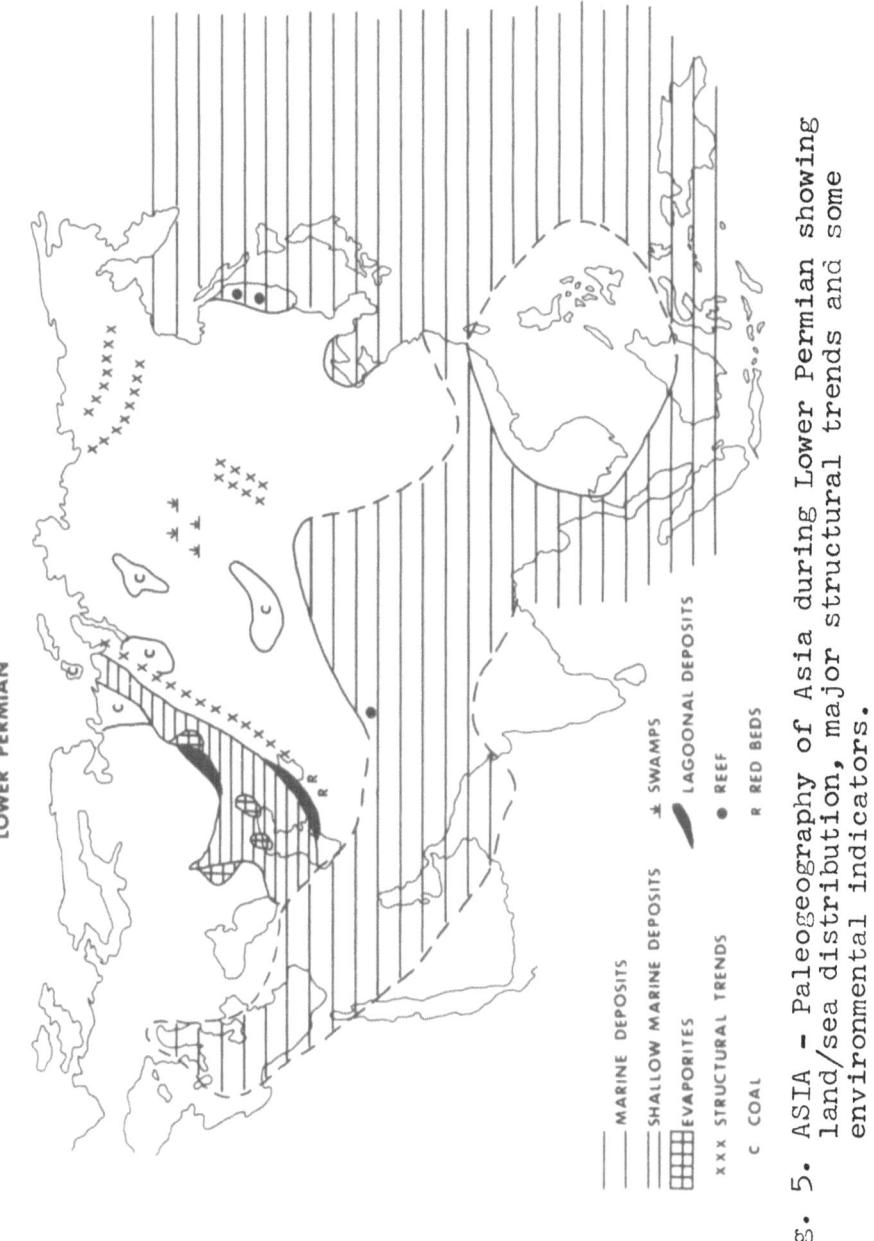

Fig. 5. ASIA – Paleogeography of Asia during Lower Permian showing land/sea distribution, major structural trends and some environmental indicators.

Fig. 6. ASIA – Paleogeography of Asia during Upper Permian showing land/sea distribution, major structural trends and some environmental indicators.

shelf deposits with glauconitic sandstone and gypsum occurring. Some submarine laval flows also are found.

2.4. Europe and Asia

The broad paleogeographic pattern of the European Permian was dictated by the tectonic events of the Hercynian orogeny, which, though continuing into Permian times, reached their apogee during the Carboniferous. The resulting chain of mountains can be traced across Europe from southwestern Ireland, through Brittany, Spain, Massif Central, Vosges and Schwarzwald to the Sudetic Mountains of Poland; its farther extent must presumably be sought beneath the Carpathians Mountains. Within the mountain belt deep, presumably faulted basins received heavy sediments, and north of the mountains lay a broad plain passing to a dessert "lake" in the region of the southern North Sea. The environments have been described by Glennie (17) in terms of modern deserts. Whether the Haselgebirge facies represents a true lake or whether it had marine connections is worthy of further study and may be elucidated from the evaporitic facies (33). Certainly during later Permian times the marine connections of the Zechstein which flooded into and beyond the area occupied by the Haselgebirge lake cannot be doubted.

To the east the broad plain over Poland passed into Russia, extended towards the Urals and southeast into the Caspian depression occupied by the Tethys sea. The same sea extended over much of southeastern Europe (14). South of the Hercynian-Variscan Mountains, the Veruccano facies probably represents restricted sedimentation.

The formation of the Ural, Verkhoyansk and Chersky ranges provided topographic relief in Asia during the Permian. The Angara shield remained a continental region, with the development of coal along the southern margin (the Angara geosyncline). A-round Vladivostok Permian reefs are recorded (25) (whose presence casts doubts on the high latitude assigned in current paleomag-netic models).

2.5. South America

Despite the enormous advances of the last years detailed knowledge of the geology of much of South America is lacking. The principal source in the reconstruction adopted here is Har-rington (22) which can be augmented by more recent publications (23, 34, 35). The broad outlines are probably reasonably well established but there remain many important problems, in particu-lar concerning the presence or absence of marine conditions in

LOWER PERMIAN

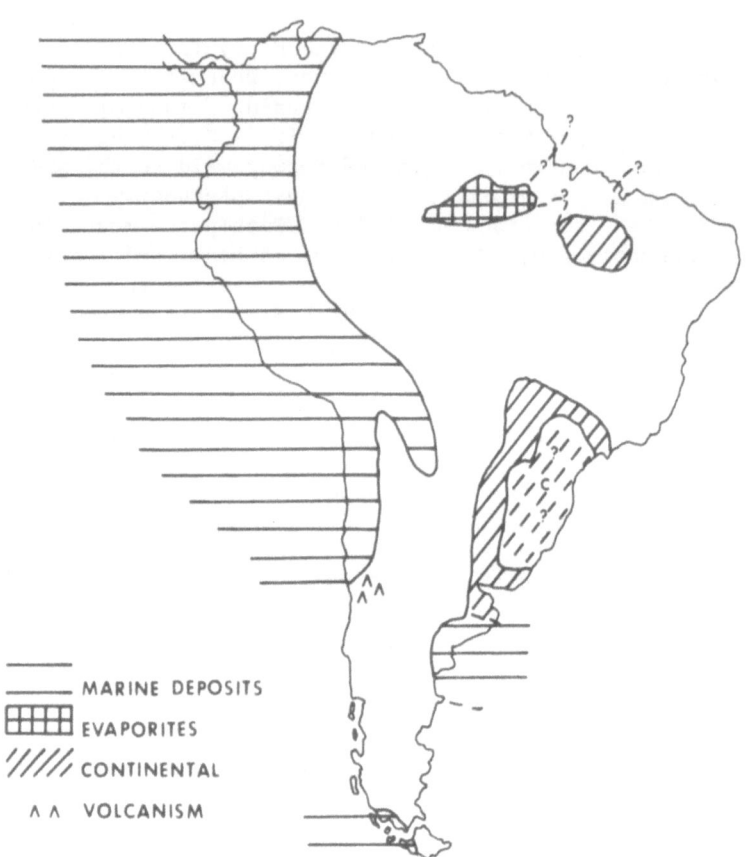

MARINE DEPOSITS
EVAPORITES
CONTINENTAL
∧ ∧ VOLCANISM

Fig. 7. SOUTH AMERICA – Paleogeography of South
America during Lower Permian showing land/
sea distribution and some environmental
indicators.

Fig. 8. SOUTH AMERICA - Paleogeography of South
America during Upper Permian showing land/sea
distribution and some environmental indicators.

the Parana Basin during Lower Permian times.

Marine conditions, generally represented by fusulinid or
brachiopod limestones, seem established for much of the Andean
region as far south as Northern Chile where the first plants are
found (accepting Harrington's 1961 age definition). Further
south in Patagonia continental conditions were established, but
in the Chilean Archipelago fusulinid limestones reccur.

Over the north half of cratonic South America relatively
little is known. According to Bigarella (23) in the middle Amazon basin the Lower Permian consists of sandstones and evaporites
followed by massive evaporites, implying marine conditions ruled
even if water access was limited. Bigarella postulates the ocean
connection was to the east rather than the west, presumably because at least temporarily seas invaded the Parnaiba basin.

The southern hills of Buenos Aires contain evidence of marine
horizons, which apparently extend also as far as the Parana basin;
however, for the greater part of the time the region appears to
be characterized by continental sedimentation. For the present
purpose the debate on the marine (22) or non-marine (34) origin
will be left unresolved. The existence of marine horizons in
Southwest Africa however, make occurrence of marine horizons
seem acceptable.

Land area appears to have been more extensive during the Upper Permian, with extensive vulcanism and redbeds. According to
Harrington (22) local glaciers contributed to fluvioglacial
deposits in Bolivia and Northern Argentina. If true this would
indicate the existence of high ground to the west.

In the Parnaiba basin both limestone and anhydrite indicate
marine conditions still persisted there, but no Upper Permian is
known in the Amazonas.

2.6. Africa

Apart from South Africa and Rhodesia, Permian rocks are relatively poorly developed in Africa. However the identification
of Ecca red shales in a borehole near Lake Victoria in Uganda
and the palynological identification of certain of the continental
beds of Gabon as Permian indicate the former wide extent of
rocks of this age.

The classical region still remains South Africa where the
Permian is represented by Ecca and Lower to Middle Beaufort beds.
One Ecca facies is well known for its economically important coals
in the Transvaal, but there are two other facies the more

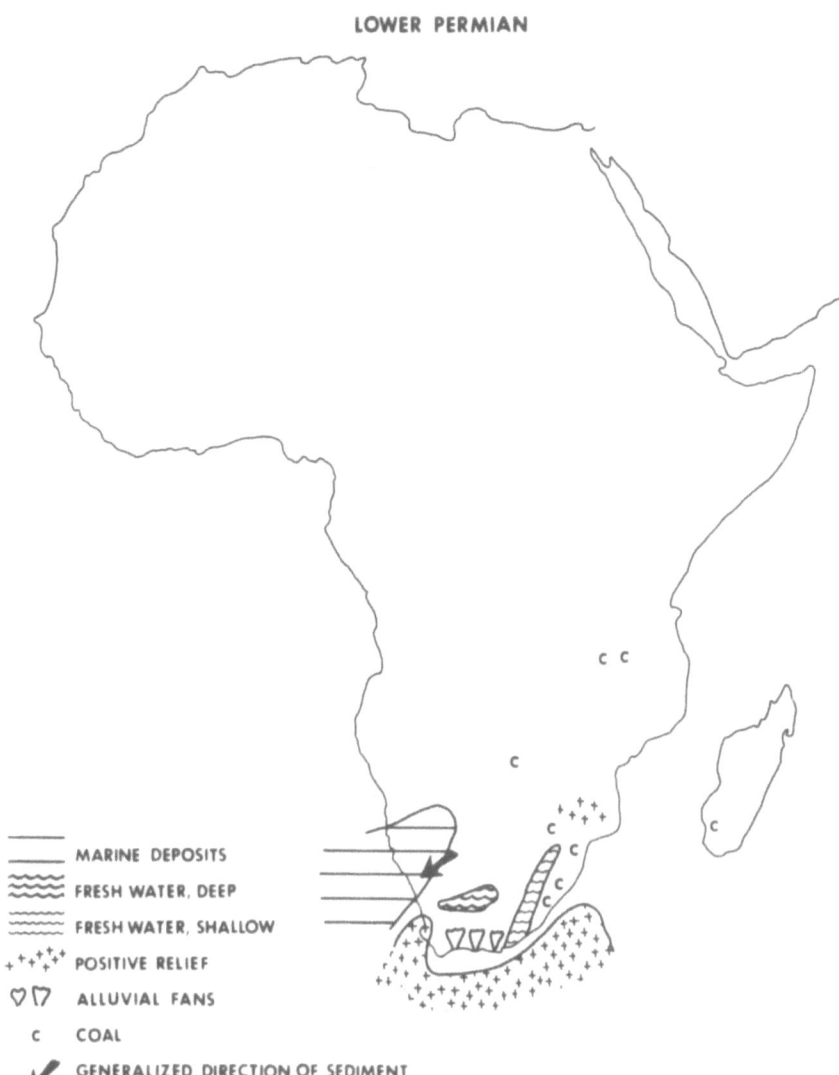

Fig. 9. AFRICA - Paleogeography of Africa during Lower
        Permian showing land/sea distribution and some
        environmental indicators.

UPPER PERMIAN

_____ MARINE DEPOSIT
≈≈≈≈ FRESH WATER, SHALLOW
°°°°° MUD AND SAND FLATS
c   COAL
XXXX BASALT FLOWS

Fig. 10. AFRICA - Paleogeography of Africa during
         Upper Permian showing land/sea distribution
         and some environmental indicators.

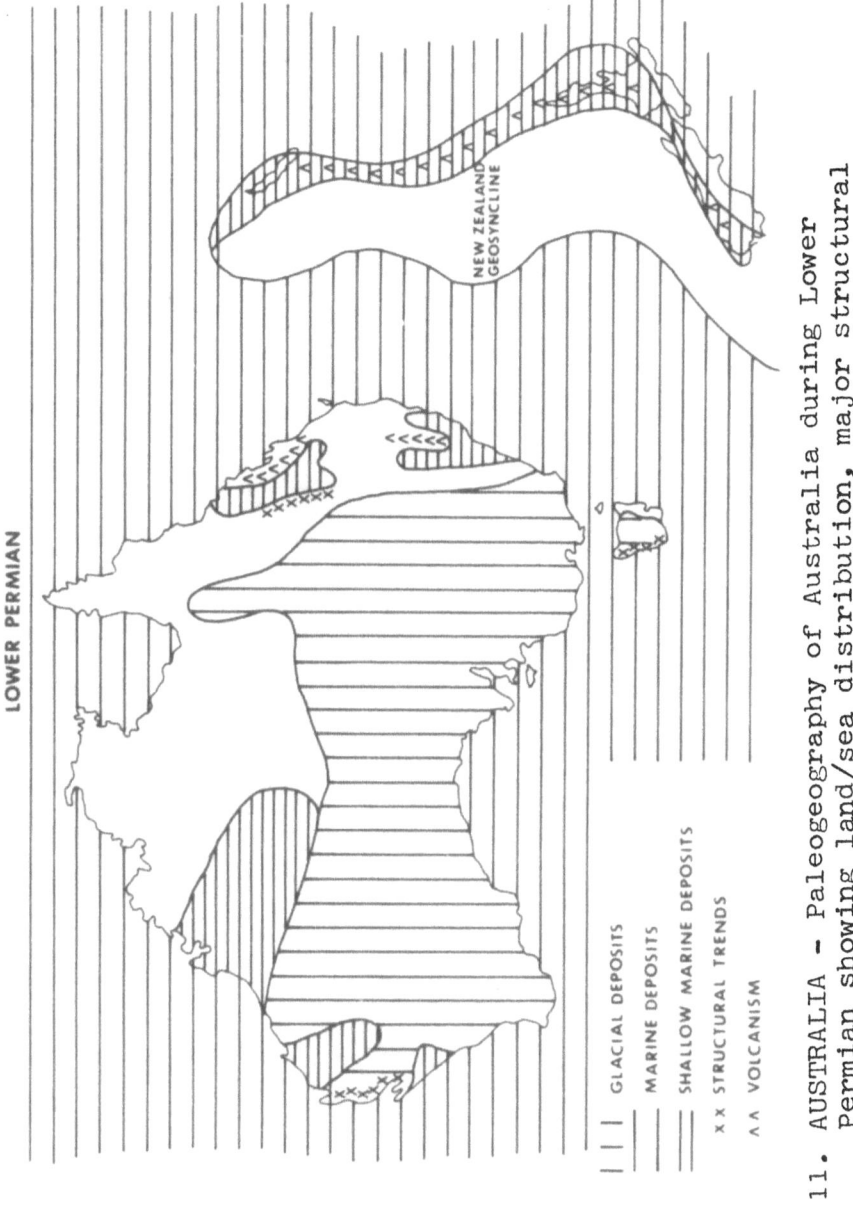

Fig. 11. AUSTRALIA - Paleogeography of Australia during Lower Permian showing land/sea distribution, major structural trends and some paleogeographic indicators.

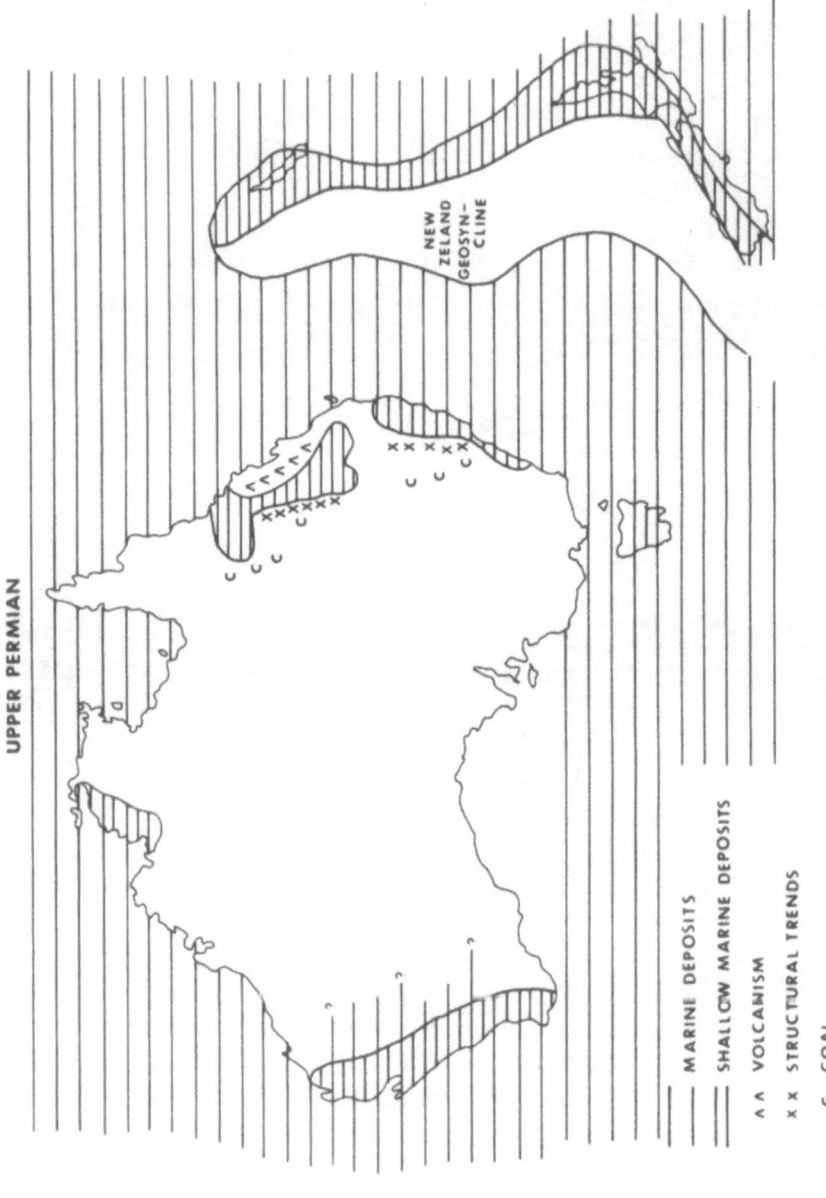

Fig. 12. AUSTRALIA – Paleogeography of Australia during Upper Permian showing land/sea distribution, major structural trends and some environmental indicators.

southerly of which was probably marine. Marine horizons have also been recorded in Southwest Africa (36) which suggest the existence of a sea in much the same area as that during Cape Supergroup times (37). The source of much of the sediment lay to the south, the region of the Agulhas bank and the Cape folds.

Wherever represented the Upper Permian is represented by a continental facies and known in South Africa for its vertebrate fauna.

In Madagascar the Lower Permian is primarily non-marine with a thin development of coal; however, towards the top, lens of limestone with brachiopods indicate a brief marine incursion. During the Upper Permian marine conditions were well established in the northern part of the island (with a fauna showing affinities with the Salt Range.

2.7. Australia

The Permian record of glaciation in Australia is well known, but the area covered by the ice sheets is perhaps still open to discussion. The arguments for the ice sheet being of continental type are reviewed by Crowell and Frakes (21). The evidence of marine invasion seems to be confined to relatively small marginal basins, the most prominent of which is the broad downwarp extending from southern Queensland (Bowen Basin) to the New South Wales coast (Sydney Basin), separated by a narrow ridge from the Gympie and Manning-Macleay Basin. In the west an Artiskian marine transgression is recorded in the Canning Basin, and the Bonaparte Gulf deposits are regarded as glacio-marine.

Tasmania was largely under the Early Permian continental ice sheet; however, in later Permian times the lowlands were on occasion marine. The deltaic sedimentation however suggests a nearshore environment sometimes emergent and sometimes submerged, and emergent phases were related to coal formation.

New Zealand is characterized by a very thick (up to 20,000 meters) section of which the lower part consists of volcanics or volcanically derived sediments, laid down in the New Zealand geosyncline in which developed a marginal facies of shelf limestones and conglomerates along western and southern and possible southeastern flanks (Hakonui facies). There was early Permian tectonic activity with thick basalts, spilites and serpentinites. The source of sediments was older regions now represented by coastal Westland and Fiordland regions of the South Island (19).

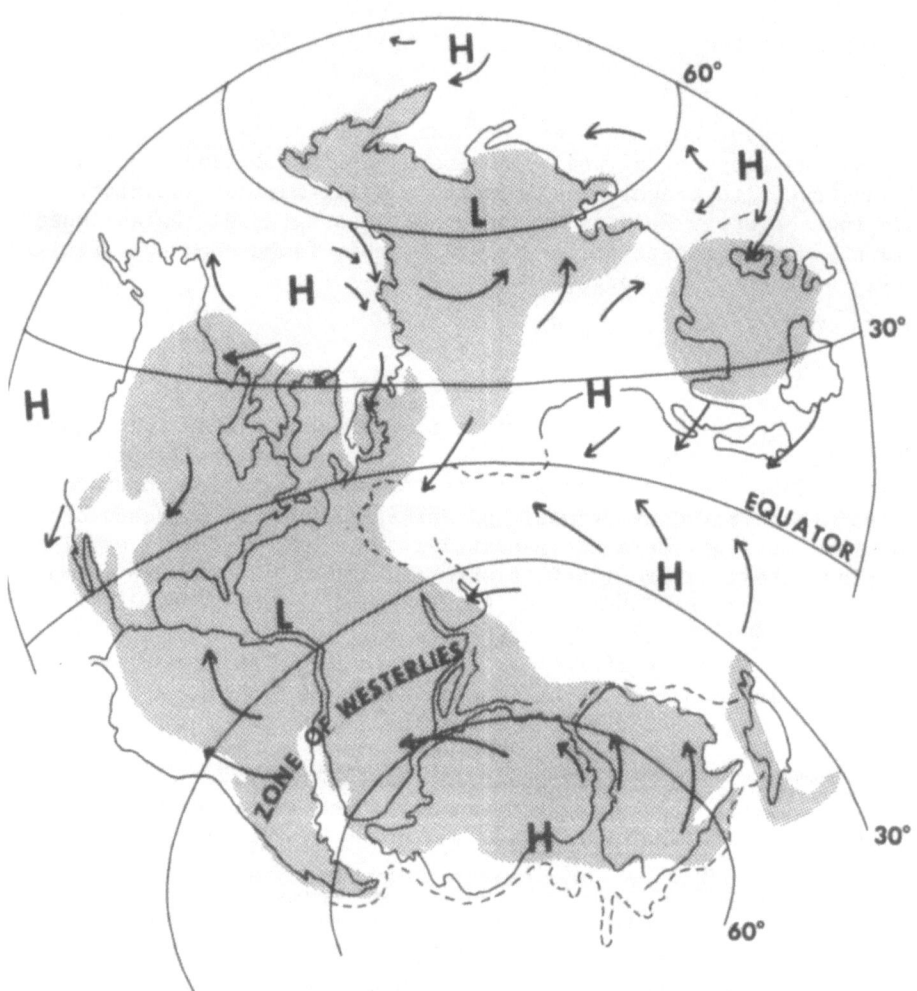

Fig. 13. Tentative surface wind circulation pattern
for the Lower Permian during the northern
summer.

LOWER PERMIAN NORTHERN WINTER

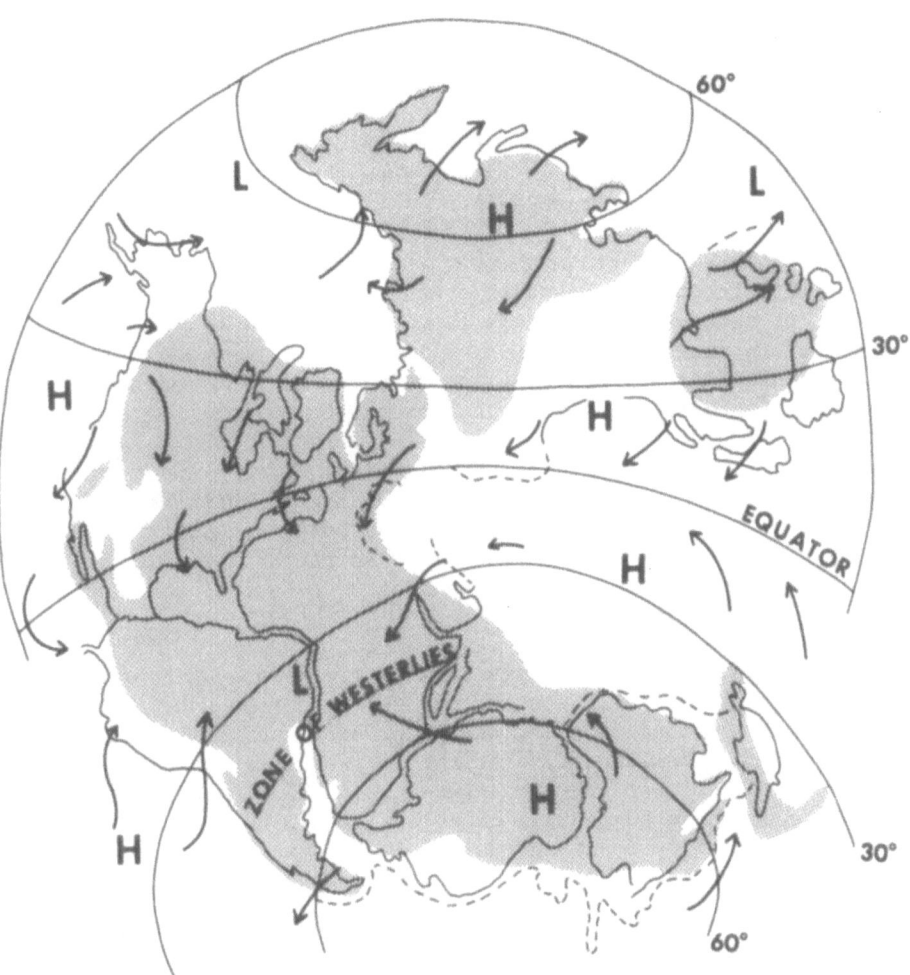

Fig. 14. Tentative surface wird circulation pattern
for the Lower Permian during the northern
winter.

3.   PALEOCLIMATES

The "paleoclimatic" maps are empirical attempts to figure
the surface wind patterns for the Upper and Lower Permian summer
and winter seasons, respectively. They are based upon a number
of assumptions. The principal assumption is that the location of
the intertropical convergence and the sub-tropical high is dominated by the earth's rotation, that the intertropical convergence
will follow the heat equator (i.e., it may diverge from the actual
equator by a few degrees), and that the sub-tropical high is unlikely to be displaced polewards more than five to six degrees
(38). Polar highs will also exist, though the location of their
limits is more difficult to define.

The paleogeographic maps show the existence of a land hemisphere with its median, Tethys, ocean (Figures 14 - 17) and an
oceanic hemisphere not figured. The chief effect of the land/sea
distribution lies in whether it will engender a monsoon effect or
not. Insufficient information is assembled here to consider the
ternary effects of mountain ranges except on a local scale.

During the Lower Permian the principal effects appear to be
the development of a monsoonal effect in the southern hemisphere
during the austral summer and the strong polar high both summer
and winter associated with a land mass in part polar (Figures 14,
and 15). The northern land masses appear to be too broken up to
generate a northern monsoon.

The Upper Permian, consistent with the paleogeographical
changes indicated, indicates changes in the northern hemisphere
where now a monsoonal effect, though probably weak, matches that
in the southern hemisphere. (At the present time, we have not
attempted to estimate pressure gradients.) In Gondwana, whose
general appearance suggests a close analogue with present Eurasia
(excluding the many mountain chains), the polar high is weaker
than during the Lower Permian, with the monsoonal effect remaining relatively unchanged (Figures 15 and 16).

In both Upper and Lower Permian over the land hemisphere
figures there is no obvious westerly wind belt. This nevertheless
does occur, much as at present day. with cyclones associated with
the polar front breaking through between the polar and subtropi-
cal highs. Over the oceanic hemisphere the pattern may be presumed similar to the southern hemisphere pattern of today.

We wish to stress that the maps given are empirical working
models, and must therefore not be considered in any way as final.
As data is improved a computor similation model may be attempted.

UPPER PERMIAN – NORTHERN SUMMER

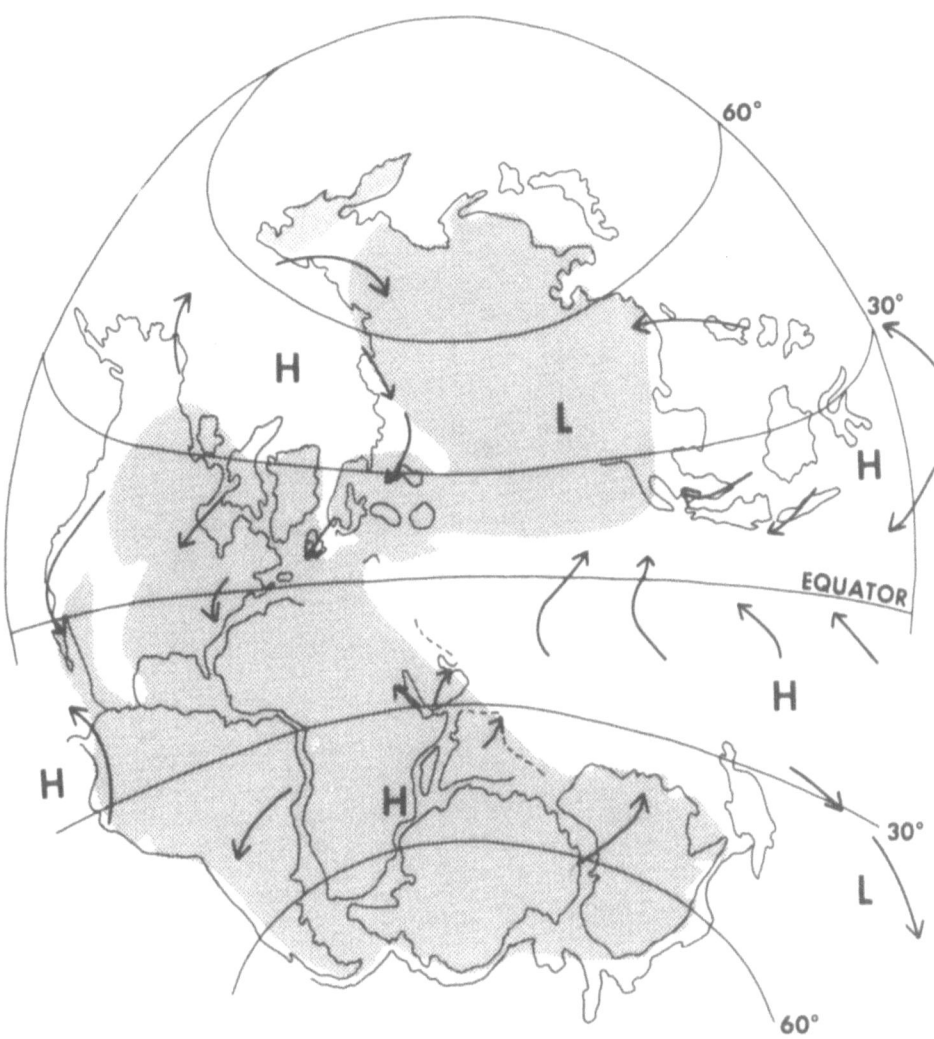

Fig. 15. Tentative surface wind circulation pattern during the Upper Permian during the northern summer.

UPPER PERMIAN - NORTHERN WINTER

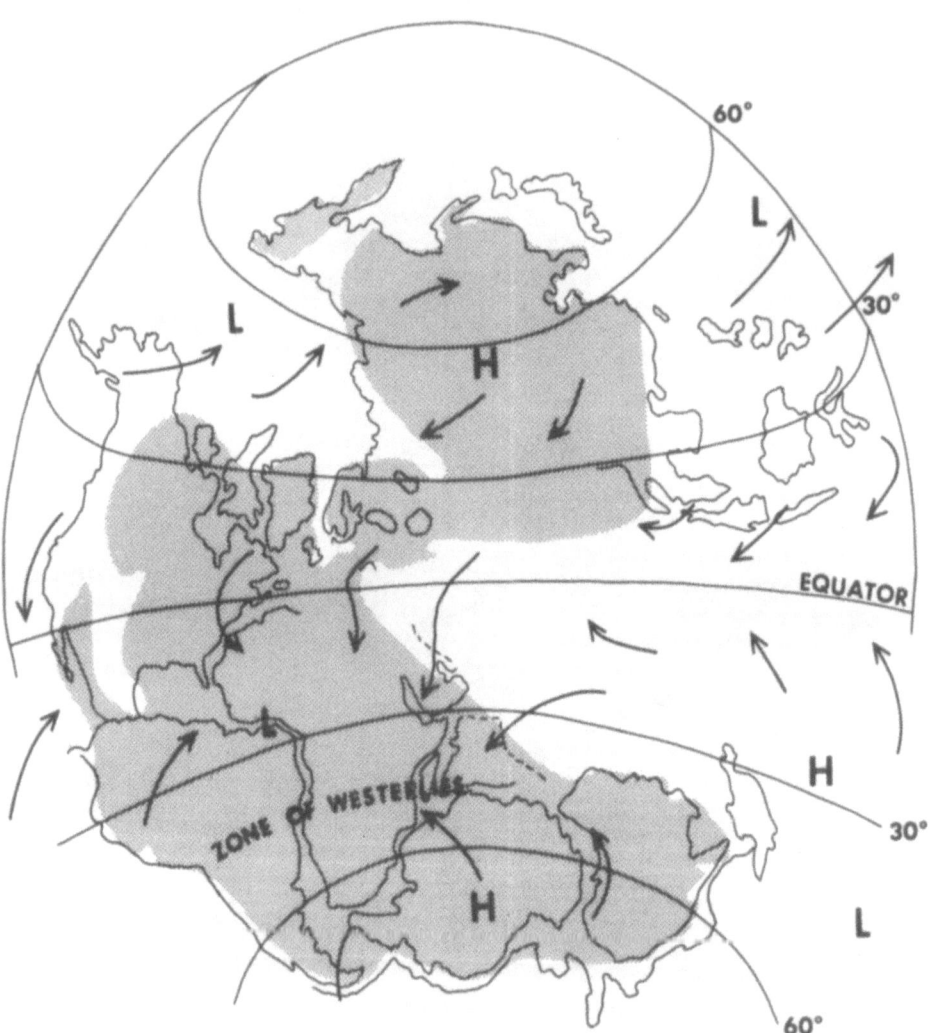

Fig. 16. Tentative surface wind circulation pattern in
the Upper Permian during the northern winter.

Given the predicted wind pattern, the main oceanic current pattern can be derived. These currents and zones of upwelling serve to stabilize the air circulation patterns above them -- an example of a feed-back mechanism. Although paleontologically of great importance, climatologically their importance is restricted.

4. DISCUSSION

Given the empirical nature of the wind circulation pattern, a detailed discussion of regional climates is premature; however, we shall consider a few regions in terms of prediction and geological observation.

4.1. Western United States

There is little difference between Lower and Upper Permian. The area lies under the high pressure cell with wind tracks continental, and this would suggest dry climate. Geologically the evidence of redbeds, evaporites and the measured wind directions are consistent.

4.2. Laurasia

During the Lower Permian northern summer a weak low pressure cell develops over the main Asian land mass, with high pressure over the "Arctic" sea and in the area of present southeastern Asia. In winter a high replaces the Asiatic low, the southeast Asia high remains, but a low dominates the sea to the northwest. The principal source of precipitation in Europe is thus likely to be from marine "Arctic" air drawn in during the summer season, as it rises over the Hercynian mountains. This is consistent with the thick sedimentation of the Saar-Nahe basin. It can also explain the relative paucity of sediment on the lee side of the mountains, where precipitation may be from occasional thunderstorms. The climate during the Upper Permian is not ameliorated by the development of the monsoon effect. Eastern Asia should experience relatively well watered conditions, with the interior dry largely because of distance from the sea.

4.3. Gondwana

The climate of Gondwana is dominated by the strong polar high, which exists summer and winter, and the summer low inducing a summer monsoon. The polar high is weaker in the Upper Permian as the land mass moves away from a polar location, and at this time Gondwana presents a strong analogy with present Asia. The

region of East Africa should have been a well watered region passing to northwestern Africa. The polar high however would ensure
that southern Africa, Australia and Antarctica would remain cold
and relatively dry, with cyclonic precipitation from the wester-
lies. The weakness of the high pressure cell in the Upper Permian
leads to a climatic amelioration. This seems born out by the
final disappearance of the Gondwana ice sheet and the development
of coals, regarded as temperate in Africa, India and South America.

In summary the circulation model, derived without respect to
geological evidence of climate, superficially seems to provide a
satisfactory enough interpretation of the climatic data to warrant
further refinement.

REFERENCES

1.   M. Schwarzbach The Climatic History of Europe and North
     America, in: Descriptive Paleoclimatology ed. by A.E.M.
     Nairn, Interscience Publishers, New York and London, 1961.
2.   F. Lotze, Steinsalz und Kalisalze, 2, Borntraeger, Berlin
     1957.
3.   F.G. Stehli, Am. Jour. Sci., 255, 607, 1957.
4.   C.E.P. Brooks, Climate Through the Ages, Benn. London, 1926.
5.   W. Koppen and A. Wegener, Die Klimate der Geologischen
     Vorzeit, Gebruder Borntraeger, Berlin, 1924.
6.   E. Bullard, J.E. Everett, and A.G. Smith, Phil. Trans R.
     Soc. London, A 258, 41, 1965.
7.   R.S. Dietz and J.C. Holden, J. Geophys. Res., 75, 4939, 1970.
8.   J.C. Briden, A. G. Smith and J. T. Sallomy, Geophys. J. R:
     Astr. Soc., 23, 101, 1971.
9.   A.G. Smith and A. Hallam, Nature, 225, 139, 1970.
10. R. Van der Voo and R. B. French. Earth-Sci. Rev., 10, 99,
     1974.
11. N.M. Strakhov, Principies of Lithogenesis 1, Oliver and
     Boyd, Edinburgh and London 1967.
12. F. G. Stehli, Geol. Soc. Am. Bull., 78, 455, 1967.
13. R. Green, Paleoclimatic Significance of Evaporites, in:
     Descriptive Paleoclimatology, ed. by A.E.M. Nairn, Inter-
     science Publishers, New York and London, 1961.
14. R. Brinkmann, Abriss der Geologie, Ferdinand Enke, Stuttgart,
     1966.
15. H. Termier and G. Termier, Histoire Geologique de La Bio-
     sphere, Masson and Co., Paris, 1952.
16. N.M. Strachow, Palaoklimatologische Weltkarten, 1963.
17. K.W. Glennie, Bull. A.A.P.G., 56, 1048, 1972.
18. P.J. Burek, Structural Deduction of the Initial Age of the
     Atlantic Rift System, in: Implications of Continental
     Drift to the Earth Sciences, ed. by D. H. Tarling and S.K.

Runcorn, 2, Academic Press, London and New York, 1973.

19. D.A. Brown, K.S.W. Campbell, and K.A.W. Cook, The Geologic Evolution of Australia and New Zealand, Pergamon Press, Sydney, 1968.

20. E. Stump, Earth Evolution in the Transantarctic Mountains and W. Antarctica, in: Implications of Continental Drift to the Earth Sciences, ed. by D. H. Tarling and S. K. Runcorn, 2, Academic Press, London and New York, 1973.

21. L. A. Frakes, J. L. Matthews and J. C. Crowell, Bull. G.S.A., 82, 1581, 1971.

22. H.J. Harrington, Bull. A.A.P.G., 46, 1773, 1962.

23. J.J. Bigarella, Geology of the Amazon and Parnaiba Basins, in: The Ocean Basins and Margins, ed. by A.E.M. Nairn and F.G. Stehli, 1, Plenum Press, New York, 1973.

24. E.D.McKee, S.S. Oriel et al., Paleotectonic Investigations of the Perm System in the United States, 515, U.S. Government Printing Office, Washington, 1967.

25. D.W. Nalivkin, The Geology of the U.S.S.R., Pergamon Press, New York, Oxford, London and Paris, 1960.

26. T. Arkle, Jr., ed., I.C. White Memorial Symposium, West Virginia Geological Survey, Morgantown, 1972.

27. D. Birkelund, A. K. Bridgwater, A.K. Higgins, and K. Perch-Nielsen, An Outline of the Geology of the Atlantic Coast of Greenland, in: The Ocean Basins and Margins, ed. by A.E.M. Nairn and F.G. Stehli, 2, Plenum Press, New York, 1974.

28. Y.O. Fortier et al., Geology of the North Central Part of the Arctic Archipelago Northwest Territories, Geol. Survey of Canada, Dept. of Mines and Technical Surveys, 1963.

29. C. K. Seyfert and L. A. Sirkin, Earth History and Plate Tectonics, Harper and Row, New York, 1973.

30. P. B. King, U. S. Geol. Survey Prof. Paper 215, 1948.

31. N.D. Opdyke and S.K. Runcorn, Bull. Geol. Soc. Am., 71, 959, 1960.

32. E. Lopez-Ramos, A.A.P.G. Bull., 53, 2399, 1969.

33. O. Braitsch, Salt Deposits: Their Origin and Compositions, Springer Verlag, Berlin, 1971.

34. J. C. Mendes the Passa Dois Group, in: Problems in Brazilian Gondwana Geology, ed. by J.J. Bigarella, R.D. Becker and J.D. Pinto, Ind. Symposium on Gondwana Stratigraphy and Paleontology, Curitiba and Parana, 1967.

35. I.W.D. Daziel and D.H. Elliot, the Scotia Arc and Antarctic Margin, in: The Ocean Basins and Margins, ed. by A.E.M. Nairn and F. G. Stehli, 1, Plenum Press, New York, 1973.

36. H. Martin, The Atlantic Margin of Southern Africa Between Latitude 17° South and the Cape of Good Hope, in: The Ocean Basins and Margins, ed. by A.E.M. Nairn and F.G. Stehli, 2, Plenum Press, New York, 1974.

37. Izak C. Rust, The Evolution of the Paleozoic Cape Basin, Southern Margin of Africa, in: The Ocean Basins and Margins, ed. by A.E.M. Nairn and F.G. Stehli, 2, Plenum Press, New York, 1974.

38. J.S. Sawyer, Possible variations of the general circulation of the atmosphere, in: World Climate from 8000 to O.B.C., Roy. Met. Soc., London, 1966.

THE PERMOCARBONIFEROUS BASIN AND RANGE PROVINCE OF EUROPE. AN APPLICATION OF PLATE TECTONICS

V. Lorenz [1] and I.A. Nicholls [2]

[1] Geologisches Institut, Johannes Gutenberg-Universität, 65 Mainz, W. Germany
[2] Department of Earth Sciences, Monash University, Clayton, Victoria, 3168, Australia

ABSTRACT. Distribution and development of European late Hercynian intermontane troughs and associated volcanicity are related to an upper mantle diapir which formed above two subduction zones during the Hercynian cycle of orogenesis. Lateral spreading of the diapir caused regional extension of the previously folded crust and enabled tholeiitic magmas from the upper mantle and rhyolitic magmas from the lower crust to reach the surface.

INTRODUCTION

The Hercynian orogeny in Europe was followed by the formation of a series of intermontane troughs in Upper Carboniferous and Lower Permian time. In addition there was widespread volcanic activity. The area of trough formation and associated volcanism extends from Britain to Poland and from Morocco to Roumania (and Caucasus). At present any analysis of this large province and its continental rocks is hindered by in part poor stratigraphic correlation between individual basins, largely due to limited or poor exposures, and the small number of reliable isotopic age determinations available. In addition, Permocarboniferous terrains in S Europe were affected by the Alpine cycle of orogenesis, and Permocarboniferous rocks in this region are in part dislocated and metamorphosed. Palaeogeographic reconstruction of Permocarboniferous Southern Europe therefore requires major rearrangement of microplates. In

H. Falke (ed.), The Continental Permian in Central, West, and South Europe, 313-342. All Rights Reserved.
Copyright © 1976 by D. Reidel Publishing Company, Dordrecht-Holland.

their analysis of the evolution of the Alpine system
Dewey et. al. (1) gave a reconstruction for the time
prior to Late Triassic. As no major tectonic events
are thought to have taken place between the early Permian and Late Triassic it seems justified to apply the
Triassic reconstruction to relationships in the Permo-
carboniferous. However, for some Permocarboniferous
troughs in S Europe, positions on the palaeogeographic
map can be assigned only very tentatively at the present
time.

Both trough formation and associated volcanism in
Permocarboniferous Europe have usually been considered
as related to the release of compressive forces within
the crust at the end of the Hercynian orogeny, and the
volcanic activity was therefore classed as "subsequent"
(2, 3). Similar features of the tectonic, sedimentary
and volcanic histories of many of the troughs suggest
that large scale processes, involving the upper mantle,
controlled events throughout this large province. This
paper presents a tentative geodynamic model for trough
formation and associated volcanism, and attempts to
relate diverse geological data to plate tectonic processes at the end of the Hercynian orogeny. The model
presented is believed to explain many of the features
of the Permocarboniferous basins, and it is hoped that
it will stimulate future research.

THE HERCYNIAN OROGENY

Plate tectonic interpretations of the Hercynian
fold belt have been presented in a number of recent
papers. Several authors (4-8) propose the former
existence of a mid-European ocean that was consumed
along a southward dipping subduction zone during the
Carboniferous, i.e. during the "Sudetan" and "Asturian"
phases of folding, metamorphism, and granite emplacement within the period 330-290 m.y. B.P. As was pointed
out for the Rhenohercynian zone (9, 10) and for SW
England (7), the zone of folding migrated with time
from S to N. According to Anderson (4), and Dewey &
Burke (6) this migration implies north-ward displacement of a trench system. Final suturing between N and
S European continental plates was completed in Westfalian time, the suture line extending along the
northern margin of the Rhenohercynian zone.

Another subduction zone, along the southern margin
of Southern Europe has been proposed by Dewey & Burke

(6) and Nicolas (11). In his analysis of Hercynian
Southern Europe, Flügel (12) recognized migration of
the orogenesis with time. If the region of the Southern
Alps is rotated back to its original orientation the
Hercynian orogenesis appears to have migrated with time
from NW to the SE, the folding of the Hochwipfelflysch
having taken place only in post-Lower Westfalian time.

Migration of the two proposed subduction zones at
the plate margins of S Europe was thus accompanied by
migration of two fold belts from a central zone out-
wards and resulted in enlargement of the S European
continent. The onset of metamorphism and initiation of
partial melting of the lower crust, yielding granitic
melts, must also have migrated outward. Northward migration with time of granitic intrusive activity near
the northern plate margin has been demonstrated in the
Black Forest region (13). The Hercynian foldbelt is
characterized by a high temperature-low pressure metamorphism (14), and by granites rich in normative Or
(15-18). According to Hall (15-18), both facts imply
a high regional geothermal gradient.

The last major orogenic event of the Hercynian
orogenesis was the "Asturian" event in SW Europe which
Laurent (7) believes to be related to the collision
of Africa and SW Europe. This collision probably led
to the termination of the activity of the southern subduction zone of Southern Europe east of the area of
continental collision.

THE CONTINENTAL PERMOCARBONIFEROUS OF EUROPE
SEDIMENTATION

The main Hercynian ("Sudetan") orogeny was followed
by retreat of the sea, regional erosion and localized
deposition of continental debris eroded from neighboring elevated areas (19). Retreat of the sea is interpreted as due to a regional uplift of Central and
Western Europe which culminated in Lower Permian time.
During the later Permian, however, this regional movement reversed and finally allowed the Upper Permian sea
to flood large areas in the northern part of the province and to a smaller extent also parts of Southern
Europe.

Simultaneously with the regional uplift a number
of troughs formed and collected large amounts of rock
debris eroded from the neighboring mountain ranges.

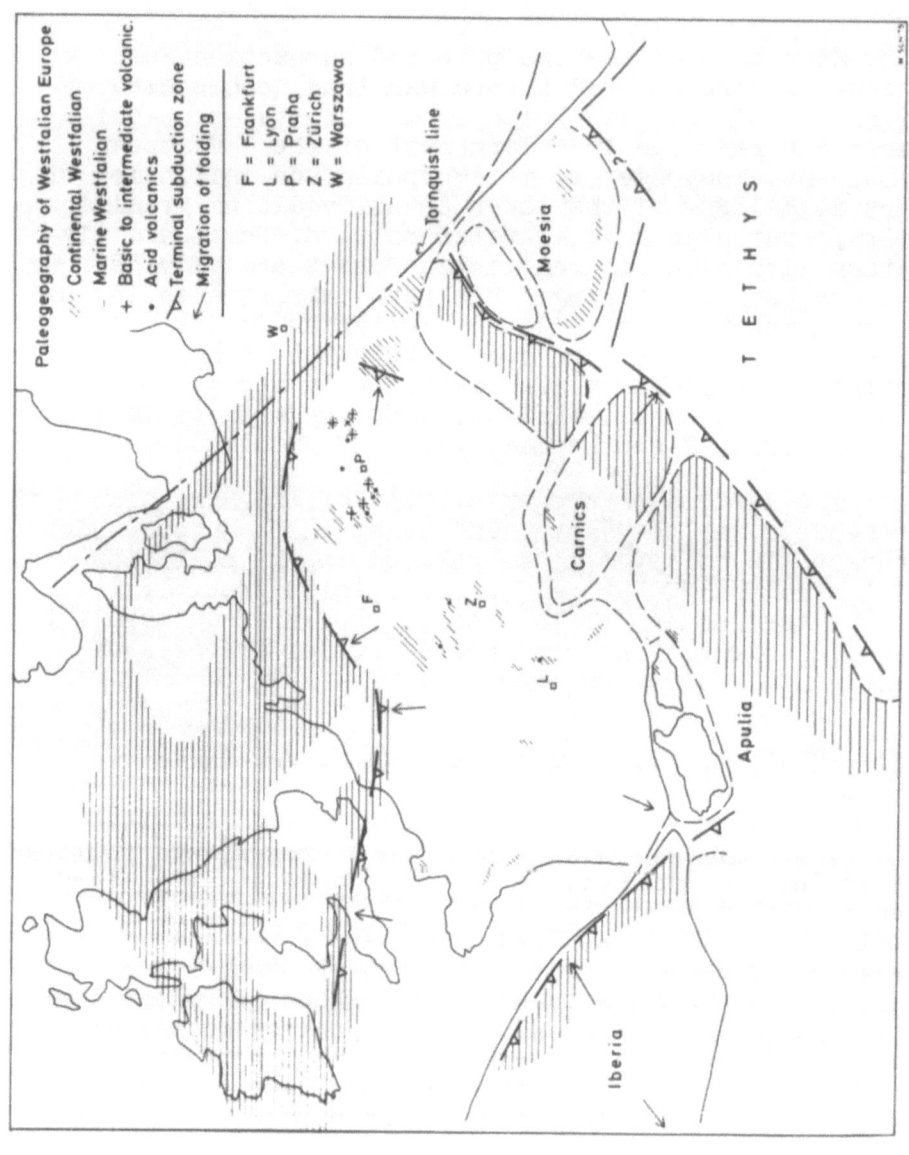

Trough formation apparently started in the uppermost
Viséan and Namurian in a central zone extending from
the Innersudetic Basin in the Sudetic Mountains (20)
through the region of Borna-Hainichen in the southern
GDR (55), the Black Forest and Vosges (21, 22) into
the Brianconnais of the Western Alps (23, 24) and the
Vendée (25). Westfalian continental deposits occur in
intermontane troughs of a much larger area extending
farther to the North (Normandy, Saar-Nahe and Mansfield
district south of the Harz-Mountains) and to the South
(Corsica, Brianconnais Ligure, Manno trough near
Lugano) (Fig. 1).

However, the Westfalian sea still covered or inter-
mittently flooded the Pyrenees, S Portugal, Britain,
Northern Germany, and coastal regions of Southern
Europe (12, 16).

During the Stefanian deposition of continental
debris extended into Britain, and Northern Germany,
pointing to a retarded regional uplift in Britain and
Northern Germany when compared to areas further South
(Fig. 2).

Lower Permian deposits are even more widespread
(27-29) in troughs which during this time increased not
only in number but also in extent. In the North, Permian
deposits occur in Southern Scandinavia and Scotland
(Fig. 3). Thus the margin of the zone of troughs in
which continental debris was collecting had migrated
far northwards by Permian time.

Southern Europe also experienced migration in the
onset of regional uplift and formation of troughs with
accumulations of continental debris from the Westfalian onwards through the Stefanian into the Lower
Permian. A 300-700km wide strip of the continental margin
acted as a shelf during the Upper Carboniferous and
Permian (12). Rau and Tongiorgi (26) have described
coastal Upper Carboniferous deposits on Elba and in
Southern Tuscany, and Lower Permian continental deposits
in these areas indicate progressive uplift, accompanied
by the retreat of the sea towards the South.

Fig. 1. Schematic palaeogeographic map of the Westfalian showing intermontane troughs and associated volcanic
rocks, migration of foldbelts and final position of former subduction zones. Base map in part after Dewey et.
al. (1).

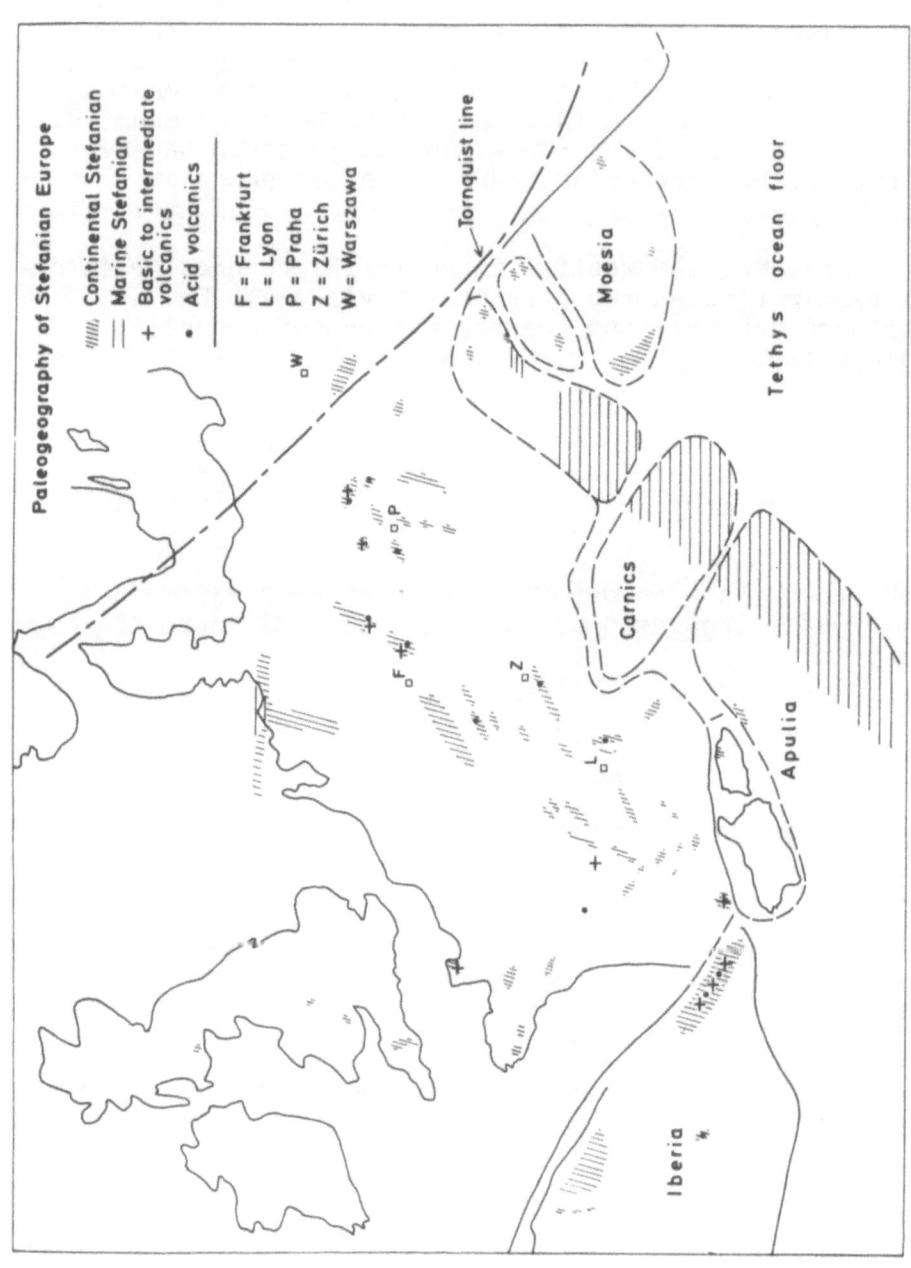

A similar development existed on Iberia where two foldbelts migrated away from the central Galicia-Castile zone and, at least in the South the zone of trough initiation also migrated southwards (30).

Depending on the rate of subsidence of the respective troughs and the amount of debris supplied from the adjoining ranges, coarse to fine-grained clastic sediments accumulated, reaching maximum thicknesses in excess of 5000 m in the Saar-Nahe basin, 4500 m in the Brianconnais, 3000 m in SW England, the Thuringian basin, and the Mescek. In a further 25 basins the thickness of the trough deposits is at least 1000 m.

The deposits frequently alternate in cyclic sequences, initiation of which suggests increased relief energy, and which are believed to indicate periods of increased subsidence of the respective basins. Associated coal seams are widespread in the continental Upper Carboniferous and to a smaller extent in the Lower Permian.

The sediments deposited late in the Lower Permian history of some troughs (North of the Alpine region) are fine-grained and non-cyclic, and indicate termination of the subsidence processes, filling of the basins and progressive levelling of the ranges (28). Steep morphological gradients were no longer present and in many areas a peneplain was established (19, 26, 31). Between the Lower Permian and Upper Permian the area then subsided to allow the Upper Permian sea to flood parts of Britain, Germany, Denmark and Poland. In Upper Permian time the sea also invaded Southern Europe including the present areas of Elba, the Apuane Alps, the Dolomites (26).

The Verrucano

The Upper Permian Verrucano deposits of the Alpine region are mostly represented by quartz-rich conglomeratic rocks and laterally equivalent sandy rocks, which in general unconformably overlie the Lower Permian sediments (23, 26, 32-35). Whereas the Lower Permian sediments are restricted to basins, the Upper Permian Verrucano was deposited on larger plains on the edge

Fig. 2. Schematic palaeogeographic map of the Stefanian showing intermontane troughs and associated volcanic rocks. Base map in part after Dewey et. al. (1).

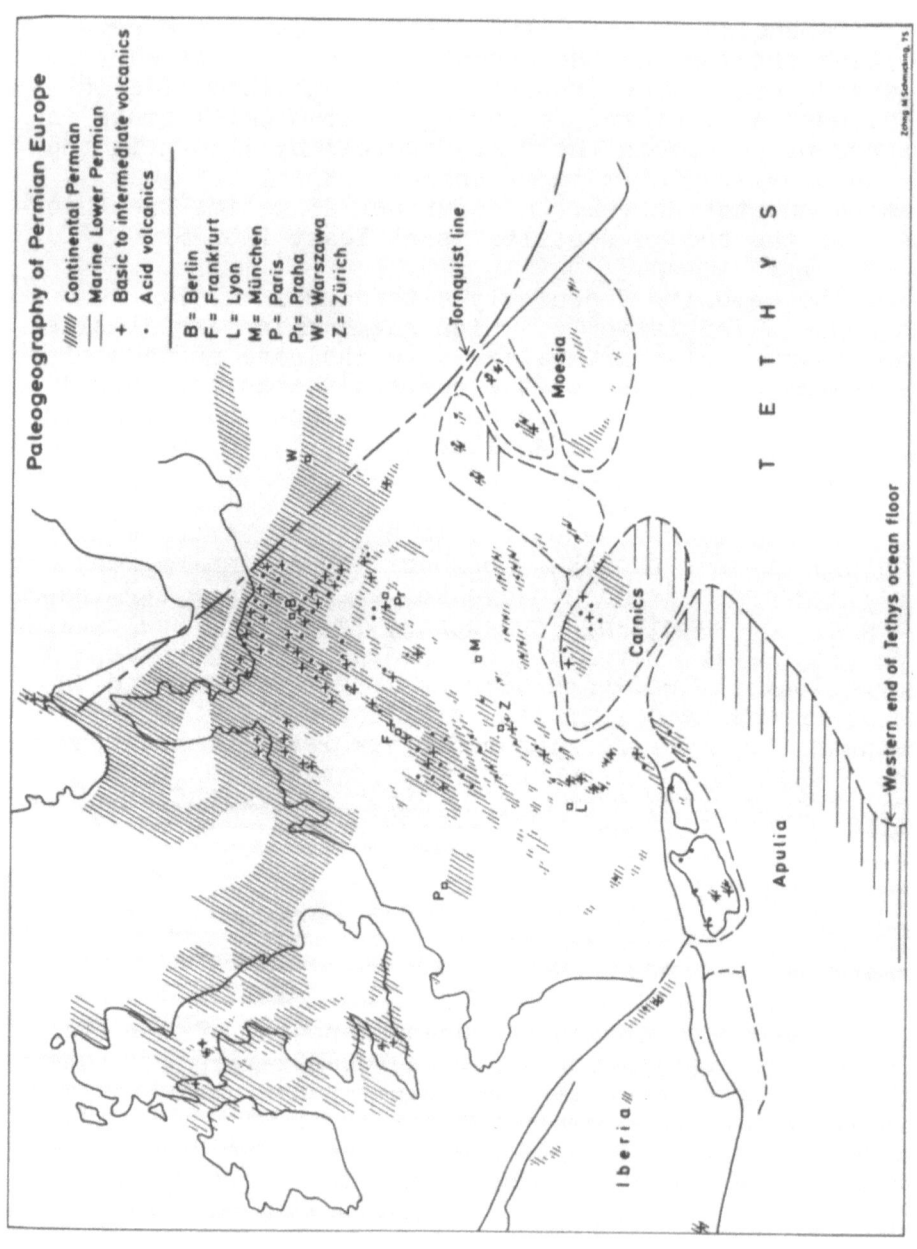

of rapidly eroding highlands (26, 35).

According to Visscher's (36) palynological studies
the Verrucano Lombardo and Verrucano Brianconnais are
confined to the uppermost Permian (Thuringian). Equivalent conglomeratic rocks of the same age do not exist
in Central and Northern Europe. This indicates the
occurrence of a new tectonic event, confined to the
Alpine region, which triggered uplift, consequent erosion and sedimentation, yielding very coarse-grained
rocks. It is suggested that this event represents the
thermal expansion and uplift of the crust prior to the
development of an embryonic rift (in the sense of
Sleep, (37)). Later opening of this rift caused formation of the ocean basin separating the Helvetic and
Penninic domain from the Austroalpine domain (38-41).

TECTONISM

Of special interest in this context is the mode of
formation of the troughs or basins and their relevant
internal structures.

Typical troughs exist only SW of the Baltopodolian
line or Tornquist line, the boundary between stable
Fennosarmatia and mobile Hercynian Europe. The northwestern margin of the zone of troughs extends from the
Oslo graben to Scotland. The southeastern margin is
given by the continental margin or more precisely by
the Southern European coast. And towards the SW the
area of trough formation extends through Spain into
Morocco. In general, Permian volcanism is also confined
to this area.

The trend of the troughs closely follows previous
structural trends (19, 27, 28), i.e. NE/SW and NW/SE in
Hercynian Europe, N/S in Britain (Caledonian and Precambrian trends (19) and NNE/SSW in the Oslo graben
region (Precambrian trend (42)).

Some troughs are known to be bounded on at least
one side by faults which were active during their formation, e.g. the Oslo graben (42), the Saar-Nahe basin,
the Wittlich basin (27, 28), the Midland basins (19),
the Glarus basin (35, 43), the Brianconnais trough (33,

Fig. 3. Schematic palaeogeographic map of the Permian
showing intermontane troughs and associated volcanic
rocks. Base map in part after Dewey et. al. (1).

44), Collio Basin (26), and the Bohemian furrows (20, 29). The great thicknesses and generally coarse grain-sizes of the deposits, the small width of basins and the general lack of internal morphology indicate that the troughs either formed as grabens, or tilted blocks as in the case of some of the troughs in Britain (19). A number of these grabens were continuously active from Westfalian to Lower Permian times.

Formation of such grabens or semigrabens requires extension of the crust. The recent investigation of the NW boundary fault of the Saar-Nahe basin shows that at present this fault extends into the upper mantle (45). Along this major fault the basin subsided through extension of the crust. The formation of a large zone of similar troughs implies spreading and thinning of the crust for a prolonged period of time on a regional scale. Intensive regional extension of the crust is also indicated by the widespread, mainly Permian basaltic volcanism which required magma con-duits extending into the upper mantle.

Since no major compressive stress field has affected the province north of the Alps since the Upper Carboniferous, the result of crustal thinning should still be measurable today. Later processes of erosion and sedimentation have probably modified crus-tal thickness only slightly, while underplating of basaltic material during Permian and Tertiary magmatism may have caused local thickening. Hence the present day Western European crustal thickness of around 30 km should be closely similar to that established at the end of the Lower Permian but slightly less than that at the end of the Hercynian orogeny.

As discussed above, the first troughs formed in the Late Viséan and Namurian, indicating an initial central zone of crustal extension running south-west from the Sudetic Mountains to the Massif Central and Vendée. Folding of Namurian age in regions to both north (9) and south of this zone (12) indicates the presence of contemporaneous marginal zones of crustal compression. The northern and southern zones of fol-ding persisted into Westfalian times, and they and the zones of trough initiation migrated northward and south-ward with time. The overall result was a near-symmetric increase in the area containing active extensional troughs. Once formed, basins within a given area under-went active subsidence over a long period, usually extending into the Lower Permian.

As indicated above, the troughs are oriented
along different Hercynian or older structural trends,
and crustal spreading apparently took place in a
number of directions simultaneously. In the areas
where trends change direction, faults displaying a
large strike slip component, and probably also com-
ressional features, should be present. Structures of
this kind are described from the Elbe line and the
Grand Sillon d'Houiller (46) both in areas where the
main structural trends of the Hercynian fold belt and
Permocarboniferous troughs change direction. On a
smaller scale such features should also exist within
some individual basins because spreading rates are
unlikely to have been uniform along the length of at
least the larger basins.

UNCONFORMITIES

In many basins unconformities within the Upper
Carboniferous and Permian deposits have been attributed to the "Asturian" and "Saalian" phases of orogenesis. In a number of basins on the S Europe plate
the "Asturian" unconformities may represent either
long-range effects of processes at the plate boundaries (continent-continent collisions of S Europe
with Africa in the SW and with Northern Europe and
N America in the N and NW). Or they may represent
short-range effects of subsidence leading to graben
formation (see below). In most cases, subsidence and
sedimentation indicative of continued extension recommen-
ced within a short time.

Unconformities within the Permian deposits are
attributed to the "Saalian" orogenic event, and define
the boundary between the Lower Rotliegende (Autunian)
and Upper Rotliegende (Saxonian). With the possible
exception of SW Europe and Morocco where post-Autunian
deformations might still have been related to suturing
of the three plates the causes of the intra-Permian
"Saalian" unconformities are not well understood and
of doubtful stratigraphic value. Three such unconformities occur in the type locality - the Saale basin -
two of which are associated with the emplacement of
the Halle rhyolite (27, 28, 47). In a number of basins,
similar Permian unconformities are present only in
marginal zones e.g. the Glarus basin (35, 43, 48), and
in others, they are completely absent e.g. the Lodeve
Basin (49) and the Vanoise-Mont Pourri Zone (23). In
several cases, there is a clear relationship between

unconformities and volcanic features, as in the Saar-
Nahe basin (50, 51) and the Glarus basin (43). In
some areas of Southern Europe, volcanic activity has
been taken as the only expression of the "Saalian"
phase (26, 33), while in SW England, GDR and CSSR,
volcanic activity predates unconformities referred
to the "Saalian" phase. And in some cases, several
widely separated volcanic events have occurred e.g.
in the Westfalian, Stefanian, and Permian of the Innersudetic basin (52).

It is clear that no concensus on the nature of
the "Saalian phase" has been reached. It is therefore
suggested that the use of the terms "Saalian phase",
"Saalian movements", and "Saalian unconformity" be
discontinued. Further grounds for this suggestion are
as follows:
1) In rapidly subsiding intermontane troughs, marginal
unconformities are to be expected due to local variation
in the rates of the basin subsidence and sedimentation.
Correlation of such unconformities with a regional
orogenic event, as in the current usage of the term
"Saalian phase" is not justified.
2) Unconformities within basins may be related to the
formation of central anticlinal uplifts, examples of
which occur in the Saar-Nahe basin, in the Massif
Central (53) and the Orlice basin in Bohemia (29). The
central anticline of the Saar-Nahe basin is known to
have had a long history, beginning as a central area
of diminished subsidence and sedimentation (54). The
formation of such structures is likely to reflect only
local processes. Large calderas and impact structures
often contain central uplifts, termed resurgent domes,
which are interpreted as due to rebound of crustal
rocks. Experiments on graben formation (55) show
similar central uplifts. These, and equivalent natural
features in Permocarboniferous basins, may well represent similar rebound structures, formed in response
to rapid rates of initial subsidence.
3) Unconformities related to volcanic activity depend
on the type of volcanic feature formed and the geological relationship to its surroundings, as noted by
Lorenz (51) for the case of rhyolite domes in the
Saar-Nahe basin.

VOLCANISM

Volcanic rocks are a characteristic feature of
the troughs throughout the province (Fig. 1-3). A range

of rock types is present, but extensive alteration and frequent lack of petrographic and petrologic data make it difficult to generalize the nature of volcanism and its products. Future investigations in this field should provide further insight into the nature of petrogenetic processes in the upper mantle and crust and the validity of the petrogenetic model presented in the following sections.

In the larger basins rhyolites/dacites and basalts/ basaltic andesites form bimodal suites, rocks of inter- mediate composition being relatively rare. In small troughs such as those of the Black Forest only rhyolitic rocks or intrusive equivalents are developed. Occur- rences of basaltic rocks thus seem to be confined to basins formed by more pronounced crustal extension.

Activity of rhyolitic magmas led to formation of large intrusive domes, lava flows, ignimbrites and extensive pyroclastic air-fall sheets. Basaltic rocks occur as dykes, sills, extensive plateau-type lavas and pyroclastic sheets. Owing to synsedimentary vol- canism in basins into which large amounts of water carrying sediment had flowed and within which con- sequently large volumes of groundwater were present, a large proportion of the pyroclastic rhyolitic and basaltic rocks is believed to be phreatomagmatic in origin (56, 57). Intrusion of basaltic sills and rhyolitic domes is also believed to point to the in- fluence of synsedimentary volcanism where magma could not break through the great thicknesses of wet and incompletely consolidated sediments (51, 57).

The beginning of volcanic activity in the Permo- carboniferous basin and range province seems to have been contemporaneous with the formation of the basins in the central zone which extended from S. Poland (through S DDR, Bohemia, Black Forest, Vosges, Brian- connais) to the Massif Central and Vendeé in France. In this zone continental volcanism was active from the Upper Viséan until the Lower Permian. Only at the Stefanian/Permian boundary did the volcanism spread over the whole province (Fig. 2, 3) pointing to a period of intensified crustal extension which allowed large volumes of magmas access to the surface.

In certain areas the onset of volcanic activity migrated with time. Thus in the S DDR the age of the earliest volcanic rocks decreases toward the NW (107). In the Glarus basin, activity migrated southward with time (35).

The Lower Permian volcanism was apparently restricted to parts of the European continent above sea level,
and no volcanic rocks have been reported from the extensive Permian shelf region of southern Europe.
Activity on the subaerial continent was confined mainly
to the basins, but less commonly it occurred on ranges,
as in the southern DDR (52,58) and the Bolzano area.

PETROLOGY

Many of the basaltic rocks of the Permocarboniferous basins were originally termed melaphyres. They
are often highly altered, and chemical analyses of fresh
examples are rare. Relatively fresh, glass-bearing basalts from two diatremes in the Saar-Nahe basin have
20-30% hypersthene, and 3% olivine to 5% quartz in the
CIPW norm, and are therefore classified as olivine
tholeiites or quartz tholeiites (59,60). However, they
have high and variable potassium contents (1.1-2.0%
K_2O), and K_2O/Na_2O (0.4-0.9) and K_2O/SiO_2 values similar to those of the calcalkaline association of island arcs and continental margins. Similar features
have been noted for basaltic rocks from other parts
of the Saar-Nahe basin (e.g. 61).

The presence of xenocrysts of quartz and sodic
plagioclase, high contents of Rb, Cs, U, Th, Pb, and
light rare earth elements, and high initial $^{87}Sr/^{86}Sr$
(0.7059-0.7069) (62) indicate contamination with sialic
crustal material, and suggest that the high K_2O contents of these basalts do not reflect the chemistry of
the mantle source regions. On the other hand, high
contents of MgO, Ni and Cr, and the presence of high
pressure phenocrysts of orthopyroxene, clinopyroxene
and olivine demonstrate that the contaminated basaltic magmas underwent little fractionation at crustal
levels.

Phase relationships for one of these basalts show
that a parental magma could have been in equilibrium
with olivine of a peridotitic mantle to a maximum
pressure of around 10kb, and Nicholls & Lorenz (59)
interpreted this result to indicate a maximum crustal
thickness of about 30km beneath the Saar-Nahe basin in
the Lower Permian. On the basis of estimated water
contents during crystallization of the basaltic rocks,
Nicholls & Lorenz also suggested that the upper mantle

was relatively rich in water (0.4-0.8% H_2O), part of
which was probably derived from dehydration of oceanic
lithosphere during a prior episode of subduction.

The few available analyses of mafic rocks of
other Permocarboniferous basins on the former southern
European continent - CSSR (Holub, 1972) including the
western Carpathians (63), GDR (64, 65), the Harz (66,
67), the Massif Central (68), and in Alpine areas of
France, Switzerland and Italy (44, 69-72) - show that
these rocks are in general too highly altered for their
original geochemical affinities to be clear. Contents
of Al_2O_3, Fe_2O_3, K_2O and H_2O are often anomalously
high, and CaO in particular is often very low. However,
contents of TiO_2, which are considered to be little
effected by hydrothermal alteration, are low (<1%), a
feature of the fresh rocks of the Saar-Nahe basin (59).
In all these areas, the volcanic rocks are neither
highly mafic, nor highly alkaline, suggesting that
they were derived from magmas of shallow mantle origin
(see e.g. 73).

On the other hand, Permocarboniferous mafic volcanic rocks of the former northern European continent
- SW England (74), Scotland (75), the Oslo province
(see review by Heier & Compston, 1969) (76), and
northern Germany (77, 78) are often much more alkaline,
characteristically more mafic, and almost always richer
in TiO_2 than their southern counterparts. They were
probably derived from basaltic magmas of deeper origin
and in the case of highly alkaline types from melting
within the mantle low-velocity zone (see e.g. Green
(73, 79)).

The large volumes of acid rocks associated with
Permocarboniferous basins are usually slightly younger
than the oldest associated mafic rocks (26, 52, 78).
Like the latter, they are often highly altered, and
available analyses frequently show extreme enrichment
in K_2O, strong depletion in Na_2O or both (see references for mafic rocks above; (33, 80, 81)). As a result
almost all have high normative Or/Ab, a characteristic of associated granitic intrusives (e.g. 15-18).

Few trace element and isotopic data are available
for the Permocarboniferous acid volcanics and associated intrusives. However, detailed studies by Hahn-Weinheimer & Ackermann (82) and Brewer & Lippolt (83) in
granitic rocks of the southern Black Forest have demonstrated high K_2O/Na_2O (1.0-1.3), high Rb and Ba and

high initial $^{87}Sr/^{86}Sr$ (0.710-0.721) for rocks post-
dating the Sudetan phase of the Hercynian orogeny
(dated at 330Ma).

The geochemical characteristics of the acid rocks
indicate derivation from magmas of crustal origin and
a number of workers have presented petrological evidence in support of this view (84, 85). In his study
of the rhyolitic rocks of the Saar-Nahe basin, Theuerjahr (85) suggested that the formation of acid magmas
took place at pressure of around 10kb, corresponding
to depths of the lowermost crust. However, Hall (15-18)
indicated shallower dephts of origin for Hercynian
acid magmas with high Or/Ab and on these grounds, proposed that partial melting of crustal material took
place under a high geothermal gradient.

COMPARISON WITH ROCKS OF YOUNG ZONES OF CRUSTAL EXTENSION

Bimodal volcanism has been observed in a number
of areas for which a tectonic environment similar to
that of Permocarboniferous Western Europe may be inferred. The most intensively studied example is that
of the western U.S.A. in the Late Cenozoic. Recently,
Lipman, et. al. (86) and Christiansen & Lipman (87)
have discussed in detail the nature of Cenozoic volcanism in this large province, and have examined the
large and small scale tectonic processes associated
with magmatic activity. A major feature is a change
from orogenic and post-orogenic calc-alkaline volcanism,
grading eastward into more alkaline and potassic suites,
in the early and middle Cenozoic, to mafic or bimodal
volcanism, associated with extensional faulting, in the
late Cenozoic. Christiansen & Lipman (87) relate these
changes in tectonic and volcanic processes to the termination of active subduction of oceanic lithosphere
beneath the American plate as the East Pacific Rise
collided with the continental margin.

The Great Basin of SW Nevada provides a good
example of a late Cenozoic bimodal volcanic province.
Here, as in much of the western U.S. the basalts of the
post-subduction phase of volcanism are mainly alkali
olivine basalts. Analyses of associated rhyolite ash-flow
volcanics (e.g. 88) show $K_2O/Na_2O > 1$. Further to the
east, in the Southern Rocky Mountains province, studies
of the basaltic rocks have demonstrated the role of
crustal contamination in producing high K_2O contents

and high $^{87}Sr/^{86}Sr$ (89). Here also, tholeiitic basalts are found in zones of major crustal extension (90).

The striking similarities between the tectonic and volcanic histories of Permocarboniferous Europe and the Western United States in the Cenozoic have previously been noted by Nicholls & Lorenz (59). For this reason, the model of tectonism and volcanism in Permocarboniferous Europe presented in the following section is based on a model for the Great Basin presented by Scholz, et. al. (91).

MODEL FOR TECTONISM AND VOLCANISM IN PERMOCARBONIFEROUS WESTERN EUROPE

A prominent feature of the geodynamic model of the Great Basin presented by Scholz, et. al. (91) is the postulated presence of a rising diapir of hot, partially molten mantle material beneath the western margin of the North American continent during most of the Cenozoic. During the early to middle Cenozoic, lithosphere subduction beneath the continental margin resulted in extensive calcalkaline volcanism and acid plutonism. As a result of the regional compressive stress field during subduction, the postulated mantle diapir, situated to the continental side of the main volcanic arc, was unable to penetrate to crustal levels. However, about 30 Ma ago, collision between the East Pacific Rise crest and the continental margin led to a progressive slowing and later cessation of subduction, and finally to relaxation of compressive stress. The mantle diapir was then able to rise into the crust and spread laterally. Resultant crustal extension triggered widespread normal faulting, and allowed large volumes of basaltic magma, separated from the diapir itself to reach the surface. Heating of the lower crust caused partial melting, and the resulting acid magmas gave rise to extensive ash-flow volcanism. There is some evidence that acid volcanism commenced within a "core area", and spread toward the margins of the Great Basin (87, 91).

It is proposed that a similar sequence of events took place in Western Europe in Permocarboniferous times, involving one or more mantle diapirs established during periods of lithosphere subduction beneath the NW and SW margins of the southern European continental plate. During and following the suturing of the northern and southern European continents, and the African continent,

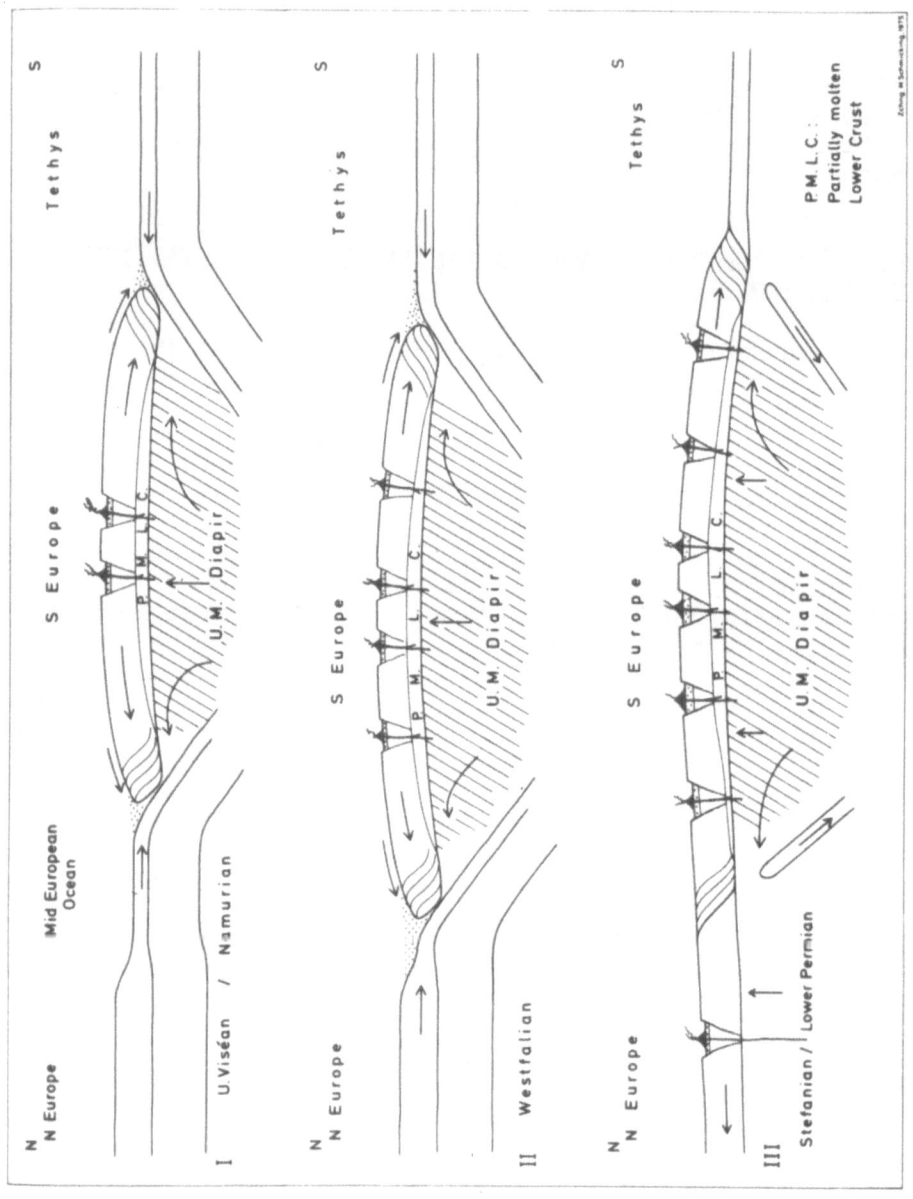

during the "Sudetan" and "Asturian" phase of orogenic movements, relaxation of compressive stress allowed uprise and spreading of hot mantle material, which in turn caused crustal extension, faulting to form graben-like basins and the initiation of widespread bimodal volcanism (Fig. 4).

Due to the release of abundant water during the subduction, dehydration and probable partial melting of former oceanic crust at the two subduction zones dipping beneath the southern European continent, the upper mantle in this region was relatively hydrous, and post-orogenic partial melting processes gave rise to basaltic magmas close to silica-saturation and poor in TiO_2, similar to those of modern intraoceanic and continental margin volcanic arcs. These magmas contrast with the more mafic and alkaline magmas, with higher TiO_2, generated beneath the northern European continent, probably within and above the low velocity zone of an upper mantle with "normal" low water content.

Uprise and lateral spreading of hot mantle material caused uplift of the overlying continental crust, and simultaneous retreat of the sea. Tensional faulting of the crust gave rise to "basin and range" topography, with consequent rapid erosion of material from the ranges and deposition of sediments in the basins. The first basins were formed in the central zone while the orogenic zones with their compressional features were still active to both north and south. Shortly afterward, the first basaltic and rhyolitic magmas gained access to the surface. Further spreading of hot mantle material progressively enlarged the zone of trough formation and volcanism toward the former marginal orogenic zones. Throughout this period, high temperature, low pressure metamorphism of crustal rocks took place within the orogenic zones (14) under a high geothermal gradient, and was accompanied by intrusion of granitic plutons. Synorogenic granites were succeeded by the emplacement of late Hercynian post-orogenic granites, the two groups of intrusives sometimes being fed from the same magma chambers (83, 84).

By late Westfalian times, the mid-European ocean had been consumed, and subduction at the southern margin of the newly-created European continent had also

Fig. 4. Schematic cross-sections showing the geodynamic development of South Europe and adjacent plates during the period U. Viséan to Lower Permian.

ceased. However, spreading of hot mantle material and associated crustal extension continued facilitated by the soft, partially molten lower crust and a rigid crust of reduced thickness. At the beginning of the Permian, these processes were intensified, and allowed the extrusion of large volumes of basaltic and rhyolitic magmas into the subsiding basins, including some north of the mid-European zone of continental plate suturing. This period of intensified regional spreading and volcanism was accompanied by establishment of conditions of high heat flow within basin structures, resulting in metamorphism of both sedimentary and volcanic rocks (see below).

During the later Permian, the supply of hot mantle material gradually diminished, and doming and extension of the crust declined. The basin and range topography was reduced to that of a near peneplain by erosion followed by basin filling. Gradual cooling of the crust led to final closure of K/Ar systems within crustal rocks (seebelow). Crustal subsidence ultimately allowed the sea to invade the continent by the Upper Permian, thus marking the end of the Hercynian plate tectonic cycle. During crustal subsidence isostatic readjustments should have led to reactivation of faults and formation of new ones and probably also the unconformities.

This model suggests a long time interval between the termination of crustal spreading and sedimentation within intermontane basins in Lower Permian time, and the Upper Permian marine transgression. During much of this period, the existence of a regional peneplain prevented the accumulation of a thick sedimentary succession and associated fossil record. Thus, the palynological studies of Visscher (36) have demonstrated that the greater part of the Permian in Western Europo is not represented in the sedimentary succession.

SOME CONSEQUENCES OF THE MODEL PRESENTED

The model described proposes that geodynamic processes operative in the late Carboniferous and early Permian were accompanied by the establishment of a regime of high crustal heat flow, which persisted until the diapiric uprise of hot material from the upper mantle declined. During this period, hydrothermal alteration of pre-existing sedimentary and volcanic rocks, and coalification of associated organic matter occurred. During the subsequent cooling of the crust, some of the palaeomagnetic and isotopic characteristics of these crustal rocks were established.

ALTERATION OF VOLCANIC ROCKS

Volcanic rocks of late Carboniferous to Permian age throughout the zone of Hercynian movements are variably metamorphosed. Basaltic to intermediate rocks in northern Germany, southern Poland, Switzerland and Italy are frequently spilitized. Eckhardt (77, 78) attributed the spilitization of north German examples to reactions during cooling of a thick lava pile. In the light of recent investigations of metamorphism in oceanic basalts (92, 93) these continental spilites might alternatively represent the products of slight thermal metamorphism and metasomatism during a period of increased geothermal gradient. The puzzling Kuselites of the Saar-Nahe area (94, 95), Thuringia and the Sudetes (96), might also be attributed to such a thermal event.

A common type of alteration of the basaltic rocks of many basins involves the formation of hematite, chlorite, pumpellyite, prehnite and zeolites (97), the development of amygdales and the alteration of plagioclase (97). The common secondary minerals in associated rhyolitic rocks are hematite, kaolinite, sericite and albite, and alteration frequently produces strongly corundum-normative compositions. The secondary oxidation of iron to form hematite and the formation of chlorites are characteristic of all groups of volcanic and pyroclastic rocks (98).

ALTERATION OF SEDIMENTS

"Red-beds" are common amongst basin filling sediments. Whether the characteristic hematite is of primary or secondary origin is uncertain, but either mode of origin requires elevated near-surface temperatures. These may reflect a warm climate (99-101). However, Lorenz (98) has described subsurface red-bed formation unrelated to either climatic factors, or to volcanism. It is probable that under the influence of a high crustal heat flow, Eh-pH conditions conducive to red-bed formation may occur within iron-bearing sediments and volcanic and pyroclastic rocks to depths of 500m or more (98).

COALIFICATION

The coalification of Upper Carboniferous coals in northern Germany and the Saar basin took place prior to the deposition of the Zechstein copper shale and the Triassic respectively. Coals of Hercynian Europe are of higher rank than those of the Carboniferous Moscow trough. These relationships have been interpreted by Teichmüller and Teichmüller (102) as indicating an elevated heat flow in Hercynian Europe during the period Upper Carboniferous (post-Hercynian) to Upper Permian (pre-Zechstein). Coals in a number of Lower Permian basins are also of high rank (103).

PALAEOMAGNETIC DATA

In their investigation of the palaeomagnetism of rocks of the Saar-Nahe area, Berthold, et. al. (104) observed a stable secondary component of magnetization which they attributed to a reheating process. This is believed to have been due to burial of the rocks under conditions of high heat flow.

K/AR RADIOMETRIC AGE DATA

Two examples of glass-bearing basalts of the Saar-Nahe basin have been dated by the K/Ar method, yielding Upper Permian ages of 239 ± 10 Ma (105), and 238 ± 4 Ma (R. Duncan, Research School of Earth Sciences, Australian National University, Canberra, unpublished data). Stratigraphic relationships demonstrate conclusively that these basalts are as old, or slightly older than associated rhyolitic intrusives which have yielded a Lower Permian Rb/Sr age of 278 m.y. (104, 106). The anomalously young K/Ar ages of the basalts are probably due to post-eruptive loss of Ar from the glassy matrix during a period of high heat flow, and represent the time at which K/Ar systems finally became closed during decline of the crustal geothermal gradient.

REFERENCES

1.   Dewey, J.F., Pitman III, W.C., Ryan, W.B.F. & Bonin, J., 1973: Plate tectonics and the evolution of the Alpine system.- Geol. Soc. Amer. Bull., 84, 3137-3180.

2. Stille, H., 1940: Zur Frage der Herkunft der Magmen.- Abh. Preuß. Akad. Wiss., Jg. 1939, Math. naturwiss. Kl., Nr. 19.
3. Stille, H., 1950: Der "subsequente" Magmatismus.- Abh. zur Geotektonik Nr. 3, Berlin.
4. Anderson, T.A., 1975: Carboniferous subduction complex in the Harz Mountains, Germany.- Geol. Soc. Amer. Bull., 86, 77-82.
5. Burret, C.F., 1972: Plate tectonics and the Hercynian orogeny.- Nature, 239, 155-157.
6. Dewey, J.F. & Burke, K.C.A., 1973: Tibetan, Variscan and Precambrian basement reactivation: products of continental collision.- J. Geol., 81, 683-692.
7. Johnson, G.A.C., 1973: Closing of the Carboniferous sea in Western Europe.- In: Tarling, D.H. & Runcorn, S.K. (Eds.): Implications of continental drift to the earth sciences, 2, 843-850, (Academic Press), London, New York.
8. Laurent, R., 1972: The Hercynides of South Europe, a model.- I.G.C., 24 sess., Montreal, sect. 3, 363-370.
9. Krebs, W., 1968: Zur Frage der bretonischen Faltung im östlichen Rhenoherzynikum.- Geotekt. Forschungen, 28, 1-71.
10. Wunderlich, H.G., 1965: Maß, Ablauf,und Ursachen orogener Einengung am Beispiel des Rheinischen Schiefergebirges, Ruhrkarbones und Harzes.- Geol. Rundschau, 54, 861-882.
11. Nicolas, A., 1972: Was the Hercynian orogenic belt of Europe of the Andean type?- Nature, 236, 221-223.
12. Flügel, H.W., 1975: Einige Probleme des Variszikums von Neo-Europa.- Geol. Rundschau, 64, 1-62.
13. Drach v., V., Lippolt, H.J. & Brewer, M.S., 1974: Rb-Sr-Altersbestimmungen an Graniten des Nordschwarzwaldes.- N. Jb. Miner. Abh., 123, 38-62.
14. Zwart, H.J., 1967: The duality of orogenic belts.- Geologie en Mijnb., 46, 283-309.
15. Hall, A., 1971: The relationship between geothermal gradient and the composition of granitic magmas in orogenic belts.- Contr. Mineral. Petrol., 32, 186-192.
16. Hall, A., 1972: Regional geochemical variation in the Caledonian and Variscan granites of Western Europe.- I.G.C., 24 sess., Montreal, sec. 2, 171-180.
17. Hall, A., 1972: New data on the composition of Caledonian granites.- Min. Mag., 38, 847-862.
18. Hall, A., 1973: Geothermal control of granite compositions in the Variscan orogenic belt.- Nature

Phys. Sci., 242, 72-75.
19. Smith, D.B., 1972: The Lower Permian in the
    British Isles.- In: Falke, H. (Ed.): Rotliegend.
    Essays on European Lower Permian.- Int. Sediment.
    Petrograph. series, 15, 1-33.
20. Havlena, V. & Tásler, R., 1966: Upper Carboniferous
    and Permian basins.- In:Svoboda et al. (Eds.):
    Regional geology of Czechoslovakia, I, 668p.,
    Geol. Surv. Czechoslovakia, Prague.
21. Burgath, K., 1973: Kulmische Lahare im Südschwarz-
    wald.- Jber. u. Mitt. oberrhein. geol. Ver., N.F.
    55, 83-93.
22. Burgath, K. & Maass, R., 1973: Die variszische
    Entwicklung im südlichen Schwarzwald.- C.R. 7 Congr.
    Strat. Géol. Carbonif., Krefeld, 1971, 2, 196-204.
23. Fabre, J., Feys, R. & Greber, C., 1960: Strati-
    graphie du bassin houiller Brianconnais.- C.R.
    4 Congr. Strat. Géol. Carbonif., Heerlen 1958, 1,
    169-173.
24. Ellenberger, F., 1966: Le Permien du Pays de
    Vanoise.- Atti del symposium sul Verrucano, 170-
    211, Pisa.
25. Stille, H., 1951: Das mitteleuropäische varis-
    zische Grundgebirge im Bilde des gesamteuropäischen.-
    Beih. Geol. Jb., H. 2, 138 p.
26. Rau, A. & Tongiorgi, M., 1972: The Permian of
    middle and northern Italy.- In:Falke, H. (Ed.):
    Rotliegend. Essays on European Lower Permian. Int.
    Sediment. Petrograph. Series, 15, 216-280.
27. Falke, H., 1972: Vergleich zwischen den Ablagerungen
    des Verrucano in den Westalpen und des Rotliegenden
    in Süddeutschland und Frankreich.- Verh. Geol.
    B. - A., Jg. 1972, 11-32.
28. Falke, H., 1972: The continental Permian in North-
    and South Germany.- In: Falke, H. (Ed.): Rotliegend.
    Essays on European Lower Permian. Int. Sediment.
    Petrograph. Series, 15, 43-113.
29. Holub, V., 1972: Permian of the Bohemian Massif.-
    In: Falke, H.(Ed.): Rotliegend. Essays on European
    Lower Permian. Int. Sediment. Petrograph. Series,
    15, 137-188.
30. Lotze, F., 1945: Zur Gliederung der Varisziden der
    Iberischen Meseta.- Geotektonische Forschungen,
    H. 6, 78-92.
31. van Waterschoot van der Gracht, W.A.J.M., 1938:
    The paleozoic geography and environment in north-
    western Europe as compared to North-America.-
    C.R. 2 Congr. Strat. Carbonif., Heerlen 1935, 3,
    1357-1429.
32. Bloch, J.P., 1966: Le Permien du domaine brian-

connais ligure. Essai de chronologie des formations anté-triasiques.- Atti del symposium sul
Verrucano, 99-115, Pisa.

33. Fabre, J. & Feys, R., 1966: Les séries bariolées
du massif de Rochachille. Leurs rapports avec le
Verrucano de Briancon et les "Permiens" de
Maurienne et de Tarentaise.- Atti del symposium
sul Verrucano, 143-169.

34. Lefèvre, R., 1966: Les formations détritique
versicolores du Néopermien de la bande d'Acceglio-
Longet (Alpes Cottiennes Franco-Italiennes).-
Atti del symposium sul Verrucano, 136-142, Pisa.

35. Trümpy, R. & Dössegger, R., 1972: Permian of
Switzerland.- In: Falke, H. (Ed.): Rotliegend.
Essays on European Lower Permian. Int. Sediment.
Petrograph. Series, 15, 189-215.

36. Visscher, H., 1973: The Upper Permian of Western
Europe - A palynological approach to chronostratigraphy.- Canad. Soc. Petrol. Geologists, Mem. 2,
200-219.

37. Sleep, N.H., 1973: Crustal thinning on Atlantic
continental margins: evidence from older margins.-
In: Tarling, D.H. & Runcorn, S.K. (Eds.): Implications of continental drift to the earth sciences,
2, 685-692, (Academic Press) London, New York.

38. Bickle, M.J. & Pearce, J.A., 1975: Oceanic mafic
rocks in the Eastern Alps.- Contrib. Mineral.
Petrol., 49, 177-189.

39. Dietrich, V., Vuagnat, M. & Bertrand, J., 1974:
Alpine metamorphism of mafic rocks.- Schweiz.
Mineralogische Petrograph. Mitt., 54, 291-332.

40. Dietrich, V.J., 1975: Plattentektonik in den Ostalpen.- Geotektonische Forschungen, (in press).

41. Förster, H., Soffel, H. & Zinsser, H., 1975:
Paleomagnetism of rocks from the Eastern Alps from
North and South of the Insubrian Line.- N.Jb.
Geol. Paläontol. Abh., 149, 112-127.

42. Ramberg, I.B. & Smithson, S.B., 1975: Gridded fault
patterns in a late Cenozoic and a Paleozoic continental rift.- Geology, 3, 201-205.

43. Fisch, W. & Ryf, W., 1966: Der Verrucano in den
Glarner Alpen.- Atti del symposium sul Verrucano.
233-245, Pisa.

44. Fabre, J., 1961: Contribution à l'étude de la zone
houillère en Maurienne et en Tarentaise (Alpes de
Savoie).- Mém. 2, B.R.G.M., 315p., Paris.

45. Murawski, H. (Berichterstatter), 1975: Die Grenz-
zone Hunsrück/Saar-Nahe-Senke als geologisch-geo-
physikalisches Problem.- Z.dt.geol. Ges., 126,
49-62.

46. Schönenberg, R., 1971: Einführung in die Geologie
    Europas.- 300 p., Rombach Hochschul Paperback,
    Bd. 18, (Rombach) Freiburg.
47. Katzung, G., 1968: Perm.- In: Grundriß der Geolo-
    gie der Deutschen Demokratischen Republik, Berlin.
48. Trümpy, R., 1966: Considérations générales sur le
    "Verrucano" des Alpes Suisses.- Atti del symposium
    sul Verrucano. 212-232, Pisa.
49. Feys, R. & Greber, C., 1966: Brève presentation
    du Permien en France.- Atti del symposium sul
    Verrucano.- 311-323, Pisa.
50. Lorenz, V., 1971: Zur Stratigraphie und Tektonik
    des Oberrotliegenden in der Umgebung von Schweis-
    weiler und Winnweiler/Pfalz.- Abh.hess.L.-Amt-
    Bodenforsch., 60, (Heinz-Tobien-Festschrift), 263-
    275.
51. Lorenz, V., 1973: Zur Altersfrage des Kreuznacher
    Rhyolithes unter besonderer Berücksichtigung der
    Stratigraphie und Überschiebungstektonik in seiner
    südlichen Umrandung (Saar-Nahe-Gebiet, SW-Deutsch-
    land).- N.Jb. Geol. Paläont. Abh., 142, 139-164.
52. Eigenfeld, F. & Schwab, M., 1974: Zur geotektonischen
    Stellung des permosilesischen subsequenten Vul-
    kanismus in Mitteleuropa.- Z.geol.Wiss., 2, 115-
    137.
53. Monomakhoff, C., 1961: La tectonique tangentielle
    dans les bassins houillers de la France et sa
    répercussion sur la continuité et comportement de
    ces gisements.- C.R. 4 Congr. Strat. Géol. Carbonif.,
    Heerlen,1958, 2, 423-435.
54. Falke, H. & Kneuper, G., 1972: Das Karbon in lim-
    nischer Entwicklung.- C.R. 7 Congr. Int. Strat.
    Géol. Carbonif., Krefeld 1971, 1, 49-67.
55. Schmidt, K. & Franke, D., 1975: Stand und Probleme
    der Karbonforschung in der Deutschen Demokratischen
    Republik. Teil I : Unterkarbon.- Z. geol. Wiss.,
    3, 819-849.
56. Lorenz, V., 1973: On the formation of maars.-
    Bull. volcanologique, 37, 183-204.
57. Lorenz, V., 1974: Die Pyroklastika des Permischen
    Saar-Nahe-Beckens.- 64. Jahrestagung Geol. Ver-
    einigung, Kurzfass. d. Vorträge, Bochum 1974.
58. v. Hoyningen-Huene, E. 1966: Entwicklungsge-
    schichte des Permokarbons im Vorfeld der Mittel-
    deutschen Hauptscholle.- Atti del symposium sul
    Verrucano, 355-380, Pisa.
59. Nicholls, I.A. & Lorenz, V., 1973: Origin and
    crystallization history of Permian tholeiites from
    the Saar-Nahe trough, SW Germany.- Contrib. Mine-
    ral. Petrol., 40, 327-344.

60. Green, D.H. & Ringwood, A.E., 1967: The genesis
    of basaltic magmas.- Contrib. Mineral. Petrol.,
    15, 103-190.
61. Jung, D. & Vinx, R., 1973: Einige Bemerkungen zur
    Geochemie der Magmatite des Saar-Nahe-Pfalz-Ge-
    bietes.- Annales Scientif. de l'Université de
    Besancon, Géol., 3 série, fasc. 18, 197-202.
62. Nicholls, I.A., Whitford, D.J. & Lorenz, V.: Geo-
    chemistry of Permian tholeiites of the Hirschberg
    and Rödern diatremes, SW Germany. In preparation.
63. Andrusov, D., 1964: Geologie der Tschechoslova-
    kischen Karpaten I. 263 p., (Akad.Verl.) Berlin.
64. Hoehne, K., 1961: Zum Alter der Porphyre im Wal-
    denburger Bergbaugebiet (Niederschlesien).- Geol.
    Jb., 78, 299-328.
65. Siegert, C., 1967: Zur Petrochemie der Vulkanite
    des Halleschen Permokarbonkomplexes.- Geologie,
    16, 1122-1135.
66. Schneider, A., 1963: Rhyolithischer Vulkanismus
    des Südharzer Rotliegenden.- Beitr. Mineral.
    Petrogr., 9, 148-174.
67. Müller, G., 1965: Beiträge zur Stratigraphie des
    Südharzer Rotliegenden.- Geol. Jb., 82, 755-770.
68. Morre, N., 1966: Etude pétrographique des laves
    carbonifères de la région de Figeac (Lot).- Bull.
    Soc. géol. de France, (7), 8, 322-327.
69. Morre, N. & Thiébaut, J., 1964: Constitution de
    quelques roches volcaniques permiennes de la
    Sierra Del Cadi (Pyrénées Catalanes).- Bull. Soc.
    géol. de France, (7), 6, 389-396.
70. Amstutz, C.G., 1954: Geologie und Petrographie
    der Ergußgesteine im Verrucano des Glarner Frei-
    berges.- Publ. Vulkaninst. Friedländer, 5, 150p.
71. Maaskant, A., 1941: De Geologie van het Gebied
    Tusschen het Val Seriana en de Mte Guglielmo.-
    68 p., Van Gorcum's Geologische Reeks, 2, (Van
    Gorcum & Comp. N.V.), Assen.
72. Pichler, H., 1957: Geologische Untersuchungen am
    Südrand der Bozener Porphyrplatte nordöstlich von
    Trento (Oberitalien).- Diss., 162 p., Univ.
    München.
73. Green, D.H., 1973: Experimental melting studies
    on a model upper mantle composition at high
    pressure under water-saturated and water-under-
    saturated conditions.- Earth Planet. Sci. Letters,
    19, 31-53.
74. Cosgrove, M.E., 1972: The geochemistry of the
    potassium-rich Permian volcanic rocks of Devon-
    shire, England.- Contrib. Mineral. Petrol., 36,
    155-170.

75. Francis, E.H., 1968: Review of Carboniferous-
     Permian volcanicity in Scotland.- Geol. Rundschau,
     57, 219-246.
76. Heier, K.S. & Compston, W., 1969: Rb-Sr isotopic
     studies of the plutonic rocks of the Oslo region.-
     Lithos, 2, 133-146.
77. Eckhardt, F.J., 1968: Vorkommen und Petrogenese
     spilitisierter Diabase des Rotliegenden im Weser-
     Ems Gebiet.- Geol. Jb., 85, 227-251.
78. Eckhardt, F.J., 1972: Vulkanismus im Rotliegenden
     Mitteleuropas.- DFG-Forschungsbericht: Unterneh-
     men Erdmantel, 269-270, (Fr. Steiner) Wiesbaden.
79. Green, D.H., 1971: Compositions of basaltic magmas
     as indicators of conditions of origin: application
     to oceanic volcanism.- Phil. Trans. Roy. Soc.,
     A268, 101-125.
80. Théobald, N. & Thiébaut, J., 1961: Les rhyolites
     permiennes du massif de Chagey (H.-S.)- Annal.
     Scientifiques de l'Université de Besancon, 2 Sér.,
     Géologie, fasc. 15, 13-23.
81. Mahel', M. et al., 1968: Regional geology of
     Czechoslovakia. Part II. The West Carpathians.-
     723 p., (Schweizerbart) Stuttgart.
82. Hahn-Weinheimer, P. & Ackermann, H., 1967: Geo-
     chemical investigations of differentiated granite
     plutons of the Southern Black Forest - II. The
     zoning of the Malsberg Granite pluton as indi-
     cated by the elements titanium, zirconium, phos-
     phorus, strontium, barium, rubidium, potassium
     and sodium.- Geochim. Cosmochim. Acta, 31, 2197-2218.
83. Brewer, M.S. & Lippolt, H.J., 1974: Petrogenesis
     of basement rocks of the Upper Rhine region eluci-
     dated by Rubidium-Strontium systematics.- Contrib.
     Mineral. Petrol., 45, 123-141.
84. Emmermann, R., 1972: Die Südschwarzwälder Granite.-
     Fortschr. Mineral., 50, Beiheft 2, 1-13.
85. Theuerjahr, A.K., 1973: Geochemisch-petrologische
     Untersuchungen an jungpaläozoischen Rhyolithen des
     Saar-Nahe-Gebietes.- 86 p., Diss., Univ. Mainz.
86. Lipman, P.W., Prostka, H.J. & Christiansen, R.L.,
     1972: Cenozoic volcanism and plate-tectonic evo-
     lution of the Western United States. I. Early and
     Middle Cenozoic.- Phil. Trans.R.Soc.Lond., A271,
     217-248.
87. Christiansen, R.L. & Lipman, P.W., 1972: Cenozoic
     volcanism and plate-tectonic evolution of the
     Western United States. II. Late Cenozoic.-Phil.
     Trans.R.Soc.Lond., A.271, 249-284.
88. Lipman, P.W., 1966: Water pressure during differen-
     tiation and crystallization of some ash-flow magmas

from southern Nevada.- Amer.J.Sci., 264, 810-826.

89. Doe, B.R., Lipman, P.W., Hedge, C.E. & Kurasawa,
    H., 1969: Primitive and contaminated basalts from
    the Southern Rocky mountains, U.S.A.- Contrib.
    Mineral.,Petrol., 21, 142-156.

90. Lipman, P.W., 1969: Alkalic and tholeiitic basal-
    tic volcanism related to the Rio Grande depression,
    Southern Colorado and Northern New Mexico.- Geol.
    Soc.Amer.Bull., 80, 1343-1354.

91. Scholz, C.H., Barazangi, M. & Sbar, M.L., 1971:
    Late Cenozoic evolution of the Great Basin, Western
    United States, as an ensialic interarc basin.-
    Geol.Soc.Amer.Bull., 82, 2979-2990.

92. Cann, J.R.: Spilites from the Carlsberg-Ridge,
    Indian Ocean.- J.Petrology, 10, 1-19.

93. Hughes, C.J., 1972: Spilites, keratophyres and the
    igneous spectrum.- Geol.Mag., 109, 513-527.

94. Koch, I., 1938: Die Kuselite des Saar-Nahe-Gebietes.-
    N.Jb.Mineral.Geol.u.Paläont., Beilage Bd. 73,
    Abt. A, 419-494.

95. Vetter, U. & Jung, D., 1971: Kuselite.- Erl.Geol.
    Kte. Rheinland-Pfalz 1:25000, Bl. Kusel, 42-46,
    Mainz.

96. Bederke, E., 1959: Probleme des Permischen Vul-
    kanismus.- Geol.Rundschau, 48, 10-18.

97. Dreyer, G., 1973: Neue Mineralien der Rheinpfalz.-
    Mitt.Pollichia, III Reihe, 20, 113-136.

98. Lorenz, V., 1972: Sekundäre Rotfärbung im Rot-
    liegenden der Saar-Nahe-Senke, SW-Deutschland.-
    N.Jb.Geol.Paläont.Mh., Jg. 1972, 356-370.

99. Walker, T.R., 1967: Formation of red beds in modern
    and ancient deserts.- Geol.Soc.Amer.Bull., 78,
    353-368.

100. Walker, T.R., 1968: Formation of red beds in modern
     and ancient deserts: reply.- Geol.Soc.Amer.Bull.,
     79, 281-282.

101. van Houten, F.B., 1961: Climatic significance of
     red beds.- In: Nairn, A.E.M. (Ed.): Descriptive
     Paleoclimatology, 89-139, New York.

102. Teichmüller, M. & Teichmüller, R., 1966: Die In-
     kohlung im saarlothringischen Karbon, verglichen
     mit der im Ruhrkarbon.- Z.deutsch.geol.Ges., Jg.
     1965, 117, 243-279.

103. Lorenz, V., Teichmüller, M. & Teichmüller, R.:
     Zum Wärmefluß in der Saar-Nahe-Senke während des
     Permokarbon.- In preparation.

104. Berthold, G., Nairn, A.E.M. & Negendank, J.F.W.,
     1975: A paleomagnetic investigation of some igneous
     rocks of the Saar-Nahe basin.- N.Jb.Geol.Paläontol.
     Mh., Jg. 1975, 134-150.

105. Negendank, J., 1969: Über permische und tertiäre
     Magmatite im Untergrund des Mainzer Beckens.-
     Geol.Rundschau, 58, 502-512.
106. Orliac, J. & Schäfer, K., 1969: Das Rubidium-
     Strontium-Alter unterpermischer Rhyolithe des
     Saar-Nahe-Gebietes.- N.Jb.Geol.Paläont.Mh., Jg.
     1969, 246-252.
107. Gallwitz, H., 1959: Die Stellung der Magmatite im
     Permokarbon der Mitteldeutschen Hauptscholle.-
     Geol.Rundschau, 48, 27-32.

INDEX

H. Falke (ed.), The Continental Permian in Central, West, and South Europe, 343-352. All Rights Reserved.
Copyright © 1976 by D. Reidel Publishing Company, Dordrecht-Holland.